WARFARE AND POLITICS IN SOUTH ASIA FROM ANCIENT TO MODERN TIMES

WARFARE AND POLITICS IN SOUTH ASIA FROM ANCIENT TO MODERN TIMES

Edited by
KAUSHIK ROY

MANOHAR
2011

First published 2011

ISBN 978-81-7304-913-2

Published by
Ajay Kumar Jain *for*
Manohar Publishers & Distributors
4753/23 Ansari Road, Daryaganj
New Delhi 110 002

Typeset at
Digigrafics
New Delhi 110 049

Printed at
Salasar Imaging Systems
Delhi 110 035

In honour of
Prof. Subhas Ranjan Chakraborty
who taught me in Presidency College, Kolkata.

Contents

Preface

This volume is the product of the panel on South Asian warfare as part of the 20th Conference of the International Association of Historians of Asia (IAHA), held at the Jawaharlal Nehru University (JNU) between 14 and 17 November 2008. For the first time to my knowledge, in the case of a history conference/workshop in India, a whole panel was devoted to the study of South Asian military history. This has been possible due to the hard work and generosity of Prof. Deepak Kumar of Zakir Husain Centre for Educational Studies at JNU. Panel 22 of IAHA was attended not only by historians from India and foreign countries but also by several students from the Centre for Historical Studies, JNU. This reflects the burgeoning interest in military history within the academic circuit. I thank some of the participants for living up to their promises and delivering essays which made this edited volume possible. Several scholars from outside India who could not attend this conference due to lack of time and funds also kindly agreed to give their articles for this collection. Besides the editor, the fifteen scholars who contributed articles for this volume reflect a balanced mixture of young and mature academicians and as well as Indian and foreign scholars. I also thank Mr. Ramesh Jain of Manohar who was not only present during the seminar but also encouraged the publication of a volume of the selected papers.

KAUSHIK ROY

Contributors

TORKEL BREKKE is a Professor at the University of Oslo. He has also worked as adviser to the Norwegian Ministry of Defence. His main interest is the interface between religious history and violent conflict. He has published ten books and a number of articles and has received several research prizes. His principal publications include *Makers of Modern Indian Religions* (2002) and an edited volume titled *The Ethics of War in Asian Civilizations* (2006).

PRANAB K. CHATTOPADHYAY is a Senior Fellow at the Centre for Archaeological Studies and Training, Eastern India, Kolkata. He is working on a project of documenting cannons of eastern India for the Indian National Science Academy, New Delhi.

ZIAUDDIN CHOWDHURY has a Master's degree in Fine Arts from the University of Chittagong. Presently, he is working as an Assistant Display Officer in the Chittagong University Museum, at Chittagong, Bangladesh.

AZHARUDIN MOHAMED DALI is a Senior Lecturer of Indian history at the Department of History, University of Malaya, Kuala Lumpur, Malaysia. Among his articles is 'The Ghadar Movement in Southeast Asia, 1914-1918', *Jati, Journal of Southeast Asian Studies*, University of Malaya, 2002.

SABYASACHI DASGUPTA is a Lecturer at the Department of History, Visva-Bharati, Santiniketan, West Bengal, India. He is currently working on the post-Independence Indian Army.

SCOTT GATES is Director of the Centre for the Study of Civil War (CSCW), International Peace Research Institute (PRIO), Oslo, and a Professor of the Norwegian University of Science and Technology (NTNU). He has several books and numerous articles and specializes on counter-insurgency and child soldiers.

SHANTHA HARIHARAN did her post-doctoral research as a scholar of the *Fundação para Ciência e Tecnologia*, Lisbon, Portugal, and worked on the revival of Portuguese trade from western India in the eighteenth century as well as on Luso-British relations in India in the same period. She is the author of *Cotton Textiles and Corporate Buyers in Cottonopolis: A Study of Purchases and Prices in Gujarat, 1600-1800* (2002).

AMARENDRA KUMAR is a Senior Lecturer at the National Defence Academy, Khadakwasla, Pune (Maharashtra), and is working on the Maratha Navy.

BERNARDO A. MICHAEL is an Associate Professor of History and Director of the Centre for Public Humanities at Messiah College, Grantham, PA, USA.

JASON MIKLIAN is a researcher at the International Peace Research Institute (PRIO), Oslo, with a focus on insurgencies in Nepal and India. Miklian has authored several PRIO publications on Nepal, including examinations of post-conflict power sharing, land reform, *hawala* trading networks, and ethnic separatism in the Terai. He has also published in *Foreign Policy Analysis, Strategic Analysis, Nepali Times* and *Outlook* (India).

PRATYAY NATH is doing a doctorate on Mughal warfare at the Centre for Historical Studies at the Jawaharlal Nehru University, New Delhi.

KAUSHIK ROY is a Senior Researcher at the Centre for the Study of Civil War (CSCW) at International Peace Research Institute (PRIO), Oslo, and Reader in the Department of History, Jadavpur University, Kolkata, West Bengal, India. He is the author of nine books and forty articles.

GAGAN PREET SINGH is a doctorate student at the Centre for Historical Studies, Jawaharlal Nehru University, New Delhi.

CHANDAR S. SUNDARAM holds a doctorate in history from McGill University in Montreal, Canada. Specializing in the history of war and society of later colonial India, he has written on the Indianization of the Indian Army's officer corps and also on the Indian National Army. He is writing a history of the Imperial

Cadet Corps, 1900-14. He is currently with the Centre for Armed Forces Historical Research at the United Service Institution of India, New Delhi.

RAM KRISHNA TANDON is a Reader in the Department of Defence and Strategic Studies, University of Allahabad, India. He holds a D.Phil. in Defence Studies.

FRANK H. WALLIS received his Ph.D. from the University of Illinois in 1987 with a speciality in post-1688 Britain. His work includes *Popular Anti-Catholicism in Mid-Victorian Britain.*

Introduction

KAUSHIK ROY

The present volume operates on the assumption that South Asia is a distinct unit. South Asia includes India, Pakistan, Bangladesh, Nepal, Sri Lanka, Burma (Myanmar) and also Afghanistan. The fifteen essays in this collection address the two interlinked themes of warfare and politics. The operative assumption of this volume is that warfare and politics are not two watertight compartments but are interlinked with each other like the double helix structure of a DNA. The concept of warfare has been defined quite broadly in this connection. Warfare means not merely activities concerning the armies, navies, battles and campaigns but also military diplomacy involving diplomatic coercion and the resultant forceful negotiations, political directions of the armed forces, etc. The nature of a state depends upon the politics followed by the armed forces of the state. Warfare in this volume comprises both inter-state war and intra-state conflicts (what is also known as insurgency, low-intensity conflict, unconventional war, sub-conventional operations, operations other than war, ethnic upheavals, ethno-national conflicts, and so on). Whether the above-mentioned approach to the historical study of warfare could be categorized as New Military History or not is debatable.

Mark Moyar while analysing recent British and American writings on warfare says that from the 1960s, numerous scholars began studying the non-military aspects of war, thus creating what became known as the New Military History. Moyar continues that to an extent the term 'New' is erroneous because both Herodotus and Thucydides integrated military history with social, cultural and political aspects.[1] Jeremy Black opines that New Military History focuses on the position, relationship and experience of the rank and file. The objective is to study the military subalterns in their social context. However, such an approach, rightly says Black, could not provide a long-term explanatory model for explaining military changes over the ages. Black claims that military changes in different societies over time cannot be explained without referring to the different political contexts.[2]

Politics in the perspective of this volume involves attempts of state building at a higher level and its interaction with tortuous negotiations among the numerous local actors with contradictory aims at the lower level. Besides elite political manoeuvrings, the activities of the local actors as well as politics of the non-state and sub-state actors have been taken into account. Further, this volume also highlights the political aspect of religion. The clear separation of politics and religion is the product of Enlightenment thinking and is not viable while undertaking historical analysis of the extra-European world.

This Introduction, instead of providing merely a summary of the fifteen essays contained in the volume, attempts to locate the essays within the broader historiographical context. Further, an attempt is made to point out the interconnections between the essays on South Asian history with the broader Eurasian history. Comparison across various cultures is another strong point of this collection. Moreover, the contributors not only come from various fields—history, political science, and engineering—but also from a wide geographical arena. Besides South Asia, some of the essayists hail from South-East Asia, Scandinavia and the USA.

The essays in this volume have been arranged chronologically. The collection of essays temporally covers the time period from the Vedic Age to recent times. Both continuities and changes throughout South Asian history have been emphasized. While two essays (R.K. Tandon's and Kaushik Roy's essay on Kautilya) focus on military theories and a military theorist of ancient India, Pratyay Nath turns the limelight on Mughal siege warfare in medieval India. Another essay of Roy covers the nature of South Asian warfare during the ancient and medieval eras. A short essay by two Bengali engineers (one Indian and another Bangladeshi) notes the technical aspects of Mughal cannons in Bengal. The maritime aspect of early modern India is addressed in Amarendra

Kumar's essay. As regards the colonial era, in general historians concentrate mostly on the doings of the British. We forget that early modern India was a playground for various European powers like the French, the Dutch and the Portuguese and in the end the British won. Shantha Hariharan's essay turns the spotlight on the Portuguese in Goa. While Bernardo A. Michael links the politics of frontier with the origins of the Anglo-Gorkha War (1814-15), Frank H. Wallis and Hariharan analyse the dynamics behind British imperialism in South Asia during the first half of the nineteenth century. And Gagan Singh's essay notes the disarmament efforts of the colonial regime in Punjab which was part of the British attempt to build up a strong state in South Asia. The essays by Chandar Sundaram and A.M. Dali cover the Raj and its British-Indian Army during Second World War. Three essays (Sabyasachi Dasgupta, Torkel Brekke, and the joint essay by Scott Gates and Jason Miklian) turn the attention to the post-colonial period. Dasgupta's essay deals with civil–military relations in post-1947 Indian Army. Dasgupta also brings to the fore the issue of colonial continuities in the state building project of post-colonial times. Three essays (that of Roy on Kautilya, Brekke's essay on Sikhism and the joint essay by Gates and Miklian) deal with intra-state war.

Spatially, two essays deal with Nepal (Michael's essay and the joint essay of Gates and Milkian), one essay (Dali) covers the actions of the expatriates from South Asia in South-East Asia during Second World War, which in turn had consequences for the political-cum-military environment of the former region.

The different regions of India are also well represented in the various essays. While Ziauddin Chowdhury and Pranab K. Chattopadhyay look at Mughal Bengal (present-day Bangladesh and West Bengal), Nath studies Mughal siege operations in north and central India, Singh and Brekke focus on Punjab, and Hariharan turns our attention towards west India. Wallis' essay covers Afghanistan and Sind. Roy's two essays and Tandon, Sundaram and Dasgupta's essays are pan-Indian in scope.

Besides the core of the subcontinent, the collection also brings into its orbit the role of the frontier. Michael, Wallis, Singh and Brekke highlight the role of the frontier in shaping the history of the subcontinent. While most of the contributors are landlubbers,

Kumar shifts our attention away from the continental landmass to the Arabian Sea and the western coast of India.

ETHICS OF WAR, POLITICS AND RELIGIOUS TRADITIONS IN SOUTH ASIA

Tandon, Brekke and Roy's essay on Kautilya discuss the issue of ethics (norms of behaviour) of warfare in the different religious traditions of South Asia. While Tandon compares and contrasts Kautilya with Sun Tzu, Roy analyses Kautilya's theory of counter-insurgency. Despite Kautilya's attempt to separate religion from *rajadharma* (statecraft), the *Arthashastra* was composed within the broad parameters of Hinduism. Similarly, one could argue that Sun Tzu's *Art of War* could not be totally insulated from Confucianism. Both Sun Tzu and Kautilya were what in modern day could be termed as strategic studies experts. Strategic studies could be defined as the study of statecraft and its relation to the use of military power.[3] There are lot of similarities as well as dissimilarities in the theoretical frames of Sun Tzu and Kautilya.

There is a debate among the historians of ancient China about the background of Sun Tzu and his real identification and dating of the text *The Art of War*. Generally, the later half of fifth century BCE is taken as the date of composition of Sun Tzu's *magnum opus*. While some regard Sun Tzu as the inhabitant of Ch'i, others think that he was a native of Wu.[4] A similar debate rages among the scholars of ancient India regarding the background of Kautilya. Both Sun Tzu and Kautilya focus on grand strategy. Grand strategy (what the Americans call national security policy) refers to the political, financial and military assets used by a state for achieving its objectives.[5] One of the key concepts of Sun Tzu's strategy is: 'subjugating the enemy's army without fighting is the true pinnacle of excellence'.[6] Both Sun Tzu and Kautilya emphasize disinformation and misinformation in order to gain victory over enemy troops.[7] Rather than fighting a decisive battle, Sun Tzu advocates that whenever possible, victory should be achieved through diplomatic coercion, thwarting the enemy's plans and alliances. And even when a campaign is launched, the focus should be on minimum risk and exposure and limiting the destruction to be inflicted and suffered.[8] Even when battle becomes necessary, emphasizes Sun Tzu, indirect attacks rather than direct confrontations should be the norm.[9] Kautilya and Sun

Tzu, unlike the Prussian military theorist Carl Von Clausewitz (1780-1831), advocate that if possible it is better to avoid a decisive set piece battle resulting in bloodshed. For Clausewitz, the key to war is battle and battle for him is *schlacht* (bloodshed).[10]

There are a lot of dissimilarities in the theoretical frameworks of Sun Tzu and Kautilya. For instance, Sun Tzu emphasizes *chi*, which can be translated as spirit or the essential vital energy of life. *Chi* is for Sun Tzu, the foundation of courage. So, *chi* is an essential ingredient of troops' morale. Such a concept is absent in Kautilya's *Arthashastra*. Ralph D. Sawyer asserts that Sun Tzu's criteria of analysing the enemy and battlefield situations consist of some forty paired, mutually defined interrelated categories that can be abstracted from the text. And this reflects the Taoist influence on Sun Tzu's frame.[11]

Unlike Sun Tzu, Kautilya introduced the concept of *asurayuddha*. It involves total destruction of the enemy kingdom, execution of the enemy ruling elite, plundering their wealth and violation of the women belonging to the ruling clan of the enemy regime. Kautilya urges the *vijigishu* (ideal world conqueror, also known as *chakravartin*) not to follow *asurayuddha* as it results in permanent enmity with the subject populace of the conquered enemy kingdom. Interestingly, the Spanish Dominican Francisco de Vitoria (1492-1546) explained that the enemy princes should not be deposed at the outcome of each and every just war. Punishment should be diminished in favour of mercy. This is a rule of not only human law but also of natural and divine law. Harm done by the enemy might be a sufficient cause for war but is inadequate to justify the extermination of the enemy kingdom and deposition of the legitimate princes. Such a practice is considered as savage and inhuman.[12] Kautilya's greatest contribution to security studies is the *mandala* (balance of power) theory. In Kautilya's framework, the various states are organized in a series of circles known as the *mandala* and the *vijigishu*'s state which aims to dominate the circle of states (international system of states) is located at the centre. In order to acquire hegemony in the *mandala*, Kautilya emphasizes various policies like *sandhi* (temporary truce), *vigraha* (war) and *dvaidhava* (dual policy of maintaining peace with one party and campaign against another state). In accordance with the doctrine of *mandala*, Kautilya advocates preventive war and preemptive strike on part of the *vijigishu* to ward off possible enemy actions in the near future. Alberico Gentili (1552-1608), an Italian Protestant and a professor of law at Oxford stated boldly that for defence of the Christian Commonwealth, actions might be taken against the dangers posed by the enemy states that are probable and possible.[13]

Roy's essay on Kautilya and Brekke's essay on Sikhism explore how ethics in warfare have been treated in the two major cultural and religious traditions (Hinduism and Sikhism) of South Asia. The concept of just war is present in most of the religious traditions of the world. Both in Hinduism and Sikhism, tensions exist regarding what type of wars could be categorized as just wars. A war can be categorized as just if its cause(s) are legitimate and its conduct is also in accordance with the prescribed rules. In medieval Christian tradition, wars for self-defence or wars to recover property were categorized as just.[14] In Judaism, territorial or religious expansion does not justify war. As regards the conduct of just war, hostages should not be taken, non-combatants if possible should be spared from the ravages of war and minimum casualties should be inflicted to attain the legitimate objectives of war.[15] Roy's essay on Kautilya shows that *Arthashastra* pushes the concept of *kutayuddha* (unjust war and the opposite of *dharmayuddha*) in a forceful manner. Kautilya's *kutayuddha* comprises deceit, treachery, sudden attacks, and so on. Brekke in an article published elsewhere asserts that Kautilya's *kutayuddha* marks a break from the deontological tradition, where actions in accordance with *dharma* are more important. The deontological tradition has been forcefully argued in the epics (*Ramayana* and *Mahabharata*). Kautilya represents the consequentialist tradition in which human actions are good or bad only in respect of their results.[16] Brekke's article in this collection claims that Sikhism did not elaborate on just war because when Sikhism emerged it was at the receiving end of aggressive Mughal imperialism and the Sikhs were on the defensive.

Hans Kung challenges the view that monotheist religions are more violent than polytheist religions. The relation between just war and holy war in Islam and Christianity has been fuzzy throughout history. Under the Roman Empire, Christians were forbidden to serve in the army. However, the ecclesiastical ideology of the Early and High Middle Ages gave rise to the Crusades.[17] The Crusades could be categorized as holy wars.

Thomas Aquinas (CE 1225-74) had written that Christians wage just wars against the unbelievers to prevent them from hindering the practice of the faith.[18] Here, we see that Aquinas is fusing the concepts of just and holy wars. In other words, holy wars (all or at least some of them depending upon the circumstances) are just. Aquinas' assertion is somewhat similar to the Islamic concept of *jihad* (literal meaning striving in the path of God). John Kelsay argues that *jihad* is equivalent to the just war of Christianity as initiating both *jihad* and just war require a legitimate authority, a just cause and righteous intention.[19]

In contrast, neither Hinduism nor Sikhism emphasizes the war against the unbelievers for their forceful conversion. This is not to suggest that Hinduism and Sikhism are pacifist religions. In fact, the pacificism of Hinduism is an invention of Mohandas Karamchand Gandhi (1869-1948) during the early twentieth century. Gandhi argued that the *Bhagavad Gita* emphasizes the eternal struggle between the force of darkness and the force of evil. Gandhi emphasized that *satya* (truth) is most important and truth is God. Gandhi derived his concept of *satyagraha* which centred around *ahimsa* (non-violence) from the *Gita.*[20] And Sikhism during late pre-colonial, colonial and post-colonial eras has been considered a martial religion. Balbinder Singh Bhogal claims that both the *Adi Granth* and the *Dasam Granth* equate *bhagati* (love) with *kharag* (violence/sword). Religious love and political violence are two movements of the same action. Bhogal continues that Guru Nanak saw the existential moment of life as one of terror and inevitable violence. The underlying message is the realization that life cannot be lived without violent encounters.[21] The absence of the concept of holy (not just) war in Hinduism and Sikhism is nothing unique of South Asian history. If we believe Norman Solomon, there is nothing in the *Bible* and rabbinic traditions about a holy war against the idolaters beyond the borders of Israel.[22]

THE COLLAPSE OF *ANCIEN REGIME* IN SOUTH ASIA AND THE RISE OF BRITISH POWER

In recent times, the dominant argument advanced for explaining pre-colonial South Asia's collapse against the British East India Company (EIC) centres round the weak state theory. In accordance with this theory, pre-British India throughout history comprised segmentary states. Such shadow polities exercised only ritual sovereignty in the peripheries away from the core. Moreover, political sovereignty in pre-colonial India comprised a variety of overlapping rights. The state in India unlike in West Europe never enjoyed a monopoly of violence. Disarming the private war bands and armed peasantry of the countryside was never an agenda for the pre-British Indian rulers. The weak states had weak armies and the latter were capable of only raiding and not serious fighting. Due to prevalence of large number of armed warriors and horses, drilled and disciplined standing armies never emerged in pre-colonial South Asia. Warfare in pre-British India comprised numerous clashes between militias and the focus was on displaying individual valour. The objective of warfare was not to destroy the enemy combat forces permanently but to defeat them and then incorporate the enemy forces within the political umbrella of the hegemonic polity. When the hegemon declined then political alliances and borders shifted. Due to the prevalence of the twin concepts of *dharma* (literal meaning way of life; due to the influence of *dharmayuddha*, the hegemon do not annex the defeated principalities. This in turn allows the defeated princes to challenge the hegemon repeatedly, resulting in continuous warfare) and *fitna* (internecine struggle within the Islamic polities), continuous political infightings and skirmishes characterized the pre-colonial Indian landscape.[23]

However, the British changed the rules of the game. Britain like Prussia was a military-fiscal regime. During the third quarter of the eighteenth century, Prussia and Britain spent on an average two-thirds of their annual revenues for military and naval purposes. The EIC established a fiscal military state with a standing bureaucratized army led by a hierarchical professional officer corps. The military entrepreneurs were no more enlisted and the practice of giving rights on the produce from land for military service was done away with. The drilled and disciplined handguns-equipped, infantry-centric army was a product of the military revolution of West Europe. The Company's *fauj* (force) following the Clausewitzian strategy of annihilating the enemy field forces was able to defeat and destroy the feudal cavalry of the 'native' regimes in a series of decisive set piece battles during the eighteenth century.[24]

Roy's article on pre-British warfare in South Asia asserts that at certain moments in history, India witnessed the emergence of strong centralized polities which maintained combat effective standing armies. In contrast, pre-modern Europe did not always maintain standing armies. During the later Roman Empire, Germanic warriors under the leadership of their clan leaders comprised the auxiliary and federated units of the imperial army. And West Europe after the decline of the Western Roman Empire had weak regimes and feudal cavalry. Modern studies show that decisive battles were quite uncommon in medieval West Europe. Medieval warfare comprised skirmishes, small-scale raids, and so on.[25]

The introduction of gunpowder weapons in warfare, which is taken as the end of medievalism and the beginning of modernity, is a tricky issue as far as South Asian history is concerned. In general, Gustav Oppert's assertion that gunpowder was known to the Vedic Aryans of ancient India is no more taken seriously. Oppert justifies his case by quoting passages from the ancient Hindu text *Nitiprakashika.* Now, we know that the passages in the *Nitiprakashika* which refers to *nalika* (hand-held firearms) were later interpolations, probably added during the sixteenth century.[26] However, the Hindus of early medieval India probably knew about *agnichurna* (saltpetre or shining salt) and a rudimentary sort of gunpowder was probably used for firing pyrotechnical devices like *bans* (rockets). Whether the Bahmanis, Vijayanagara Empire and pre-Mughal Gujarat Sultanate of Bahadur Shah had access to gunpowder weapons or not is still debated. Chowdhury and Chattopadhyay assert that pre-Mughal Bengal had a tradition of using gunpowder but whether this region had access to cannons or not cannot be established at the present state of our knowledge. The Mughals in Bengal made use of both the *Rumi* (Turkish) expertise of cannon manufacturing as well as *firingi* methods (probably referring to the naval guns of the Portuguese).

Nath agrees with Jos Gommans' formulation that artillery played an insignificant role in the Mughal conduct of sieges under Babur and Humayun. Unlike the Mughals, the Ottomans during the first half of the sixteenth century made extensive use of siege artillery. During the Siege of Rhodes (1522), the Turks fired 85,000 iron and stone shots.[27] Nath shows that Babur and Humayun while conducting sieges did not use gunpowder mines and tunnelling to a great extent. In Europe, gunpowder mines were first used in 1439 to undermine the defence of Belgrade.[28] However, unlike Gommans who regards Mughal warfare as a sort of political spectacle and theatrics geared for intimidating and incorporating the enemy rather than destroying him,[29] Nath reaches the conclusion that the Mughal Army was able to conduct decisive sieges. As regards the issue whether pre-colonial South Asian warfare could be categorized as theatrics and spectacle, at least one could confidently say that the jury is still out.

British domination of South Asia was partly possible due to British maritime supremacy. British supremacy in the Indian Ocean and in the Bay of Bengal and the Arabian Sea was part and parcel of maritime supremacy of the West *vis-à-vis* the rest during the early modern era. A revolution in the design of the sailing ship occurred in west Europe during the fourteenth and fifteenth centuries. The result was a square rigged sea-going vessel which had much greater endurance and seaworthiness compared to its predecessor.[30] Throughout the sixteenth, seventeenth and eighteenth centuries, west European rulers paid for and encouraged maritime exploration, mapping and reporting that generated extensive and systematic knowledge about global geography. In contrast, the Ming emperors rejected maritime exploration and commerce and turned their society inwards after CE 1433. West Europe's discovery and exploitation of reliable sea passage throughout the globe gave it a commercial, diplomatic and military edge over other societies. By the late 1700s, west European mariners had reliable techniques for measuring latitude and longitude.[31] The Portuguese used *caravels*. Each *caravel* weighed less than 100 tons and was about 70-80 feet in length. The *caravel* drew little water and had a flat bottom.[32]

The indigenous naval response was sporadic and ineffective. Aurangzeb proposed to build a powerful navy in order to protect the Haj pilgrims from the attack of the *firingi* pirates. However, lack of expert sailors and navigators forced the emperor to abandon the project. In 1670, Siddi Yaqut Khan (an Abyssinian slave of Fath Khan, the Bijapuri governor of Konkan) was enrolled in the imperial service. The Siddi was ordered to protect the Muslim pilgrims and merchants in return for an annual subsidy of Rs. 300,000. The Siddi's fleet was, however, no match for the west European battleships in the high seas. In 1734, the British ships blocked the Mughal port of Surat.[33] After the collapse of Mughal power in west India during the first half of the eighteenth century, the Maratha Confederacy emerged as the premier power in that region. Kumar's essay argues that the Maratha Navy was no match to the west European navies in the high seas but gave a good fight in the coastal water. Kumar's argument is in tune with Anirudh Deshpande's assertion that in the coastal waters of west India, the so-called west European Naval Revolution did not prove to be significant. Finance and politics played a more important role in the collapse of the coastal Maratha navy.[34]

The principal competitors of the EIC in India at least till the early decades of the nineteenth century were the Portuguese, the Dutch and the French. The Portuguese were never very successful in the land. The Portuguese army in India remained small. In 1821, they had 6,920 troops and 296 sepoys.[35] However, Lord Mornington during the Fourth Anglo-Mysore War (1799) feared that the French troops might enter India from Egypt through Goa. So, he sent British troops to occupy Goa. Hariharan's essay shows how between 1799 and 1815, the Portuguese authorities in Goa had to tackle continuous British pressure of sending troops in Goa to garrison this enclave ostensibly to protect the Portuguese from other 'interlopers' like the French and Marathas.

The EIC's fiscal military machine introduced in India the concept of a linear border which is a straight line in the map. The concept of a clear linear frontier separating the two sovereign states is a modern Eurocentric concept. In contrast, the indigenous polities' concept of border comprised a shaded hazy buffer zone with overlapping contradictory rights by different power brokers at both the local and regional levels.[36] Michael's essay argues that the differing concepts of border resulted in the clash between the EIC, whose power was radiating out from its client state of Awadh Nawabi and the dynamic and aggressive Gorkha Kingdom which had spilled into the Terai region.

The fiscal-military Company state (derived from John Brewer's concept and applied in case of eighteenth-century South Asia by C.A. Bayly and Douglas M. Peers) continued to expand till it had gobbled up the whole subcontinent by 1849. The emergence of military-fiscalism was the product of the early modern Military

Revolution which first occurred in north-west Europe and then spread to various parts of the world. The demands of raising and maintaining drilled and disciplined gunpowder-equipped infantry armies resulted in the expansion of the state. The polity's managerial expertise rose by leaps and bounds as the 'new' armies were costly and to maintain and organize these forces, written codes and records were required. The net result was the rise of a technical culture and an expanding bureaucracy which replaced the previous household governments of the princes. The end result was the emergence of the 'modern' state.[37]

The Old Cambridge School view of H.H. Dodwell was that the decline of the Mughal *imperium* and the failure of the predatory Mughal successor states to establish a stable political environment resulted in the expansion of the EIC's holdings in India. The New Cambridge History pushes the argument that a combination of the working of international capitalism, designs of an aggressive nation state (Britain) and collaboration of certain social groups in India (mostly merchants) made imperialism possible.[38]

Long before the publication of the volumes of *New Cambridge History*, W.H. McNeill in 1982 argued that the prevalence of free market economy in the West resulted in continuous development of technology (especially military technology) and commercial growth. These two factors in turn resulted in the genesis of powerful polities in West Europe.[39] In contrast, the command economy of China resulted in social stagnation. Here, one can smell Capitalism–Socialism rivalry of the Cold War world of the 1980s.

Michael Geyer and Charles Bright in a long ranging essay assert that the Cold War which was occasionally punctuated by skirmishes between *czarist* Russia and imperial Britain in Central Asia was not a Great Game as explained by a older tradition of diplomatic historiography but the end game in a long-standing struggle over empire along the Eurasian heartland with China, India, Russia and the Ottoman Empire as the main protagonists. The British wars on the Afghan border like the Russian wars along Caucasus were not directed by London or St. Petersburg but by the imperial regional sub-centres. With its base in India, the maritime British Empire transformed itself into a continental land power and tried to dominate the geopolitical heartland of Eurasia at least till the mid-nineteenth century.[40] Wallis in his essay claims that rather than economic greed, it was the hunger for power which led the governor-generals to invent security threats at the frontiers of the British-Indian Empire which in turn resulted in continuous westward expansion of the frontiers of the colonial state. The EIC's claim in the 1830s that they were defending the Bengal Presidency at Afghanistan is somewhat equivalent to the State Department's claim in the 1960s that they were defending the democracy of United States of America in the paddy fields of Vietnam.

The British in India in accordance with the ideology of the strong nation state (those which had emerged in post-Enlightenment west Europe) established a monopoly of violence in the public sphere. In pursuit of this objective, the British EIC and later the Raj launched intensive pacification campaigns (the Raj's theorist termed it as Small War) and demilitarized the rural society. Singh's essay in this volume shows that the Raj's attempt to demilitarize Punjab and the North-West Frontier during the late nineteenth century ran into trouble.

Besides Small War, one article in this collection focuses on spy war. During Second World War, Britain's Secret Intelligence Service (SIS) operated from Calcutta. The SIS also cooperated with the intelligence section of General Head Quarter India and with Special Operations Executive for gathering intelligence in South-East Asia. The SIS besides collecting tactical battlefield intelligence for immediate wartime use also acquired information about long-term strategic intelligence which had social, cultural and political dimensions. After the war, Ho Chi Minh cooperated with the SIS in rounding off the Indian National Army (henceforth INA) leaders who had taken refuge in Hanoi and were thinking of escaping to China or Russia.[41]

FROM RAJ TO *SWARAJ*

The principal prop of the British-Indian Empire was the British-Indian Army. More than 30 per cent of the Raj's budget was consumed by its military establishment. Moreover, the army had enormous impact on indigenous society. The British-Indian Army was a long-term volunteer force. Britain lacked the manpower for policing the subcontinent. Moreover, Indians were cheaper and more suited for deployment in the terrain and climate of

South Asia compared to the British soldiers. About 20,000 small peasants were recruited in the British-Indian Army and the average size of the British-Indian Army was about 120,000 during the second half of the nineteenth century. During the two world wars, the size of the British-Indian Army crossed the mark of 1 million. Moreover, pay, pension and gratuities in cash monetized the agrarian society of large parts of north and north-west India. In the post-colonial era, India maintains a million strong army, whose rank and file are recruited mostly from the small peasants. While in the colonial era, the army was the largest employer, in independent India, army is the second largest employer after the railways.[42]

Dasgupta tackles the issue of recruitment and its effect on the loyalty of the Indian Army towards the political establishment. Dasgupta evaluates the dominant argument that civil supremacy over the military is the gift of British colonialism. A group of scholars argue that the British crafted the structure of civil domination over the military and the British legacy still continues in India. But they tend to forget that during the colonial era, the military challenged the supremacy of the civilians several times. The most famous military challenge to civil supremacy was the Kitchener-Curzon controversy.[43] Even before Lord Kitchener's (Commander-in-Chief of India) disobedience of Lord Curzon (Viceroy/Governor-General of India), the land forces in India (both British and Indian) several times challenged civilian supremacy of the Raj's officials. Just after the 1857 Uprising, the Bengal Europeans (private white army of the EIC) rebelled when they heard that as part of the post-Mutiny reorganization, the EIC's private white army would be amalgamated with the line units of the British Army. The Bengal Europeans were afraid that this amalgamation would adversely affect their prospects of promotion as the promotion slots after amalgamation would be dominated by the officers of the regular units of the British Army.[44] Even in the pre-1857 Uprising era, military mutinies rocked the EIC state. In 1824, during the First Anglo-Burma War, some Bengal infantry regiments mutinied at Barrackpur.[45] Again, the overemphasis on British legacy to explain civil supremacy over the military could not explain why in Pakistan the army dominates the civilians though Pakistan as a successor state also inherited the traditions of governance of the Raj. From the 1950s onwards, unlike India, Pakistan devoted 50 per cent of the annual budget for its defence. The Pakistan Army's influence in governance increased with time. In 1954, the army took over the defence portfolio and in 1958, a military coup occurred. A civilian government came to power in Pakistan after 1971. In 1977 another military coup occurred when General Zia-ul-Haq removed Zulfiqar Ali Bhutto from power.[46] Till this date, unlike India, Pakistan for most of the time has lived under military dictatorships.

The question of civil–military relations is linked with the issue of recruitment in the army. Recruitment in the British-Indian Army was guided by the Martial Race theory. While Stephen P. Cohen, David Omissi and Roy[47] argue that the Martial Race theory emerged in the 1880s in response to the Russian threat across north-west India, Gavin Rand, a young British historian opines that the Martial Race theory emerged in response to the 1857 Uprising. In Rand's view, the Martial Race discourse was part of a wider discourse of imperial power and government. The theory for Rand justified the British assumption that its Indian subjects were not modern citizens and legitimized the subjection of the Indians to the rule of the white men.[48]

Are we still following the Martial Race theory? Data regarding the social and regional composition of the post-colonial Indian Army is very difficult to acquire. Omar Khalidi in a pathbreaking article shows that not theoretically but practically the Indian Army is still recruiting in accordance with the Martial Race theory. The favoured 'martial races' of the Raj, like the Sikhs and the Gurkhas who constituted a disproportionate share of the British-Indian Army, are over-represented in the Indian Army. However, there has been a change. The Muslims were highly represented in the British-Indian Army but they are underrepresented in the Indian Army. Khalidi asserts that this is because majority of the Indian civilians (who belong to the majority community) and the 'Hindu' officers of the Indian Army do not trust the Muslims of India.[49]

Another reason for the decline of Muslim representation in the Indian Army could be the fact that the most favoured Muslim 'martial races' of the Raj joined the British-Indian Army from west Punjab (especially the Salt Range) and North-West Frontier Province. And these two regions had gone to Pakistan during the

1947 Partition. The Muslims from west Punjab dominate the Pakistan Army, and the Sindhis and the Baluchis are underrepresented. It seems that the Pakistan Army is also following the Martial Race theory. Again, not only the Muslims, but several regional groups like the Hindus of West Bengal and the Oriyas are also heavily underrepresented in the Indian Army. So can we argue that the Indian Army and the Indian government is anti-Bengali or anti-Oriya? However, Khalidi's overall argument that for a balanced civil–military relationship, the Indian Army needs to be representative of all the communities inhabiting the different regions of India is acceptable.

Civil–military relationship could be conflictual or based on concordance (symbiotic). In case of a conflictual civil–military relationship, the political authorities maintain their dominance through systematic suppression of lateral communication among the officer corps, purges, promotions based on personal loyalty rather than merit, and other such techniques. Stephen Biddle and Robert Zirkle argue that under Saddam Hussein, the civil–military relationship in Iraq was pathologically conflictual. The loyal Ba'thist party members were made officers. From 1970, Ba'thist political commissars and morale officers were appointed down to the battalion level (a move reminiscent of the Red Army under Josef Stalin in the aftermath of Tukachevsky purge), and micro-management of military affairs by the government was practised. Further, the government discouraged independent initiatives on part of the officer corps. The net result was a loyal military but with low combat effectiveness.[50] In India, under Defence Minister Krishna Menon, the civil-military relationship became conflictual and the Indian Army's low combat effectiveness was amply brought forward during the 1962 China War.

In contrast, concordance theory emphasizes partnership and dialogue between the civilian institutions and the military organizations. In Rebecca L. Schiff's view, concordance means agreement among the three partners: political elites, the military and the citizenry. She rightly says that for most of the time, India's civil–military relationship has been characterized by concordance.[51] Even within the democracies, the quantum of power enjoyed by the military varies from case to case. For instance in Israel, since the 1967 Arab–Israeli War, high ranking officers participate in the government meetings and enjoy considerable influence in the decision-making process. And senior officers after retirement become ministers. This trend, claims Eva Etzioni-Halevy, is detrimental to the operation of democracy.[52] In India, the three service chiefs occasionally participate in Cabinet meetings on invitation. And retired military officers rarely become governors and ambassadors. Rather, the Indian Administrative Service (IAS) officers more frequently become governors and ministers. This shows domination of the civilian bureaucracy in the body fabric of post-1947 Indian state.

The real break between the colonial army and the post-colonial army occurred in the realm of military culture. Stephen Wilson defines military culture 'to designate the military way of life and the value system associated with it; and also such military values as seem to be adopted by civilian society'.[53] Wilson continues that military culture originates in and with the military and is diffused from there to the rest of the society, but being so diffused it is radically changed.[54] The Anglo-Indian culture of colonial India was definitely militaristic. Military officers enjoyed more prestige and power than the civilian counterparts like the collectors and superintendents of police. The biggest item of expenditure in the Raj's budget was the army. And military culture dominated the upper echelons of the civil society of Britain and Anglo-Indian society in British-India. Sundaram's essay portrays the partly 'Orientalist' culture of the British officers who came to serve in India. Sundaram agrees with Douglas M. Peers that Edward Said's concept of Orientalism is inadequate to capture the totality of British military culture in India. Both Peers and Sundaram accept that racial stereotypes influenced British image of India. Sundaram's piece shows that Paul Scott was the authorized and favourite reading for the young British officers who arrived in India during Second World War. Peers argues that Walter Scott's historical romances were considered as suitable readings for the European rank and file during the mid-nineteenth century. Scott's works were available in the regimental libraries. The popular culture in Britain during the 1890s was saturated with nationalist and militarist ideas. And nationalism, militarism and empire were closely interrelated.[55]

The culture of the officer corps changed drastically in post-1947 India. Post-colonial India witnessed decline in the prestige and status of military officers. Their pay and perquisites were reduced *vis-à-vis* the IAS officers. The net result was that the social composition of the officer corps changed from being an upper class and upper middle class to lower middle class members from the small towns. In fact, the aversion of the university-educated upper middle class from the metropolitan cities to serve in the army resulted about 13,000 vacancies in the officer corps of the Indian Army in the 1990s. There cannot be any question of military values dominating the civil society of post-1947 India.[56]

INSURGENCIES, COUNTER-INSURGENCIES AND STATE BUILDING IN SOUTH ASIA

The end of Cold War has resulted in the proliferation of insurgencies. This in turn has encouraged a group of scholars to advocate aggressively that the age of inter-state war is over and the stage has been set for intra-state war, which is going to be the dominant trend of the twenty-first century.[57] Terrorism and insurgency are distinct. According to Ivan Arreguin-Toft, terrorism is mainly an urban phenomenon. Terrorism aims to inflict casualties on the non-combatants so that the latter will pressurize their government to accede to the terrorists' political demand or to delegitimize a government in order to coerce or replace it. Hence, terrorism is most effective in case of a democracy where the people have a say in the government.[58] James D. Fearon and David D. Laitin define insurgency as 'a technology of military conflict characterized by small, lightly armed bands practising guerrilla warfare from rural base area'.[59] Insurgency in this paper is defined as a civil war in which the military is deployed.

In recent times, South Asia is experiencing two types of insurgencies. While the Sikh insurgency and the ongoing insurgency in Kashmir and north-east India are partly political and partly religious, the Maoist insurgency in Nepal is secular and based on the premises of Mao's theory of revolutionary warfare. In fact, the Maoist capture of power in Nepal and the spreading of Maoist insurgency in several states of India including West Bengal in recent times prove that it is too early to write off Communism. The Maoist insurgency in large parts of South Asia and the long-drawn Liberation Tigers of Tamil Ealam movement which was finally crushed in mid-2009 prove that it would be wrong to link all the insurgencies in the post-Cold War era with religion in general and Islam in particular.

Monica Duffy Toft writes that religious outbidding results in religious war. The elites attempt to outbid each other to enhance their religious credentials and thereby gain the support they need.[60] This results in crossing the threshold and religious violence. This model could be applied in case of the Sikh insurgency in Punjab during the 1980s. The Akali Dal's position gradually hardened and it ultimately resulted in the rise of Sant Jarnail Singh Bhindranwale and consequent bloodshed. The twelve-year long Sikh insurgency (whose turning point was the Indian Army's action at the Golden Temple of Amritsar against Sant Jarnail Singh Bhindranwale in 1984) resulted in the death of 250,000 people.[61] Operation Blue Star in June 1984, which aimed at flushing out Bhindranwale's terrorists from Amritsar's Golden Temple, involved the use of six tanks, four infantry fighting vehicles, three armoured personnel carriers and heavy artillery with assault infantry. Along with Bhindranwale, about 4,172 civilians were also killed. The Sikh community viewed this operation as an assault upon the Sikh religion.[62] Duffy Toft's other argument is that Islamic theory failed to separate religion from state and this made Islam vulnerable to use by the insurgents.[63] The same could be applied to the case of Sikhism. In Sikh theology, the Gurus somewhat like the Caliphs are both spiritual and temporal heads of their followers.

Gates and Miklian's paper assesses the political aspects of the Maoist insurgency in Nepal. According to one strand of thought among the political scientists, semi-democracies (regimes intermediate between a democracy and an autocracy) are prone to insurgencies. Autocratic countries do not suddenly become democracies. They travel through a period of tortuous transition, during which mass politics mixes with authoritarian elite politics in a volatile way. Such political changes heighten the risk of civil war. Especially when authoritarianism collapses and inefficient attempts are made to establish democracy. Such an interim period of anarchy is ripe for ethnological and ideological leaders for organizing rebellions against the state.[64]

Nepal from late twentieth century onwards, when the Maoist insurgency spread, may be classified as a semi-democracy. Prior to 1991, Nepal was an absolute monarchy. The monarch devoluted power but the democratic institutions are yet to strike deep roots. Again, Nepal's political institutions are not strong and the monarch till recent times retained control over the army. The Maoist insurgency in Nepal started in 1996.[65] Between 1996 and 2006, the insurgency claimed 13,000 lives.[66] Prachanda, the leader of the Nepali Maoists, emerges in Gates-Miklian's analysis as an ideologue with mass mobilizing capacity. S. Mansoob Murshed and Gates in their article argue that not greed for capturing resources but grievances, i.e. poverty, was the principal factor behind the rise of Maoist insurgency. The class struggle of the Maoists was an extension of political struggle against the elite (Bahun-Chetri-Newari) domination of political and economic life of Nepal.[67] Fearon and Laitin agree that low per capita income favours the generation of insurgency. Recruiting young men to the life of a guerrilla is easy when the economic alternatives are worse.[68] This is applicable in case of young men joining the Maoist rank in Nepal. George Graham says that besides poverty, rising aspirations to gain political power and prestige and a spirit of vengeance against oppression by the security forces encouraged the young generation to become *Maobadis* (supporters of the Maoist insurgency).[69] Halvard Buhaug and Gates in a joint article assert that insurgency movements that endure year after year tend to encompass a broad territory. Probably influenced by Mao Tse Tung's writings, Buhaug and Gates continue that rough terrain which is difficult for the government to control is ideal for guerrilla warfare. Mountains and forest regions give advantage to rebel troops and allow them to expand their base of operations.[70] Gates and Miklian's joint article in the present volume shows that the Maoist insurgency gradually spread from the mountains of central Nepal to the forested Terai regions at the southern border of Nepal.

In recent times, American experts have rediscovered Sun Tzu and are trying to use his teachings for fighting the 'shadow warriors'. For instance, Caleb M. Bartley asserts that the terrorists use Sun Tzu's basic battlefield strategies to harass and fatigue their larger and more cumbersome enemies. Again, Al-Qaeda in accordance with Sun Tzu's teachings emphasizes the use of spies. Bartley continues that Sun Tzu could teach a lot to the Americans about how to conduct unconventional war.[71] Kautilya is not popular among such Western commentators. Roy's essay tries to build-up a case that not only Kautilya's prescriptions have influenced India's counter-insurgency policy but the *Arthashastra* is also relevant for understanding the root causes behind insurgencies in modern times. Kautilya is criticized by Brekke in an essay for failing to distinguish between internal campaigns and external violence against foreign states.[72] Actually, the *Arthashastra* argues that inter-state war and intra-state warfare are interrelated. In cases of domestic rebellion, foreign potentates are more willing to attack the neighbouring kingdoms. Using modern-day terminologies, one could argue that Kautilya unlike Clausewitz is challenging the clear-cut division between conventional war (inter-state conflict) and unconventional war (*kleinkrieg* in Clausewitz's theory). Roger Boesche rightly says that in Kautilya's framework, the policy of social justice in the long run is the best insurance against insurgencies.[73] Stuart Kinross writes that for penetrating a terrorist organization, human intelligence (HUMNIT) is more significant than signal intelligence (SIGNIT).[74] Kautilya, as Roy's essay shows, also emphasizes on HUMNIT rather than that technological superiority for fighting *kopa* (internal rebellions). Now, let us turn the focus on the essays in this volume.

NOTES

1. Mark Moyar, 'The Current State of Military History', *Historical Journal*, vol. 50, no. 1 (2007), pp. 227-8.
2. Jeremy Black, 'Military Organizations and Military Change in Historical Perspective', *Journal of Military History* (*JMH*), vol. 62 (October 1998), pp. 871, 892.
3. C. Dale Walton, 'The Strategist in Context: Culture, the Development of Strategic Thought, and the Pursuit of Timeless Truth', *Comparative Strategy*, vol. 23, no. 1 (2004), p. 93.
4. Sun Tzu, *The Art of War*, tr. with a Historical Introduction and Commentary by Ralph D. Sawyer, with the collaboration of Mei-Chun Lee Sawyer (Boulder: Westview, 1994), pp. 156-7.
5. Ivan Arreguin-Toft, 'How the Weak Win Wars: A Theory of Asymmetric Conflict', *International Security*, vol. 26, no. 1 (2001), p. 100.
6. *The Art of War*, tr. Sawyer, p. 129.
7. Roger Boesche, 'Kautilya's *Arthashastra* on War and Diplomacy in Ancient India', *JMH*, vol. 67, no. 1 (2003), p. 36.
8. *The Art of War*, p. 129.
9. Caleb M. Bartley, 'The Art of Terrorism: What Sun Tzu can Teach Us about International Terrorism', *Comparative Strategy*, vol. 24 (2005), p. 250.

10. Carl Von Clausewitz, *On War*, ed. and tr. Michael Howard and Peter Paret (1976, rpt., Princeton: Princeton University Press, 1989), pp. 226-44.
11. *The Art of War*, pp. 130, 142.
12. Gregory M. Reichberg, 'Preventive War in Classical Just War Theory', *Journal of the History of International Law*, vol. 9 (2007), p. 15.
13. Ibid., p. 18.
14. G. Scott Davis, 'Introduction: Comparative Ethics and the Crucible of War', in Torkel Brekke (ed.), *The Ethics of War in Asian Civilizations: A Comparative Perspective* (London: Routledge, 2006), p. 5.
15. Norman Solomon, 'The Ethics of War in Judaism', in Brekke (ed.), *The Ethics of War in Asian Civilizations*, p. 68.
16. Torkel Brekke, 'Between Prudence and Heroism: Ethics of War in the Hindu Tradition', in Brekke (ed.), *The Ethics of War in Asian Civilizations*, pp. 132, 135.
17. Hans Kung, 'Religion, Violence and "Holy Wars"', *International Review of the Red Cross*, vol. 87, no. 858 (2005), pp. 255, 264-5.
18. Davis, 'Introduction', in Brekke (ed.), *The Ethics of War in Asian Civilizations*, p. 7.
19. John Kelsay, 'Islamic Tradition and the Justice of War', in Brekke (ed.), *The Ethics of War in Asian Civilizations*, p. 103.
20. Katherine K. Young, 'Hinduism and the Ethics of Weapons of Mass Destruction', in Sohail H. Hashmi and Steven P. Lee (eds.), *Ethics and Weapons of Mass Destruction: Religious and Secular Perspectives* (Cambridge: Cambridge University Press, 2004), pp. 287-90.
21. Balbinder Singh Bhogal, 'Text as Sword: Sikh Religious Violence Taken for Wonder', in John R. Hinnells and Richard King (eds.), *Religion and Violence in South Asia: Theory and Practice* (Abingdon, Oxon: Routledge, 2007), pp. 110, 112, 114.
22. Solomon, 'The Ethics of War in Judaism', in Brekke (ed.), *The Ethics of War in Asian Civilizations*, p. 41.
23. Dirk H.A. Kolff, 'A Millennium of Stateless Indian History?', in Rajat Datta (ed.), *Rethinking a Millennium: Perspectives on Indian History from the Eighth to the Eighteenth Century: Essays for Harbans Mukhia* (Delhi: Aakar, 2008), pp. 51-67; Kolff, 'End of an *Ancien* Regime: Colonial War in India, 1798-1818', in J.A. DeMoor and H.L. Wesseling (eds.), *Imperialism and War: Essays on Colonial Wars in Asia and Africa* (Leiden: E.J. Brill, 1989), pp. 22-49; Burton Stein, *A History of India* (1998, rpt., New Delhi: Oxford University Press, 2004); Stephen Peter Rosen, 'Military Effectiveness: Why Society Matters', *International Security*, vol. 19, no. 4 (1995), pp. 5-31.
24. G.J. Bryant, 'Asymmetric Warfare: The British Experience in Eighteenth-Century India', *JMH*, vol. 68, no. 2 (2004), pp. 431-69; Stewart Gordon, 'Zones of Military Entrepreneurship in India: 1500-1700', in Gordon, *Marathas, Marauders, and State Formation in Eighteenth Century*

India (1994, rpt., New Delhi: Oxford University Press, 1998), pp. 182-208; Martin van Creveld, 'Thoughts on Military History', *Journal of Contemporary History*, vol. 18, no. 4 (1983), p. 551.

25. Dennis E. Showalter, 'Caste, Skill and Training: The Evolution of Cohesion in European Armies from the Middle Ages to the Sixteenth Century', *JMH*, vol. 57 (July 1993), pp. 407-10.
26. *Nitiprakasika*, ed. Gustav Oppert (1882, rpt., New Delhi: Kumar Brothers, 1970).
27. Christopher Duffy, *Siege Warfare: The Fortress in the Early Modern World: 1494-1660* (1979, rpt., London and New York: Routledge, 1996), p. 192.
28. Duffy, *Siege Warfare: 1494-1660*, p. 11.
29. Jos Gommans, *Mughal Warfare: Indian Frontiers and High Roads to Empire, 1500-1700* (London and New York: Routledge, 2002).
30. John Law, 'On the Social Explanation of Technical Change: The Case of the Portuguese Maritime Expansion', *Technology and Culture*, vol. 28, no. 2 (1987), p. 235.
31. John F. Richards, 'Early Modern India and World History', *Journal of World History*, vol. 8, no. 2 (1997), pp. 198-9.
32. Law, 'On the Social Explanation of Technical Change', p. 238.
33. Naim R. Farooqi, 'Moguls, Ottomans, and Pilgrims: Protecting the Routes to Mecca in the Sixteenth and Seventeenth Centuries', *International History Review*, vol. 10, no. 2 (1988), pp. 200-20.
34. Anirudh Deshpande, 'Limitations of Military Technology: Naval Warfare on the West Coast, 1650-1800', *Economic and Political Weekly*, vol. 27, no. 17 (1992), pp. 900-4.
35. L.A. Rodrigues, 'The Portuguese Army of India', *Journal of Indian History*, vol. LVII, pt. I (April 1979), p. 82.
36. Jeremy Black, *Rethinking Military History* (London: Routledge, 2004), p. 57.
37. Marshall Poe, 'The Consequences of the Military Revolution in Muscovy: A Comparative Perspective', *Comparative Studies in Society and History* (*CSSH*), vol. 38, no. 4 (1996), pp. 603-18.
38. *The New Cambridge History of India*, II: I; C.A. Bayly, *Indian Society and the Making of the British Empire* (Cambridge: Cambridge University Press, 1988).
39. Alex Roland, 'Technology and War: The Historiographical Revolution of the 1980s', *Technology and Culture*, vol. 34, no. 1 (1993), pp. 117-19.
40. Michael Geyer and Charles Bright, 'Global Violence and Nationalizing Wars in Eurasia and America: The Geopolitics of War in the Mid-Nineteenth Century', *CSSH*, vol. 38, no. 4 (1996), pp. 632-3, 649-51.
41. Richard J. Aldrich, 'Britain's Secret Intelligence Service in Asia during the Second World War', *Modern Asian Studies* (*MAS*), vol. 32, no. 1 (1998), pp. 179-217.

42. Kaushik Roy, *Brown Warriors of the Raj: Recruitment and the Mechanics of Command in the Sepoy Army, 1859-1913* (New Delhi: Manohar, 2008).
43. Stephen P. Cohen, 'Issue, Role, and Personality: The Kitchener-Curzon Dispute', *CSSH*, vol. 10 (1967-8), pp. 337-55.
44. Peter Stanley, '"Dear Comrades": Barrack Room Culture and the "White Mutiny" of 1859-60', *Indo-British Review*, vol. 21, no. 2 (1996), pp. 165-75.
45. Douglas M. Peers, '"The Habitual Nobility of Being": British Officers and the Social Construction of the Bengal Army in the Early Nineteenth Century', *MAS*, vol. 25, no. 3 (1991), pp. 546-7.
46. Shirin Tahir-Kheli, 'The Military in Contemporary Pakistan', *Armed Forces and Society*, vol. 6, no. 4 (1980), pp. 639-53.
47. Stephen P. Cohen, *The Indian Army: Its Contribution to the Development of a Nation* (1971, rpt., New Delhi: Oxford University Press, 1991), pp. 32-56; David Omisi, *The Sepoy and the Raj: The Indian Army, 1860-1940* (Houndmills, Basingstoke: Macmillan, 1994), pp. 1-46; Kaushik Roy, 'Recruitment Doctrines of the Colonial Indian Army: 1859-1913', *Indian Economic and Social History Review*, vol. 34, no. 3 (1997), pp. 321-54.
48. Gavin Rand, '"Martial Races" and "Imperial Subjects": Violence and Governance in Colonial India, 1857-1914', *European Review of History*, vol. 13, no. 1 (2006), pp. 1-20.
49. Omar Khalidi, 'Ethnic Group Recruitment in the Indian Army: The Contrasting Cases of Sikhs, Muslims, Gurkhas and Others', *Pacific Affairs*, vol. 74, no. 4 (Winter 2001-2), pp. 529-52.
50. Stephen Biddle and Robert Zirkle, 'Technology, Civil–Military Relations, and Warfare in the Developing World', *Journal of Strategic Studies* (*JSS*), vol. 19, no. 2 (1996), pp. 171-212.
51. Rebecca L. Schiff, 'Concordance Theory: A Response to Recent Criticism', *Armed Forces and Society*, vol. 23, no. 2 (1996), pp. 277-83.
52. Eva Etzioni-Halevy, 'Civil-Military Relations and Democracy: The Case of the Military-Political Elites' Connection in Israel', *Armed Forces and Society*, vol. 22, no. 3 (1996), pp. 401-17.
53. Stephen Wilson, 'For a Socio-Historical Approach to the Study of Western Military Culture', *Armed Forces and Society*, vol. 6, no. 4 (1980), p. 528.
54. Wilson, 'Western Military Culture', p. 529.
55. Douglas M. Peers, '"Those Noble Exemplars of the True Military Tradition": Constructions of the Indian Army in the Mid-Victorian Press', *MAS*, vol. 31, no. 1 (1997), pp. 109-42.
56. P.R. Chari, 'Civil-Military Relations in India', *Armed Forces and Society*, vol. 4, no. 1 (1977), p. 25.

57. Stuart Kinross, 'Clausewitz and Low-Intensity Conflict', *JSS*, vol. 27, no. 1 (2004), p. 45.
58. Arreguin-Toft, 'How the Weak Win Wars', p. 103.
59. James D. Fearon and David D. Laitin, 'Ethnicity, Insurgency and Civil War', *American Political Science Review* (*APSR*), vol. 97, no. 1 (2003), p. 75.
60. Monica Duffy Toft, 'Getting Religion? The Puzzling Case of Islam and Civil War', *International Security*, vol. 31, no. 4 (2007), p. 103.
61. Arvind Mandair, 'The Global Fiduciary: Mediating the Violence of Religion', in Hinnells and King (eds.), *Religion and Violence in South Asia*, p. 220.
62. Namrata Goswami, 'India's Counter-Insurgency Experience: The "Trust and Nurture" Strategy', *Small Wars and Insurgencies*, vol. 20, no. 1 (2009), p. 78.
63. Duffy Toft, 'Getting Religion?', p. 109.
64. Havard Hegre, Tanja Ellingsen, Scott Gates and Nils Petter Gleditsch, 'Towards a Democratic Civil Peace? Democracy, Political Change, and Civil War, 1816-1992', *APSR*, vol. 95, no. 1 (2001), pp. 33-4.
65. S. Mansoob Murshed and Scott Gates, 'Spatial-Horizontal Inequality and the Maoist Insurgency in Nepal', *Review of Development Economics*, vol. 9, no. 1 (2005), p. 121.
66. George Graham, 'People's War? Self-Interest, Coercion and Ideology in Nepal's Maoist Insurgency', *Small Wars and Insurgencies*, vol. 18, no. 2 (2007), p. 231.
67. Murshed and Gates, 'Spatial-Horizontal Inequality and the Maoist Insurgency in Nepal', p. 124.
68. Fearon and Laitin, 'Ethnicity, Insurgency, and Civil War', p. 80.
69. Graham, 'People's War?', pp. 234-46.
70. Halvard Buhaug and Scott Gates, 'The Geography of Civil War', *Journal of Peace Research*, vol. 39, no. 4 (2002), p. 422.
71. Bartley, 'What Sun Tzu can Teach Us', pp. 237, 245.
72. Brekke, 'Between Prudence and Heroism', in Brekke (ed.), *Ethics of War in Asian Civilizations*, p. 121.
73. Boesche, 'Kautilya's *Arthashastra* on War and Diplomacy in Ancient India', p. 30.
74. Kinross, 'Clausewitz and Low Intensity Conflict', p. 51.

CHAPTER 1

Asian Writings on Warfare: A Comparative Study of Sun Tzu and Kautilya

R.K. TANDON

INTRODUCTION

The Art of War by the Chinese thinker Sun Tzu and the *Arthashastra* of Kautilya, the Indian strategic thinker, are the two classic works on the study of war and statecraft. The period of both these thinkers was the period characterized by constant warfare in China and India. Both of them realized the importance of the study of war as a major element of statecraft. However, the backgrounds of these two theorists of war are not entirely similar. Sun Tzu did not play any role in establishing a new dynasty by destroying the old but instead played an important role in strengthening the state where he was leading the army. Kautilya on the other hand played a major role in the downfall of the Nanda dynasty and in the establishment of the Mauryan dynasty.

Sun Tzu considers war a matter of vital importance to the state and takes it as a road to survival or to ruin and therefore gives more importance to the details of methods/techniques of winning the battle. He takes into account every possible element to win the battle such as military organization, leadership and battle tactics. The thirteen chapters of the *Art of War* almost leave no aspect of battle tactics untouched.

Kautilya's *Arthashastra* is a study of politics, wealth, and ways of acquiring and maintaining power. He advocates the use of indirect methods to gain one's objectives and propagates the theory of silent war by adopting and using secret agents and women as weapons of war. He also discusses different methods to demoralize the enemy and gain the confidence of conquered subjects. He considers the use of force as the last resort to achieve the goal. Kautilya emphasizes on the different elements of statecraft including inter-state relationship. Kautilya's *Arthashastra* is often described as a practical manual of statecraft.

This paper addresses the suitability of approaching these two writers from the Asian strategic affair perspective in addition to studying their historical backgrounds. The primary objective of this paper is to have a comparative study of two thinkers of the ancient period, in the light of different aspects relating to war. The secondary objective is in line with the broader understanding of history, i.e. it is to be studied in order to gain knowledge for the future. Sun Tzu's writings have been utilized in modern management, however, Kautilya's work has not gained that much attention in terms of modern management application.

Fourth century BCE witnessed an unstable political environment, both in India and China. It was a period of constant wars in both countries. Sun Tzu in China and Kautilya in India are two thinkers of the era who wrote treatises of statecraft and warfare. Though their treatment of the subject was different, at places they had many commonalities. Today their work is considered as classic work in the area of warfare and statecraft. This paper analyses the subject in three sections. The first deals with Sun Tzu's thoughts, second with Kautilya's contribution and the third section attempts to compare the thought patterns of the two thinkers.

SUN TZU'S THEORY OF WARFARE

Sun Tzu, whose first name was Wu, was a general of King Ho-lu of Wu state of China, during Spring and Autumn period which is taken as the golden period of Ancient China. This was the period of many famous Chinese philosophers like Lao-Tzu and Confucius whose philosophies of Taoism and Confucianism, respectively, influenced Chinese society. China during this period had many small states. In 564 BCE, the eleven major states of China signed a non-aggression pact and temporary peace was established. This peace was broken by the king of Wu in 506 BCE, who was not a signatory to this pact and Sun Tzu was at the head of the army of Wu. China very soon witnessed the rise and fall of the Wu state and entered into the Warring States period, which lasted from 403-221 BCE.[1] This period witnessed constant warfare in China.

Sun Tzu was a native of Wu. He had a good education and composed military treatises which were noticed by the ruling class. On the recommendation of Wu Tzu-Hsu, the advisor to King Ho-lu of Wu state, he was appointed head of the army, after he performed a test on the ladies of the palace to demonstrate his theory of managing soldiers. Sun Tzu fought many battles as head of the army of Wu state and his name spread among the feudal lords of China.

Sun Tzu wrote his book in 13 chapters. The book present his overall concept about how to win a war. He begins by stating that the art of war is of vital importance to the state. It's a matter of life and death and hence cannot be neglected. The art of war is governed by five constant factors, i.e. Moral Influence (confidence in the ruler); Weather (day, night, and season); Terrain; Command (wisdom, sincerity, courage, etc.) and Organization doctrine, supply line, provision of principal items used by the army).[2] Sun Tzu emphasizes that the commander who knows these five factors will get victory.

Sun Tzu, while discussing the techniques of waging war, talks about the numerical strength of the army and about military expenditure. He says that one thousand pieces of metallic currency per day may be required to raise an army of one hundred thousand men.[3] He recommends that a commander may take equipment from own country but feeding his force from the enemy territory not only make his force sufficient in both arms and provision but also drastically reduces military expenditure of the state by putting the burden of logistics on the enemy territory.[4] According to him the spoils of war should be used to reward the troops. He condemns the prolongation of war and says that the important thing in a military operation is victory.[5] He recommends subduing the enemy without fighting and says that the supreme importance in war is to attack the enemy's strategy. In his opinion, the highest form of generalship is to disrupt the enemy's plan and prevent the junction of the enemy forces.[6] He further advocates that the most successful general is one who achieves his ends without battle or with minimum losses.[7]

While talking about the tactical disposition of army, he opines that the skilful commander takes up a position in which he cannot be defeated and utilizes every opportunity to master his enemy.[8] He says that order or disorder depends on the commander's management of warfare. He talks about the indirect methods to win a battle. He advocates two methods of war—direct and indirect; and says that the combination of these two methods will results in a series of manoeuvres.[9] He also discusses the various types of manoeuvres. Sun Tzu advocates methods to impose one's own will on the enemy, or in other words, to force the enemy to fight on one's own terms. He warns the commander not to repeat the same tactic, which initially had generated victory, but instead adopt a new one each time. Discussing the importance of manoeuvring, he says that war is based on deception. Move when it is advantageous and create changes in the situation by dispersal and concentration of forces.[10] Manoeuvring with an army is advantageous; with an undisciplined multitude, it is most dangerous.[11]

Sun Tzu's *Art of War* teaches not to rely on the likelihood of enemy not coming but on one's own readiness to confront him.[12] He also discusses the different situations regarding war on mountainous terrain, riverine valleys, high ground, and so on; and advices to be on the favourable battleground before the enemy and guard one's supply line. He gives more emphasis on the importance of the battleground. He mentions nine varieties of grounds and discusses the advantages and disadvantages of each type of terrain. Attacking using fire is also described by him and five different methods of attacking with fire are given by him (burn the soldiers in their camp; burn the stores; burn the baggage train; burn the arsenals and magazine; hurling/dropping the fire [firebombs?] among the enemy). Great emphasis has been given in his work on the use of secret agents. He mentions five types of spies namely, the local, inside, reverse, dead and living. He says only a brilliant ruler or a wise general, who can use highly intelligent agents for espionage, is sure of great success. This is essential for military operations and the armies depend on the functioning of an effective spy system during campaigns.[13]

KAUTILYA ON WAR AND DIPLOMACY

Kautilya or Vishnugupta or Chanakya are the three names given to a man who lived in fourth century BCE and was considered by historians as responsible for the overthrow of the Nanda dynasty in India and the establishment of the Mauryan dynasty. He became the prime minister of Chandragupta Maurya and wrote a book named *Arthashastra* between 321 BCE and 300 BCE.[14] The author in his book describes his philosophy for society and the state. He discussed civil and military administration, warfare, and diplomacy. Kautilya's *Arthashastra* is considered as the classical treatise on polity.[15] The history of ancient India is incomplete without the study of Kautilya's *Arthashastra*. Apart from the social and economic aspects of society, the *Arthashastra* gives detailed instructions on the control of the state, the organization of national economy and the conduct of war.[16]

He advocates a philosophy of state having seven elements, i.e. the king, the minister, the country, the fort, the treasury, the army and the allies. Kautilya's study of the interrelationship of all these seven elements gives an insight into his state-centric philosophy. Kautilya is of the opinion that the king is the pivot of the whole state and every important affair of the state revolves around him. He then goes on to discuss the qualities of the king, his duties, his daily routine, and the proper qualities of who is most qualified to be the king. Then, he discusses about the minister; mentions his qualification and duties and describes him as the main executor of the state's policies. While describing the country he tells what type of state could be developed and analyses the basic elements of the polity. The fort according to him is a place where both the money and army can be placed. He gives a detailed description about construction of the fort. The treasury is the next and most important element of the state. Kautilya says that the king should always keep an eye on his treasury and must have enough money to meet any situation.

Kautilya elaborates his theory of inter-state relationship which is known as *mandala* or the theory of circle of states. The king (*vijigishu*) is the centre of the *mandala*, i.e. the ruler with a desire to expand his empire. The *vijigishu*, his ally state and his allied state's ally are the three primary kings constituting an important aspect of the *mandala*. Besides the enemy of the *vijigishu*, the *madhyama* (middle king) and the *udasina* (neutral king) are other vital elements of the *mandala*. The *madhyama*'s territory lies between the *vijigishu* and his enemy. The *madhyama* and the *udasina's* attitudes are important for the *vijigishu* pursuing expansionist policy. The four primary circles within the *mandala* are:

- Circle I: *vijigishu*, his ally, and his ally's ally
- Circle II: the enemy, his friend, and his friend's friend
- Circle III: *madhyama*, his friend, and his friend's friend
- Circle IV: the *udasina*, his friend, and his friend's friend.[17]

To formulate his foreign policy the king, says Kautilya, should adopt the formula of six *gunas* or policies, i.e. *sandhi* (making a treaty), *vigraha* (war), *asana* (neutrality), *yana* (marching on an expedition), *samsaya* (seeking shelter with another king), and *dvaidhibhava* (double policy of *sandhi* with one king and *vigraha* with another at the same time).[18] Kautilya emphasizes the use of *sandhi* when the *vijigishu* is weaker than the enemy; and when stronger than enemy then follow *vigraha*; when equal in strength with the enemy then *asana*; when stronger than enemy then *yana*; when helpless in front of enemy then *samsaya* and when you expect help from another source then to deceive the enemy and wage the policy of *dvaidhibhava*. To implement the six *gunas* in different circumstances, Kautilya recommends adopting the four *upayas*, i.e. *sama* (negotiations), *dana* (gift/bribery), *bheda* (divide and rule) and *danda* (use of coercive power through military assets).[19] Next, Kautilya discusses the tools of *danda* and *bheda*.

He talks about the army in great detail. He describes the four wings of army—the elephant, the chariot, the cavalry and the infantry. He gives the duties and functions of the superintendents of each one of them. He discusses in detail the weapons and classifies them according to their use whether in battles, in forts or in destruction of strongholds of the enemy.[20] Kautilya also talks about the battleground and advices 'even' ground with water reservoirs for use of chariots; ground having small stones and trees and small pebbles not suited for deployment of horses and ground having big tree trunks, stones, trees, free from thorns for deployment of the infantry.[21] He also discusses different form(s) of battle arrays and their formations. The army in

Kautilya's format can be of six types: *maula bala* (standing army); *bhrata bala* (territorial army raised for a particular campaign or mercenaries); *sreni bala* (organized militia); *aatvika bala* (tribal contingents provided by the allied tribal chief); *mitra bala* (friendly troops or those of allies) and the *amitra bala* (alien forces, who happen to fight with the king for their own reasons. Probably, they were deserters from the enemy army).[22] Kautilya continues that the different constituents of the army had their own advantages as well as disadvantages.[23]

Kautilya mentions three types of wars namely, the *prakasya yuddha* (open fight); *kutayuddha* (concealed war) and *tusnim yuddha* (silent war).[24] When a king is superior than the enemy in all aspect(s), then the king, advises Kautilya, should go for open war, i.e. declare a war and then confront the enemy with the regular force in a battlefield. When the king is not superior than his enemy and the terrain and season are unfavourable to him, then the king should wage concealed war, i.e. use deceptive methods to fight. Finally, when the king is extremely weak *vis-à-vis* the hostile power, then the former should wage a silent war, i.e. use secret methods to win the war. Some of the methods of silent war are making the enemy weak by the use of secret agents, cutting the line of communication of the enemy, destroying the stores of the enemy, creating unstable conditions in enemy area, killing the enemy king, his *senapati* (general/commander) or important ministers, creating mistrust among the king, *senapati*, ministers or even between king and the queen and his sons by use of secret agents or even by the use of women.

Kautilya discusses at great length about secret agents. There may be two types of spies—the agents based at one place and roving agents. The spies are dressed as monks, householders or merchants. Kautilya advocates the use of secret agents for assassination, poisoning the enemies of the state. Kautilya's greatest breakthrough is to advocate the use of women mendicants as secret agents. He is in the favour of the use of beautiful women as *vishakanyas*, i.e. secret agents, and also using them as a weapon to kill the enemy king and other important officials. Kautilya says any information confirmed by different spies shall be considered as of value by the king.[25] He also mentions about double agents. He says that the family of the spies should be kept as hostages in order to prevent the former from betraying their state's secrets to the enemy. He talks about the establishment of small secret cells and continues that the members of the different cells should remain unknown to each other. Hence, if any one cell is busted by enemy counter-espionage, the other cells would continue to function in the enemy kingdom.

SUN TZU AND KAUTILYA: A COMPARATIVE ANALYSIS

Both Sun Tzu and Kautilya were products of a period when constant war occurred in China and India. Both were strategic thinkers and emphasized the importance of statecraft. Both had the objective of how to win wars cheaply without a lot of bloodshed. Kautilya played an important role in destroying the Nandas and establishing the dynasty of Mauryas, whereas Sun Tzu played no such role in initiating dynastic changes. The *Arthashastra* of Kautilya covers a very broad aspect of the art of war and inter-state relations whereas Sun Tzu confined himself mostly to matters related to the art of war. Both the thinkers give importance to the art of war but their treatment of the subject is different. Sun Tzu says that the art of war is of vital importance to the state and then goes into its details. Kautilya gives his philosophy of the state and art of war is one part of it, he then concentrates on the interrelationship between these two concepts. However, both of them agree that the goal of war is to get 'victory' with minimum bloodshed.

Kautilya and Sun Tzu give much importance to espionage activity. According to Kautilya the king must give high priority for creating a secret service with spies, secret agents and other specialists because they are necessary for ensuring the security of the kingdom and for furthering the objectives of expansion by conquest.[26] Sun Tzu also emphasizes the importance of spies by advocating that foreknowledge about the opponents enables a military leader or the government to overcome the other, and achieve extraordinary accomplishments.[27] Sun Tzu mentions five kinds of spies—the local spy, i.e. those hired from the enemy's country; inside spy, i.e. the enemy officials whom the *vijigishu* employs; double or reverse spym, i.e. those hired from enemy spies; dead or expendable spy, i.e. the ruler's own spy who was given fabricated information to be transmitted to the enemy. Such a spy was made to fall deliberately into enemy hands so that the false information he carries would misinform/disinform the enemy. Since the enemy would kill such a spy sooner or later, he is regarded as an expendable element; and living spy, i.e. those who return to the establishment after the conclusion of a mission.[28] Sun Tzu is of the opinion that when all five types of spies work simultaneously and none knows the others' methods of operation, then the sovereign is most successful.[29]

Kautilya mentions two categories of spies—those based at one place, i.e. *sanstha*, and roving spies, i.e. *sanchar.*[30] The *sanstha* are of five types: *kapatik* (the student as spy); *udasthita* (the recluse as spy); *grihapatik* (the householder as spy); *vaidehaka* (merchants as spy) and *tapas* (ascetic as spy). *Sanchar* spies can be of many types: *satri* spy means an orphan who is looked after by the state; *tikshma* refers to a brave agent who for money fights with anybody without any concern for his own safety or life; *rasada* is one who is cruel and is recruited as a poisoner; *parivrajike* is a wandering nun with shaven head.[31] In other words Kautilya is of the opinion that spies should be recruited from all sections of the society. All these roving spies in Kautilya's paradigm, have access to all the nooks and corners of a kingdom and hence it is easy for them to collect information. The information thus gathered from roving spies is collated together in the establishment of the spies located at one place and then transmitted by code to the *vijigishu.*[32]

Sun Tzu is of the view that if any information is known before a spy reports then one who told about it and the spy should both be killed. Kautilya says if wrong information is given by the spy, he should be punished severely. He is also of the view that if the same information is given by three unknown spies, then it should be considered as correct; and if these three spies differ frequently then they be punished in secret or dismissed.[33] Both Kautilya and Sun Tzu talk about the double agent. Sun Tzu says that double agents be given huge bribes and treated well. Kautilya says servants of the enemy be made double agents and the *vijigishu* should hold their families as hostage.

Overall, Kautilya's treatment about the use of spy is more detailed than Sun Tzu. The former gives much emphasis on employing women as spy and using them as weapon to gather information and kill the important persons of enemy, including the enemy king. He discusses in detail about the use of *vishkanyas.*

He says women and prostitutes should be used to gather information from enemy officers. He talks about counter-espionage and advocates employing roving and non-roving spies to counter-enemy spies. The spies should be used to find out the loyalty of own officers and common man of the society. Sun Tzu only advocates the use of spies to gather information about the enemy to ensure proper planning, whereas Kautilya goes further and says to use the spies to gather information about the enemy and also use them as agents to wage a psychological war in enemy area and to paralyse the enemy before the actual battle takes place by creating mistrust among higher officials such as between the king, his sons, his brothers, even his wife, minister, *senapati*, etc. Both Sun Tzu and Kautilya to a greater extent, want to take full advantage of the fifth column.

Sun Tzu discusses the qualities of a commander and also tells how a commander can command the loyalty of his men. He says that if a commander views the soldiers as his beloved children then the latter would willingly die for the commander.[34] But Sun Tzu remains silent on the issue of how to test the integrity of commanders. Kautilya on the other hand prescribes tests to be administered on the ministers and *senapatis* to test their integrity and loyalty towards the state. He suggests that the ministers and *senapatis* be humiliated by the *vijigishu* and then a bribe be offered to them by secret agents in order to shatter their integrity and honesty. The officers who pass these tests will be allowed to remain in office.

Both Sun Tzu and Kautilya have somewhat similar views on sieges. Sun Tzu is of the view that sieges are wasteful in both cost and lives of soldiers, and being time consuming results in abdication of the initiative of campaigns.[35] He also says that siege of a city is to be undertaken only as a last resort.[36] Kautilya also says that sieges are costly exercises, entail losses of men, heavy expenditure and long absence from home. But he also adds that besides laying siege, the commander of the besieging force may wage a psychological war with the aim of frightening the people inside the fort and simultaneously boosting the morale of his own soldiers.[37] Both Sun Tzu and Kautilya agree that fire be used in siege warfare. Sun Tzu gives five different purposes for which fire can be used—to burn the people, the store, the supplies, the equipment and the weapons. He also highlights the precautions to be taken while attacking with fire. Kautilya also recommends the use of fire in capturing the fort and describes in detail how fire balls be prepared but also says that when a fort can be captured by other means then fire should not be used, as fire cannot be trusted and it can also destroy the people, grain, cattle, gold, and raw materials. The capture of a enemy fort with its property totally destroyed is a source of further loss for the *vijigishu.*[38]

Sun Tzu mentions the names of some weapons like bows and arrows, spears and shields, protective mantles, breast-plates,[39] helmets, arrows and crossbows, lances, hand and body shields, etc.[40] But he remains silent on the organization of armoury and the duties of the various officials of that department. Kautilya tells us that the armoury should be under the charge of an *adhyaksha* who is responsible for the manufacture and upkeep of the weapons and maintains full record of different types of weapons. He is also to be responsible for keeping balance between the production and supply of weapons.[41] He also says that each weapon must have the king's emblem on it.[42] He classifies the equipment as weapons with sharp points, bows and arrows, swords, cutting weapons, stone weapons, armours and other protective weapons, fixed equipment for the defence of forts, and so on.[43] As part of biological warfare, Kautilya also describes the use of many chemicals/recipes to be used as weapons of mass destruction.[44]

There is no reference in Sun Tzu's work about the training of soldiers whereas Kautilya emphasizes the training of different wings of army; at sunrise every day to be inspected by the king as part of his daily routine.[45] Sun Tzu and Kautilya give much importance to terrain and weather. According to Sun Tzu the terrain is to be assessed in terms of distance, difficulty or ease of travel, and safety.[46] He describes six types of terrains and is of the opinion that all positive and negative points of the terrain are to be taken into account by the commander. He says know your ground, know your weather; your victory will then be total.[47] His opinion is that a general who is unable to use ground properly is unfit to command.[48] Kautilya describes different types of ground suitable for different wings of the army and for the construction of the military camp. He is of the view that the battleground be selected after proper survey and suitable for the type of array to be adopted during battle. He also describes different types of battleground suitable for different arrays.[49] Kautilya mentions that there should be a mountain or a forest fort in the rear as place of retreat and for holding reserves for the fighting forces.[50]

Sun Tzu and Kautilya both recommend use of deception in war. Sun Tzu is of the view that a military operation involves deception. Sun Tzu advises the commander that even though competent, he must appear incompetent in order to deceive the enemy. Though his force is effective, make it appear as ineffective in order to make the enemy careless.[51] He advocates using all possible methods to confuse the enemy and thus demoralize him. Kautilya also advocates the concept of *kutayuddha*, and gives details about how to wage it. He says that the enemy be deceived by all possible methods in the battle. Sun Tzu advocates attacking the mind of the enemy before the actual battle.[52] The silent war of Kautilya is designed to achieve this objective. He not only recommends creating mistrust among the top brass of the enemy state but even kill the enemy king/minister/*senapati* before the battle and cut the supply line to demoralize the enemy.

Both Sun Tzu and Kautilya display humanitarian attitude towards the enemy in open battle. Sun Tzu supports giving a way of escape to a surrounded enemy force and recommends not obstructing defeated enemy army returning home.[53] Kautilya's *Arthashastra* mentions that when victory is certain in a *prakasya yuddha*, then those enemy soldiers who have fallen, those who have turned their back on combat, those who have surrendered and those who have abandoned their weapons and armour should not be killed.[54]

CONCLUSION

The works of both Sun Tzu and Kautilya are the established classics of art of war and statecraft and are essential part of military training institutions all over the world. Apart from this, Sun Tzu has become one of the pioneer gurus of modern business management planning. Sun Tzu gives much importance to detailed planning. This principle is very much applicable to business planning today and the larger companies are setting up separate corporate planning departments.[55] As victory in war, to a very large extent, depends on the capabilitics of the commander, the success of a business house depends upon the capability of its chief executive officer/managing director or the like. The strategies given in the work of Sun Tzu are very much applicable to the modern business/market strategies. Sun Tzu today is studied by scholars trying to understand and provide consultancy to large and small businesses acting in new environments and a quite a few of them take recourse to Sun Tzu's basic model of situation appraisal and also of his views on evaluating strategies. Kautilya's, on the other hand, has received less patronage from business scholars, as is evident from lack of business management books based on his principles. Kautilya too deserves a higher place in the modern Indian business management education as his views on leadership and man-management are definitely at par with Sun Tzu's. Kautilya's views on leadership qualities, conduct of state and economy and the organization and the tests for loyalty and counter-measures for checking sabotages are a treasure for any management student/researcher. The *Arthashastra* apart from covering topics related to war, politics, inter-state relations, also covers in detail law, accounting system taxation, trade, sale and purchase. Planning, leadership, training, awards and punishments are some of the other topics in *Arthashastra* that provide guidelines and principles of modern (business) management education and practice.

To conclude, the theories propagated by both Sun Tzu and Kautilya more than two millennia ago are still relevant in present age of Information Technology. All our activities of life including warfare are governed and depend on information technology, a concept which these two thinkers have mentioned long ago. The silent war of Kautilya and Sun Tzu's unorthodox mechanisms for winning war are to a great extent similar to the theory of Indirect Approach as propounded by Captain Basil Liddell-Hart, a prominent British thinker of armoured warfare during the first half of twentieth century. The guerrilla operations in history (Sun Tzu and Mao's guerrilla warfare and Kautilya and *ganimi kava* of the Marathas) have their roots in the writings of these two strategic thinkers of ancient Asia.

NOTES

1. *Dictionary of World History* (London: Brockampton Press, 1994), p. 124.
2. S.B. Griffith, *Sun Tzu: The Art of War* (London: Oxford University Press, 1963), p. 63.
3. Ibid., p. 72.
4. T. Cleary, *Mastering the Art of War* (Boston: Shambhala South Asia Edition, 2004), p. 111.
5. T. Cleary, *The Art of War* (Boston: Shambhala Publication, 2005), p. 30.
6. L. Giles, *On The Art of War*, The Project Gutenberg eBooks, 1994, p. 42.
7. A. Toffler and H. Toffler, *War And Anti-War: Survival at the Dawn of the 21st Century* (Dehradun: Natraj Publishers, 2006), p. 41.
8. Griffith, *Sun Tzu: The Art of War*, p. 87.
9. Giles, *On The Art of War*, p. 56.
10. Ibid., p. 106.
11. Ibid., p. 71.
12. Ibid., p. 85.
13. Cleary, *The Art of War*, p. 170.
14. R. Shamasastry, *Kautilya's Arthashastra* (Mysore: Mysore Printing and Publishing House, 1961), p. vii.
15. A.L. Basham, *The Wonder That Was India* (London: Sidgwick & Jackson, 1985), p. 50.
16. Ibid., pp. 79-80.
17. B.A. Saletore, *Ancient Indian Political Thought and Institution* (New Delhi: Asia Publishing House, 1963), p. 476.
18. R.P. Kangle, *The Kautilya Arthashastra*, pt. III (Bombay: University of Bombay, 1965), p. 252.
19. Ibid., p. 255.
20. R.K. Mookerji, *Chandragupta Maurya and His Times* (Delhi: Motilal Banarsidass, 1960), p. 170.
21. L.N. Rangarajan (ed.), *Kautilya: The Arthashastra* (New Delhi: Penguin Books, 1992), p. 702.
22. Ibid., p. 684.
23. B.K. Majumdar, *The Military System in India* (Calcutta: Firma K.L. Mukhopadhyay, 1960), p. 61.
24. Kangle, *The Kautilya Arthashastra*, vol. III, p. 258.
25. Rangarajan, *Kautilya: The Arthashastra*, p. 503.
26. Ibid., p. 499.
27. Cleary, *The Art of War*, p. 165.
28. Griffith, *Sun Tzu: The Art of War*, pp. 145-6.
29. Giles, *On The Art of War*, p. 151.

30. Rangarajan, *Kautilya: The Arthashastra*, pp. 503-04.
31. V. Gairola, *Kautilya's Arthashastra* (Varanasi: Chaukhambha Vidya Bhawan, 1962), pp. 21-3.
32. Rangarajan, *Kautilya: The Arthashastra*, p. 505.
33. Shamasastry, *Kautilya's Arthashastra*, p. 21.
34. Cleary, *The Art of War*, p. 137.
35. Griffith, *Sun Tzu: The Art of War*, p. 41.
36. Cleary, *The Art of War*, p. 38.
37. Rangarajan: *Kautilya: The Arthashastra*, p. 728.
38. Shamasastry, *Kautilya's Arthashastra*, p. 434.
39. Giles, *On the Art of War*, p. 40.
40. Ibid.
41. Shamasastry, *Kautilya's Arthashastra*, pp. 112-13.
42. Kangle, *The Kautilya Arthashastra*, pt. III, p. 247.
43. Rangarajan, *Kautilya: The Arthashastra*, pp. 810-12.
44. Shamasatry, *Kautilya's Arthashastra*, pp. 442-5.
45. Rangarajan, *Kautilya: The Arthashastra*, p. 701.
46. Cleary, *The Art of War*, p. 5.
47. Griffith, *Sun Tzu: The Art of War*, p. 129.
48. Ibid., p. 43.
49. Rangarajan, *Kautilya: The Arthashastra*, p. 703.
50. Kangle, *The Kautilya Arthashastra*, pt. III, p. 259.
51. Cleary, *The Art of War*, p. 11.
52. Griffith, *Sun Tzu: The Art of War*, p. 41.
53. Ibid., p. 109.
54. Kangle, *The Kautilya Arthashastra*, pt. III, p. 260.
55. Lee Wee, and Hidajat, *Sun Tzu: War and Management* (Delhi: Pearson Education, 2004), p. 17.

CHAPTER 2

Kautilya on COIN: A Classical Indian Thinker on Counter-Insurgencies

KAUSHIK ROY

INTRODUCTION

About one-fifth of the world's population inhabits South Asia. From 1947 onwards, India has been the premier power in this region. In the new millennium, India is a rising global power. On an average, India's economy is expected to grow at the rate of 8 per cent until 2020 and will probably overtake United States' economy by 2042.[1] The Indian Army is the fourth largest in the world and has been engaged in more or less continuous counter-insurgency (COIN) duties from 1947 onwards. The various regions where the Indian Army is deployed to combat the insurgents are somewhat bigger than many European states. The Indian military officers use the terms Low Intensity Conflicts (LIC), insurgency, and so on interchangeably to designate internal rebellions which require the deployment of army units. Both India and the world would gain by analysing the roots behind the Indian Army's COIN doctrine.

Rajesh Rajagopalan asserts '. . . there is a strong conventional war bias that reduces the effectiveness of the Indian Army's counter-insurgency campaigns.'[2] This is indeed a sweeping statement. Most of the scholars working on the Indian Army harp on the British military tradition.[3] D.B. Shekatkar asserts that in the late 1950s, the Indian Army's COIN in Nagaland was influenced by British COIN policy of resettlement of villages and communities in Malaya which started after 1948. And this policy proved to be a total failure in Nagaland.[4] In addition to colonial legacies, this paper asserts that Classical Hindu political and military tradition as exemplified especially by Kautilya in the *Arthashastra* plays an important role in shaping the Indian Army's COIN doctrine. This article also suggests ways in which Kautilya's recommendations could be used fruitfully by India's strategic managers in particular and by the Western COIN forces in general in combating the insurgencies which are ravaging the subcontinent as well as a large part of the world.

COIN BY THE INDIAN ARMY IN RECENT TIMES

A brief glance at the physical geography of India is necessary before moving on to analyse the varied types of threats posed by the different insurgent groups. India's land frontiers exceed 15,000 km and India shares land frontiers with seven countries. India's coastline is 7,600 km long and its Exclusive Economic Zone is over 2 million sq km. The island territories in the east are 1,300 km away from the mainland. India shares the maritime boundary with five countries.[5]

In 1989, about 400,000 personnel (from the Indian Army and the various paramilitary forces) were deployed in Kashmir.[6] By 1998, about 44 per cent (156 out of 356) of the Indian Army's infantry battalions were engaged in different COIN campaigns around the country. In 2001, about 117 battalions of the Indian Army were engaged in COIN duties in Jammu & Kashmir.[7] Some examples will suffice the intensity of combat between the insurgents and the security forces. In Jammu & Kashmir between September and October 2000, 368 insurgents were eliminated by the army. Overall in 2000, about 1,432 terrorists were killed, 274 were caught and 76 surrendered.[8] The lower numbers captured and surrendered compared to the numbers killed show that the insurgents are highly motivated and are willing to fight and die for their cause.

In north-east India, more than 40 insurgent groups are operating.[9] In 1982, about 200,000 military personnel were deployed there.[10] Some examples of insurgent actions in the north-east are necessary to portray the gravity of the situation. On 20 October 2000, the People's Liberation Army (PLA) in Manipur raided the police station at Jiribam and took away 36 weapons. On 21 November 2000, the UNLF attacked a column of 17 Para Field Regiment and the 21st Regiment. In the ensuing encounter three *jawans* were killed and five were injured.[11] Thus we see that rather than being hunted, the initiative lies often with the insurgents. In order to deal with 800 guerrillas in Mizoram, the Indian Army has deployed 20,000 combat troops. The ratio of insurgents *versus* security forces appears to be 1:25.[12] Nagaland is located on the east of Bangladesh and its eastern border touches Myanmar. It is a mountainous state with the average altitude ranging from 900 to 1,500 m above sea level. Most of Nagaland's 16,488 sq km is covered with thick tropical forest and the climate is very humid.[13] Helicopters are used to move troops at short notice in order to conduct cordon and search operations.[14] Between 1986 and 1996, the Indian Army suffered a total of 2,467 deaths and 14,359 injuries in its various peacekeeping missions.[15]

The biggest overseas peacekeeping operation by the Indian Army was to send the Indian Peace Keeping Force (IPKF) to Sri Lanka in 1987 in order to maintain peace between the Sri Lankan armed forces and the Liberation Tigers of Tamil Ealam (LTTE). Two Indian infantry divisions (36th and 54th) were inducted into the island for conducting COIN against the LTTE.[16] The LTTE was equipped with the insurgents' favourite AK-47 and Soviet RPG-7 rocket launchers.[17] In March 1990, the Indian Army units started to pull out from Sri Lanka. As a point of comparison, in 2007, the US had about 140,000 troops in Iraq and in three years till 2006 had suffered 21,000 casualties.[18] Let us glance at the traditional roots of the Indian Army's COIN doctrine.

ORIGIN AND SCOPE OF KAUTILYA *ARTHASHASTRA*

The theory of polity, i.e. the genre known as *arthashastra*, emerged around 600 BCE.[19] Buddhism mainly focused on *moksha*. The *arthashastra* school emerged as a reaction to Buddhism. The *arthashastra* tradition emphasizes *lokayita* (materialism) rather than morality. The *lokayita* philosophy believes in the analysis of concrete facts. They deduced their conclusions from human behaviour and attempted an inductive investigation of the polity.[20] In P.V. Kane's view, *arthashastra* is narrower in scope than *dharmashastra* but broader than *dandaniti*. Overall, *arthashastra* in Classical Hindu discourse comprises politics, economics, law and justice.[21] *Arthashastra* means *labha* (the theory of politics for acquisition) and *palana* (good governance, i.e. the protection of material goods and subjects).[22]

The *arthashastra* writers divide the goals of human life into *chatuvarga* (four categories): *dharma* (morality), *artha* (wealth and other material requirements necessary for living properly), *kama* (desires) and *moksha* (salvation). And of these four, *artha* occupies the prominent place. Kautilya himself says that material well-being alone is supreme because spiritual well-being and sensual pleasures depend on material well-being. The *arthashastra* means the *shastra* (theory) of *artha*. The meaning of *artha* broadly refers to wealth and territory with human population. Kautilya's *Arthashastra* which was composed around 300 BCE does not really deal with the theory of generation of wealth but is a treatise on statecraft.

Kautilya's work contains fifteen *adhikarana*s (books). The first five books deal with internal administration of the polity and the next eight books mostly deal with foreign relations. The last two books are miscellaneous in character. Book 7 especially deals with foreign policy and the use of stratagems and force for gaining the objectives of a state. Books 9 and 10 deal with military preparations of the *vijigishu* (the ideal conqueror of the subcontinent). Book 12 shows how a weak king should survive against a strong king. And Book 13 mainly describes siege warfare.[23] The *Arthashastra* is in prose with verses scattered in the middle or end of the chapters.[24] The geographical scope of Kautilya's theory encompasses the whole subcontinent.

Kautilya's work does not refer to any particular historical event. This is because in ancient India according to the tradition of the *shastra*s, great work expounding timeless principles are compiled by some mythical sage. In such works, any reference to any historical individual or event is inconceivable as any such reference would reduce the value of the work.[25]

Kautilya attempts to put forward the timeless laws of politics, economy, diplomacy and war. He not only attempts to conceptualize the nature of war both within and among the states but also tries to formulate a strategy of power. He attempts to create a systematic and universal theory of power and warfare which will be applicable in all environmental contexts. The lynchpin behind Kautilya's paradigm is the assumption that polities crave power which is the most vital component for survival in the big bad world. This is same as Thucydides' view that the strong do what they can and the weak suffers as a result of it.[26] Torkel Brekke, a Norwegian scholar, criticizes Kautilya by saying that the power and influence of the kings in the international order described by Kautilya overlap and interpenetrate in ways with internal enemies that make it impossible to distinguish between external and internal affairs. Brekke in another article continues that Kautilya fails to distinguish between policing and war.[27]

Kautilya highlights the linkage between *kopa* (internal rebellion) and the shifting power structure in the international arena. He authorizes the *vijigishu* to pursue an expansionist design as regards foreign affairs and total suppression of the civil society to weed out any possibility of *kopa*. The *Arthashastra* discusses *kopa* from the perspective of state security. In Clausewitz's paradigm, diplomacy and the conduct of warfare constitute two different watertight compartments, but for Sun Tzu and Kautilya these two activities fuse to comprise a continuous seamless activity.[28]

In a recent work, Victoria Tin-Bor Hui notes the linkages between external militarism and internal insurgencies in polities. A world of self-interested actors is a world of dog-eat-dog competition not just in inter-state relations but also in state-society relations, writes Hui. If a ruler betrays his allies and breaks his word in the international realm, he is also likely to subjugate his citizens in the domestic realm. The state becomes something like a predatory mafia. Hui continues that a dynamic theory should view politics—both international and domestic—as processes of strategic interaction between the domination seekers and targets of domination.[29]

KAUTILYA ON COIN

In this section, we will analyse Kautilya's concept of internal rebellions, the factors behind their origin and the various *upayas* (means) to deal with them. We will see that many of Kautilya's *upayas* are similar to the techniques advocated by the Indian Army commanders. However, to avoid the taint of being a militarist Hindu in a multi-religious country, the military officers did not name Kautilya even while following his policy. We will also compare and contrast the views of the Western theoreticians to see how similar or dissimilar Kautilya and the Indian Army's policy are from the techniques advocated by the Western commentators on COIN.

Kautilya taking a statist perspective writes that 'The king, the ministers, the kingdom, the fortified cities, the treasury, the army and the ally, are the constituent elements of the state'.[30] These seven elements interact with each other to generate what Kautilya defines as *vigraha* or *yuddha*. These two terms refer to war (both inter-state and intra-state). The seven *prakritis* (elements of state) generating warfare are to be contrasted with the trinity of Carl von Clausewitz which comprises primeval violence, hatred, enmity (these three constitute one element), the play of chance and probability (these two constitute another element) and policy plus reason (these two constitute the third element). For some scholars, Clausewitz's trinity for war comprises the government, people and the army.[31] Like the Namierites, Kautilya does not believe in ideology behind human actions.[32] He writes that often due to ambition and treacheries of the ministers, royal family, vassal chiefs and frontier officials, internal rebellions occur.[33] In Kautilya's format, the principal threat to the *rashtra* (state) encompassing the whole subcontinent comes from internal rebellions. This is the view of various Indian military officers also.[34]

James D. Fearon and David D. Laitin define insurgency as '... a technology of military conflict characterized by small, lightly armed bands practicing guerrilla warfare from rural base areas.'[35] In their views, economic underdevelopment often encourages poor men to join the insurgent ranks.[36] Fearon and Laitin neglect the phenomenon of urban insurgency. Urban guerrilla operations emerged as a policy option among the Latin American revolutionaries during the late 1960s. The urban insurgents focused their attacks on the cities which are the nodes of political, economic and military power. The urban operations involve attacks on banks, abduction of businessmen and foreign tourists and assassination of political leaders. All these acts generate greater publicity.[37] Kautilya's *Arthashastra* notes that *kopa* could occur both in the principal urban centres as well as in the *janapada*s (rural regions of the state which are quite far away from the capital) as well as in the frontier regions.

Lieutenant-Colonel Vivek Chadha (who was associated with the Army Headquarter, New Delhi during 2005) asserts that there are two forms of LIC: terrorism and insurgency. Terrorists unlike the guerrillas did not enjoy widespread popular support. Insurgency is more dangerous than terrorism and if not handled properly, the former might escalate into civil war.[38]

Thomas A. Marks writes that neither ideology nor societal conditions but breakdown of the negotiated relationship between the leaders and their followers results in insurgencies. And not all insurgencies need to be revolutionary.[39] Kautilya unlike Chadha but like Marks focuses more on the relationship between the rulers, ruling elite and the citizens of a state rather than on social, cultural and political conditions while explaining the emergence of internal rebellions. Halvard Buhaug and Scott Gates write that the nature of a civil war depends on the objectives. For instance, a war for control over the state will be different from a war of secession. The role played by the natural resources in financing the purchase of arms is an important determinant of the nature of civil wars.[40] Kautilya observes that often pure greed on part of the state's officials, greed of the *antapalas* (warden of the marches) for acquiring mining rights, etc., generate *kopa.*

A group of Western scholars argue that civil wars occur more frequently in countries with substantial population belonging to different ethnic, linguistic and religious groups.[41] India has 18 officially recognized languages, 12 ethnic groups and 7 religious groups which are further subdivided into various sects, castes and sub-castes.[42] One aspect of the rebellion in Mizoram, claim the insurgents, is protection of the Christian religion against the 'Hindu' Indian state despite the post-independent Indian government's professed secular approaches to politics.[43] Chadha says that countries which are ethnically diverse and have a multiplicity of religions and languages are vulnerable to LICs. In fact, economic disparity along ethnic lines encourages internal rebellions. In Chadha's paradigm, the causes behind the LICs are political, social, economic and external agents. Like Kautilya, Chadha writes that with external support an ongoing LIC could escalate into a regular war between the states.[44]

Kautilya links up *kopa* within the *rashtra* with interventions by the foreign powers. He advises that it is best to wage war against an unjust king (read government) who has no public support. And it is wise to avoid war with a righteous king whose subjects will fight vigorously on his behalf. In other words, a king should march only against an enemy with disaffected subjects. Such a king's subjects weary of the unjust ruler will not help him and might even join the war against him.[45] So, Kautilya is implying that morally and practically it is easier to conquer an internally divisive kingdom which is rocked by *kopa.*

K. Santhanam claims that non-state actors indulging in insurgencies require the support of a friendly state for proper operation.[46] Between 1993 and 1994, Pakistan provided 1 million US $ to the militant outfit named National Socialist Council of Nagaland–Issac-Muviah Faction (NSCN-IM) of Nagaland.[47] China through the Inter Service Intelligence of Pakistan in collusion with Bangladesh aids the United Liberation Front of Asom (ULFA) of Assam and the NSCN-IM.[48] From late 1977, the Indian Army trained the Tamil militants of Sri Lanka.[49] Lieutenant-General Depinder Singh who commanded the IPKF at Sri Lanka during 1987-8, claims that without Indian support from the state of Tamil Nadu, the LTTE could not have survived against the Sri Lankan security forces during the 1980s.[50]

The German scholar Herfired Munkler asserts that credit is due to Mao Tse-tung for raising the guerrilla warfare from the status of secondary operation in service of conventional operation to the level of a full-fledged grand strategy for achieving political power. Further, Mao realized that a guerrilla strategy is going to be a time-consuming attritional warfare.[51] Kautilya and the Indian military officers note that initiating or destroying *kopa* is a time-consuming affair. Walter C. Ladwig III writes that analysis of India's COIN policies show that India has the patience, determination and resources to outlast the insurgents.[52]

Both the *Mahabharata* and the *Arthashastra* dislike tyrants. Kautilya says that while tyrants are interested in self-aggrandizement, efficient 'just' monarchs are more concerned with the interests of the *rashtra* (polity) rather than their personal interests. Kautilya warns the king to use *danda* (coercive force, i.e. military assets) with a sense of discrimination and by steering the middle course. Kautilya repeatedly emphasizes good governance to prevent *kopa.* He urges that if necessary then righteous customs should be initiated and unrighteous customs should be abolished.[53] Kautilya notes that the government should care about the cultural sensibilities of the people inhabiting troubled regions. The state policies should take care of the dress, language and cultural behaviour of the people in order to win and retain their loyalty.[54] Proper respect should be shown by the government to the fairs and festivals of the people in the disturbed zone.[55] And punishment should be moderate. Kautilya is for replacement of corporal punishment with monetary fines and is against imposing exorbitant monetary fines which would alienate the subjects who had erred slightly.[56] During natural calamities, in order to prevent the anger of the people from crossing the threshold and resulting in *kopa*, Kautilya warns that the state officials must initiate large-scale relief measures to alleviate the sufferings of the people in the disturbed zone.[57]

The Indian Army frequently goes for aid to civil operations during natural calamities. Two examples will suffice. On 29 March 1999, an earthquake occurred in the Garhwal region. In response, the Indian Army distributed food packets, blankets, tents and about 492 civilians were also treated by the army's medical units. During 17-18 October 1999, a cyclone from the Bay of Bengal caused devastation in the coastal areas of Andhra Pradesh and Tamil Nadu. In response, more than 5,000 army personnel were deployed in the affected areas. They rescued marooned civilians, distributed food packets and provided medical aid. About 22,288 civilians were evacuated and 33,722 were medically treated, 4,259 tons of food items distributed and 2,48,000 litres of drinking water was provided.[58]

The Indian Army believes that no insurgency could be solved by military force alone. Rather, the application of military force should prepare the ground for holding elections which would result in the formation of a democratic government.[59] Colonel Harjeet Singh (who served in the Sikh Light Infantry and also in the Army Training Command before he took premature retirement in 1998) opines: 'Low intensity conflicts cannot be won or even contained by military power alone.'[60] Depinder Singh notes in his autobiography: 'I was quite clear in my mind that no insurgency has ever been or can ever be settled militarily. Therefore, a political solution had to be found. . . . On the military plane we had to mount unrelenting pressure against the insurgents to force them to negotiate at some point in the future.'[61] In fact, Chadha claims that the only solution to insurgencies is a decentralized federal system in the spirit of self-governance.[62] Walter C. Ladwig III claims that India's flexibility and willingness to redefine internal borders and political arrangements in order to satisfy the preservationist as well as the reformist goals of the insurgents is praiseworthy.[63]

Depinder Singh claims:

> Insurgency is a consequence of political ineptitude, where alienation has been allowed to grow from minor, unfulfilled demands to a stage where alienated feel that there is no other recourse except to take up weapons. Therefore, this stage must never be allowed to be reached; if it has been reached, a political solution must be swiftly found. . . . What such sections seek is not promises of good governance, but prompt and sincere implementation. . . . Where the valves of democracy permit ethnic pressures to escape, differences will be settled by debate.[64]

Depinder Singh emphasizes on winning the minds and hearts of the people of the disturbed areas in order to prevent them from supporting the insurgents. He also suggests separation of the hardcore insurgents from the common people who should be brought within the mainstream.[65]

Kautilya and Sun Tzu warn against excessive dependence on violence.[66] The first step in COIN strategy for Kautilya is to persuade the seditious people to get back within the mainstream. Kautilya emphasizes the role of secret agents who should try to eliminate grievances among the rebelleous subjects through persuasion. If persuasion did not work, the next step in Kautilya's format is limited coercion. In this stage, the state should resort to *bheda* (divide and rule policy) in order to play off the different rebel leaders against each other and occasionally the leaders against their followers. If these two *upaya*s do not work, then the polity should resort to secret assassination of the troublesome rebel leaders. For implementing secret punishment, the secret agents should try to establish links with the family members (sons and wives as well as relatives) of the troublesome rebel leaders.[67]

As part of the minimum force to be employed in COIN operations, during October 1987, the IPKF while advancing towards Jaffna did not use artillery and tanks either to soften LTTE's defence or to provide fire support to the Indian infantry advancing into the heavily built-up urban landscape.[68] The objective was to avoid extensive non-combatant and collateral damage which would later obstruct peace building measures.

In 1993, Brigadier R.K. Nanavatty asserted that the concept of minimum force should guide India's COIN operations.[69] For the COIN forces, Harjeet Singh notes what he calls the 10 Commandments:

1. No Rape.
2. No Molestation.
3. No Torture resulting in death or maiming.
4. No military disgrace.
5. No meddling in civil administration.
6. Competence in platoon/company tactics.
7. Willingness to conduct civic actions.
8. Develop interaction with the media.
9. Respect human rights.[70]
10. Only fear God and uphold *dharma* and enjoy serving the country.

As regards Commandment no. 4, Harjeet Singh elaborates that military disgrace means loss of arms, surrender, abandonment of posts, and so on. Commandment no. 8 is detailed to use the media as a force multiplier rather than a force degrader. Harjeet Singh defines *dharma* as ethical mode of life which leads to the path of righteousness.[71] Here, Harjeet Singh is obliquely referring to *dharmayuddha* or just war concept inherent in Hinduism. And the latter part of the last Commandment refers to *nishkama karma*, i.e. doing one's own duty without looking for any tangible reward. This is a concept lifted from the *Bhagavad Gita.* Harjeet Singh's Commandment no. 3 finds support in the *Arthashastra.* The *Arthashastra* warns that prison officials should not harass and torture the prisoners and especially as regards female prisoners there should not be any sexual harassment, as such a policy is destructive for the legitimacy of the state in the long run.[72]

Stathis N. Kalyvas notes that terror is not synonymous with mass violence. Successful terror implies low level of violence. Violence fails if it destroys the subjects whose compliance is sought.[73] In Vietnam, indiscriminate use of military firepower by the United States' troops resulted in an upward spiral of civilian alienation.[74] Rod Thornton asserts that the British battalion in

Bosnia in 1992 emphasized minimum necessary force which in turn required enormous restraint. This generated a framework of trust and confidence, which was a key feature in humanitarian aid operations.[75]

However, right intentions and good policies are not always adequate for preventing *kopa*. In *The Prince*, Machiavelli writes: 'The fact is that a man who wants to act virtuously in every way necessarily comes to grief among so many who are not virtuous. Therefore if a prince wants to maintain his rule he must learn how not to be virtuous.'[76] Both Kautilya and Machiavelli argue that ends justify the means. Both agree on the utilization of wine, women, poison and spies for the attainment of the objectives.[77] Kautilya advocates that at times it is necessary to wage *kutayuddha* in order to cope up with *kopa*. The word *kuta* stands for evil and crooked. *Kutayuddha* means unjust war or war involving deceptive measures and treacheries.

Kautilya's recommendations in the *Arthashastra* are that a king should sow dissension among his enemies and always guard against fratricide. Kautilya writes about the mission of the envoy: 'Fight with the weapon of diplomacy, assassination of the enemy's army chiefs, stirring up the circle of kings, secret use of weapons, fire and poison . . . overreaching the enemy by trickery.'[78] Kautilya's *kutayuddha* also comprises winning over the enemy's commanders and councillors. He says that deserters should be encouraged from the enemy and loyal troops in the ranks of the enemy should be made hostile. Further, all sorts of tangible and non-tangible incentives should be offered in order to create treason in the ranks of the enemy.[79] Kautilya says that treaties are pieces of paper to be torn away at the most opportune moment. Lull the enemy into security by a peace pact. And then launch sudden commando attacks to destroy his brave troops.[80] The *Arthashastra* elaborates that by the divide and rule policy the state should create dissensions among the ranks of the enemy leadership. Some of the hostile chiefs should be won over and encouraged with moral and material help to fight their erstwhile commanders. If they refuse to fight then they should be secretly assassinated and the false news should be spread that they were assassinated by their comrades in arms since the former had agreed to make peace with the ruling regime.[81]

In accordance with Kautilya's dictum, the Indian state follows *bheda* against the insurgents. In October 1968, due to encouragement from the Indian state, the Sema tribal Nagas broke from the Naga Federal Government of Z.A. Phizo and made peace with the central government.[82]

During civil wars, writes Kalyvas, political actors promote amnesty programmes to encourage insurgent defection and rewards civilians who defect and collaborate with them.[83] The RAND research team also focuses on the use of rewards as an incentive for surrender or informing on insurgents.[84] On 27 February 1966, 'Operation Jericho' was launched by the Indian Army against the Mizo insurgents. By December 1966, India declared a general amnesty for those Mizo rebels who surrendered with their arms.[85]

Kautilya emphasizes the use of spies to gather knowledge about treasonable elements within the state and if necessary to use silent punishment against them.[86] Kautilya speaks about *vaidehakantevasinah*, who is an assistant to the trader spy.[87] He is for using *bhikshuki* (mendicant women) spies also.[88] Kautilya discusses *dvahsthaparampara* which refers to roving spies (carrier of information) disguised as acrobats, beggars, jugglers who come at the houses at intervals to beg. The stationary spies deployed in enemy territory in these houses take advantage of their appearance to communicate the information to them. Occasionally, the spies in these houses get orders from the secret service establishment through the above-mentioned roving spies who claim to be relations of the servants and come to visit the latter in the houses. Another concept in the *Arthashastra* is *samjnalipibhih* which means code language used by the spies for communication with each other as well for sending message to their master.[89]

In 1970, Brigadier S.K. Sinha noted: 'Sound intelligence is the bedrock for success in counter-insurgency operations.'[90] Rajagopalan asserts that long range patrols by small units are necessary in order to gather real time intelligence about insurgents.[91] Similarly, Depinder Singh asserts that good and secure intelligence functions as a force multiplier in COIN campaigns.[92] Kautilya repeatedly emphasizes the necessity of integrating the views generated by different sorts of spies (roving spies, stationary spies, double agents, etc.) with the state bureaucracy in order to generate a clear unified picture about the intelligence landscape. Integration about the various intelligence agencies is something which the Indian military officers demand but the Indian state is yet to construct an unified integrated machinery for collating intelligence acquired from different intelligence agencies. As a result, the Indian COIN strategy often suffers. For instance, one reason behind the Sri Lankan imbroglio was the fact that the Research and Analysis Wing and the Ministry of External Affair's intelligence agencies did not cooperate with the military intelligence agency of the Indian Army. The net result was that the IPKF remained in dark as regards the strength and intention of the LTTE and the Sri Lankan armed forces.[93]

The Western commentators accept the cardinal importance of intelligence for conducting COIN successfully. Thomas R. Mockaitis writes that sound intelligence remains the key to victory. Gathering accurate and timely information requires winning the trust of the general populace who support the insurgents tacitly or actively. Trust and cooperation in turn require recognizing and as far as possible addressing the real needs and the legitimate grievances on which insurgency feeds. Good intelligence allows the security forces to use force against the insurgents in a limited and focused manner to prevent further alienation of the general population.[94]

KAUTILYA'S RECOMMENDATIONS AND THEIR PRESENT RELEVANCE

The new millennium is witnessing intra-state rather than inter-state warfare. Colonel Thomas X. Hammes of United States Marine Corps categorizes it as Fourth Generation Warfare (4GW). He writes: 'It is an evolved form of insurgency. . . . Unlike previous generations, it does not attempt to win by defeating the enemy's military forces. Instead, via the networks, it directly attacks the minds of the enemy decision makers to destroy the enemy's political will.'[95] Hammes continues that Fifth Generation Warfare (5GW) is arriving. About its characteristics, he writes that such insurgent attacks could be launched by a small group and perhaps even an individual could attack the enemy country.[96]

Pakistan seems to be waging this sort of warfare against India in recent times. On 13 December 2001, the Parliament building in New Delhi was attacked by five Kashmiri insurgents backed by Pakistan. In the ensuing hour-long gun battle, all the five insurgents were killed along with seven security personnel and 22 were injured. In October 2001, a similar assault was carried out against the Assembly building of Jammu & Kashmir by the Jaish-e-Mohammad (Army of Mohammad) and in this confrontation 38 people were killed.[97] However, in both these two cases, the insurgents came very close to decapitating the political managers of India.

Instead of naming the new millennium's insurgencies as 4GW, two scholars term the post-Cold War era intra-state conflicts as New War. Domestic factors generating identity politics are the prime movers in this sort of warfare. One of the characteristics of New War is privatization of warfare. It means that non-state and sub-state agents have seized the initiative and demilitarized warfare. The latter refers to the dissolution of the distinction between combatants and non-combatants and targeting of the civilian population and the non-military infrastructure. Munkler claims that the rise of private companies which provide security is an essential characteristic of the New War. The advent of private security companies is due to the collapse of the state and also for the fact that a state finds using them economical.[98] Munkler continues that for the warlords, warfare is a business in order to make money. So, they have no interest to settle the conflict. And this distinguishes the New War from the Old insurgencies. Since, the warlords of the new era make hay in unsettled societal scenarios their wars are not state building projects. The private wars of the warlords are demodernizing in all aspects.[99]

To an extent, Munkler's hypothesis is applicable in case of certain insurgencies within India. The NSCN-IM's stated objective is to establish a sovereign greater Nagaland known as Nagalim which besides Nagaland would also comprise certain areas from Assam, Manipur, Arunachal Pradesh and Myanmar. In reality, the NSCN-IM and National Socialist Council of Nagaland–Khalpang (NSCN-K) are interested in collecting taxes from the government employees and the merchants. In 1998, every village household had to pay a house tax of Rs. 110 annually. The NSCN-IM's annual budget was about Rs. 250 million. Besides collection of taxes, the NSCN-IM is also involved in gun running and narcotics in the Golden Triangle. The Golden Triangle comprises 38 million hectares of rainforest-covered mountains in Laos, Thailand and Myanmar. These insurgents are not interested in peace as that will result in the termination of their lucrative underground extortion and smuggling operations.[100]

While the guerrillas have to depend on moral and material support of the populace among whom they operate, the terrorists who are conducting the New War do not depend on backing from the people but conduct the war by subsisting within the general populace and simultaneously by targeting them.[101] David Lonsdale writes that the irregular conflicts are protracted, attritional and people intensive.[102] Mockaitis claims that previously the insurgents operated in a local arena and now they operate at the global stage due to the development of print media, access to internet and global airwaves.[103]

One analyst claims that Clausewitz did not merely equate politics with state but defined politics as human intercourse. And this is relevant for understanding the present-day scenario when the initiative lies with the stateless marginal groups conducting insurgencies.[104] Kautilya besides indulging in politics of the *rashtra* also discusses the armed politics conducted by the stateless marginal groups like the forest chiefs and, semi-autonomous tribal chieftains within a state.

Thomas M. Kane's assertion that postmodernism includes renewed appreciation for factors which remain forever unquantifiable is acceptable. Modernism for Thomas Kane is the school of thought which is unwilling to accept the idea that nature sets inherent limits on what human beings can know and do. Modernists always emphasize progress and are unwilling to rely on intuition, inchoate personal knowledge or allegedly innate knowledge. The postmodern military thinkers focus more on human judgement, creativity and holistic thought.[105]

The *rashtra* suffers, says Kautilya, due to *vyasanas* (calamities). *Vyasanas*, are of two types: those caused by human errors and those caused by divine factors. Divine factors do not mean religious factors but those extraneous factors which are beyond the control of human beings. Divine calamities also occur due to bad luck and adverse natural conditions. Some examples of divine calamities are flood and drought. Human calamities on the other hand occur due to bad policies of the ruling elite.[106] Kautilya continues that war is like *maya*, i.e. it could take different forms in different circumstances. This statement of Kautilya is somewhat similar to Clausewitz's assertion that warfare is like a chameleon.

In recent times, Alan Beyerchen has attempted to interpret Clausewitz in a postmodernist manner. Beyerchen asserts that for Clausewitz, war is a non-linear phenomenon characterized by organized complexity. Organized complexity means a number of variables are interrelated into the organic whole of any biological or social system. Systems of organized complexity involve factors that are interdependent. Over time they change in ways that affect how and even which variables interact. The exact behaviour of such a system is almost impossible to predict. In order to emphasize the non-linear nature of warfare characterized by a large number of variables, Clausewitz focuses on friction which generates chance and makes war almost like the random roll of dice.[107]

Christopher Bassford notes that one element of Clausewitz's trinity is the play of chance and probability within which the creative spirit is free to roam.[108] In Kautilya's paradigm, society is always in chaos. Exercise of power is the only route for survival. He uses the term *matsyanyaya* which means the law of fishes. In a pond, the bigger fishes consume the smaller ones.[109] Kautilya similarly highlights the importance of *utsahashakti* (personal dynamism and vigour) of the *vijigishu*. The *vijigishu* in accordance with the circumstances has to use *buddhi* (intelligence) to overcome difficult situations.[110] The *vijigishu*'s *utsahashakti* is somewhat similar to Clausewitz's genius for war. The latter's genius for war includes all the facets of a personality, both rational and emotional.[111] The genius for war should both be able to act and to know. His action is influenced by his power of judgement.[112]

Ulrike Kleemeier asserts that Clausewitz focused on courage of the soldiers which means more than the bravery to face death in the battlefield. Courage in a positive sense for the Prussian military theorist means the capability to think in dangerous situations in order to come up with solutions. And this is the basis of the mission command system (*Auftragstaktik*) which is required more than ever in the war against terror, especially when junior officers have to undertake operations for quite a long time without receiving detailed orders from above.[113]

Rupert Smith, the retired British General, categorizes the new form of insurgency as war amongst the people. Smith influenced by the French philosopher Michel Foucault writes that power is a relationship between the rulers and the ruled. He says:

> And that is now the problem of the West, or those who use conventional armies against 'insurgents,' 'terrorists,' 'asymmetric opponents' and so forth: for if the opponent has moved amongst the people it is extremely difficult to establish this relationship to advantage since the underpinning idea of the strategies of provocation and so forth is to establish the relation to the disadvantage of conventional military force.[114]

Like Rupert Smith, Depinder Singh asserts that the COIN against the LTTE represented a war amongst the people. He elaborates that ordinary men and women and also LTTE cadres dressed as ordinary men and women mingled among the general populace and threw grenades at the IPKF personnel.[115] The LTTE used civilians as human shields when the IPKF advanced towards Jaffna in late 1987. The LTTE men wore *lungis* and fought from within the common people. The IPKF personnel could not distinguish a LTTE cadre from the ordinary crowd. Against such insurgent tactics, the IPKF had no effective operational doctrine.[116]

Kautilya offers some clue to deal with these sorts of internal rebellions. The *Arthashastra* introduces the concept of *tusnim-dandena* which means getting rid of the enemy leaders by secret assassination, poisoning, etc.[117] Kautilya speaks of silent punishment for eliminating powerful persons whose actions are responsible for injuries to the *rashtra*.[118] As part of *kutayuddha*, Kautilya notes that special cash payments and prizes should be announced for encouraging soldiers to kill the opposing side's commander-in-chief and ruler.[119] Other means of secret punishment are setting up an ambush and giving poison through physicians.[120] Kautilya is aiming to eliminate the enemy leadership/brain by deploying small cells rather than a mass deployment of force. In case of foreign powers supporting hostile elements within the *vijigishu*'s state, the latter state's diplomatic envoys in cooperation with the secret agents should abduct some of the power elites of the foreign state sponsoring terrorism in order to bring pressure on the latter to prevent it from supporting *kopa* in the *vijigishu*'s *rashtra*.[121]

In case of popular uprising (similar to Kautilya's *kopa*), the personalities of the leaders and public opinion constitute, for Clausewitz, the centre of gravity.[122] Public opinion, is an integral part of democracy especially in a country like India. In 2005, Lieutenant-Colonel Vivek Chadha asserted that in insurgencies (or LIC) ideas rather than arms are more important.[123]

Beatrice Heuser writes that Clausewitz implies in case of resistance movement, the centre of gravity is the leader who has to be wiped out.[124] Kautilya speaks of waging biological warfare against the enemies by poisoning the wells, blinding the enemy soldiers through poisonous smoke, etc. Kautilya details the various poisons to be manufactured from herbs, drugs and animals.[125] These poisons are to be administered to the top leadership of the enemy by secret agents.[126] The *Arthashastra* details that the secret service establishment should guard against counter-espionage by the enemy secret agents. So, various sorts of spies should operate in the disturbed zones and they should be unaware of each other's existence, so that even if a small cell is burst by the enemy counter-espionage department, the other spies would be able to operate.[127]

The Al-Qaeda and other successful terrorist networks around the world heavily utilize spies and the focus is on HUMNIT.[128] Rather than technology (SIGNIT in modern terminology), Kautilya focuses on HUMNIT and urges that the spies should be conversant with the culture of the region in which they are deployed.

Rajagopalan writes that successful COIN requires small highly mobile offensive patrolling units moving deep inside guerrilla territories. Large-unit cordon and search operations are useless. In fact, moving large number of security forces to sensitive areas alert the insurgents and they escape into the wilderness.[129] In 2004, an American analyst as regards COIN in Iraq emphasized small-unit operations and careful intelligence work.[130]

The *Arthashastra* tells us that military operations should be conducted taking into consideration *desha* (terrain) and *kala* (season).[131] Kautilya notes that the government troops should be ready to fight in the mountainous or forest regions and they should conduct operations with adequate flank guards and a reserve force stationed behind the attacking units.[132] Nocturnal commando attacks, says Kautilya, are to be launched in order to surprise the rebels.[133] The Indian Army has recently accepted the doctrine: 'Fight the guerrilla like a guerrilla'.[134] But a lot has to be done yet to translate this doctrine into practice.

Gavin Bulloch warns that the nature of environment in which insurgencies are conducted is changing. Now, COIN is conducted under the critical scrutiny of the law, media, human rights organizations and other international bodies.[135] Alice Hills writes that when regime change or reconstruction is the objective of COIN campaigns, then in order to justify it, the Western powers conducting COIN has to initiate humanitarian relief operations also. The incorporation of air strikes and airdrops within a single operation in Afghanistan in October 2001 is an example of this trend. Along with lethal weapon strikes, propaganda leaflets were also dropped in order to undermine the support for the Taliban. Hills continues that both Afghanistan and Iraq show that a mix of air strikes, airdrops and convoys of humanitarian aid is typical of the West's approach to conflict. The devastation of a state is followed by its reconstruction and the whole affair is legitimized by reference to humanitarian objectives or the 'war on terrorism'.[136]

Antulio J. Echevarria II points out that propaganda offering alternatives to *jihadi* lifestyle, financial assistance, job training, etc., should be used to undermine the commitments of the insurgents.[137] Christopher Dasse warns that resistance and possibility of defence are two cardinal points while conducting unconventional warfare.[138] Andreas Herberg-Rothe warns that creation of a just peace after the conflict is the most important measure in order to prevent any further insurgencies. Construction of just peace should emphasize removal of poverty, oppression and recognition of pluralism in cultures.[139]

Kautilya writes that ensuring *yogakshema* of the subjects is the principal duty of the *rashtra*. For this it is essential for the *rashtra* to maintain security of the person and his/her property. *Yogakshema* is made difficult due to the activities of anti-socials like thieves, robbers and also corrupt officials and deceitful merchants.[140] Kautilya warns that the *vijigishu* should never resort to *asuravijayin* which means behaving like a demon and seizing land, money and women of the conquered people. Rather, the state should treat the defeated parties with honour and respect in order to establish a long-lasting just peace. Kautilya notes that *lovavijayin*, i.e. extracting money and land from the defeated people would result in further *kopa* in the near future.[141]

Depinder Singh focuses on psychological warfare and public relations as part of the COIN operations.[142] E.N. Rammohan, an ex-director of the Border Security Force, argues that unemployment of the Manipuri youth encourages them to join the insurgent ranks. He urges the government to sponsor forestry, poultry, horticulture, piggery and fishery in order to absorb the unemployed youth into sustainable economic schemes.[143]

CONCLUSION

In recent times, there is a forced attempt by pro-Clausewitzian Western scholars to speculate that Clausewitz implied a lot about little war which are relevant for understanding the West's on-going war on terror against stateless groups. Both Sun Tzu and Clausewitz, unlike Kautilya, assume that the states are the primary actors in war. However, intra-state war is more important than inter-state war in Kautilya's paradigm. As this essay shows, consciously or unconsciously, the Indian Army's COIN doctrine to a great extent has been shaped by the *Arthashastra*. The Maoist armed threat in West Bengal, Madhya Pradesh, Maharashtra, Orissa and Andhra Pradesh in the near future might take the form of an aggressive large-scale armed insurgency which might require New Delhi to deploy military units. And in Kashmir, the Islamic insurgency is still going on. The Indian Army would do better to try to cull further lessons from the *Arthashastra* rather than looking at the new-fangled Western COIN theories about New War.

NOTES

1. Breena E. Coates, 'Modern India's Strategic Advantage to the United States: Her Twin Strengths in *Himsa* and *Ahimsa*', *Comparative Strategy*, vol. 27, no. 2 (2008), p. 143.
2. Rajesh Rajagopalan, *Fighting like a Guerrilla: The Indian Army and Counterinsurgency* (New Delhi: Routledge, 2008), p. 17.

3. Stephen P. Cohen, 'The Military and Indian Democracy', in Atul Kohli (ed.), *India's Democracy: An Analysis of Changing State-Society Relationship* (1988, rpt., New Delhi: Orient Longman, 1991), pp. 99-143.
4. D.B. Shekatkar, 'India's Counterinsurgency Campaign in Nagaland', in Sumit Ganguly and David P. Fidler (eds.), *India and Counterinsurgency: Lessons Learned* (London: Routledge, 2009), p. 17.
5. *Ministry of Defence Government of India Annual Report (MODAR), 2000-2001*, p. 2.
6. Sumit Ganguly, 'Explaining the Kashmir Insurgency: Political Mobilization and Institutional Decay', *International Security*, vol. 21, no. 2 (1996), p. 76.
7. Rajesh Rajagopalan, 'Innovations in Counterinsurgency: The Indian Army's Rashtriya Rifles', *Contemporary South Asia*, vol. 13, no. 1 (2004), p. 26.
8. *MODAR: 2000-2001*, pp. 20-1.
9. R.S. Grewal, 'Ethno Nationalism in North Eastern India', *Journal of the United Service Institution of India (JUSII)*, vol. CXXXIII, no. 552 (2003), p. 268.
10. Jerrold F. Elkin and W. Andrew Ritezel, 'Military Role Expansion in India', *Armed Forces and Society*, vol. 11, no. 4 (1985), p. 495.
11. *MODAR: 2000-2001*, p. 22.
12. Anil A. Athale, 'Insurgency and Counter-Insurgency in Modern India: An Overview', in S.N. Prasad (ed.), *Historical Perspectives of Warfare in India: Some Morale and Material Determinants* (New Delhi: Centre for Studies in Civilizations, distributed by Motilal Banarsidass, 2002), p. 419.
13. Shekatkar, 'India's Counterinsurgency Campaign in Nagaland', in Ganguly and Fidler (eds.), *India and Counterinsurgency*, p. 9.
14. Pradeep P. Barua, *The State at War in South Asia* (Lincoln/London: University of Nebraska Press, 2005), p. 233.
15. Colonel Harjeet Singh, *Doda: An Insurgency in the Wilderness* (New Delhi: Lancer, 1999), p. 245.
16. Lieutenant-General Depinder Singh, *Indian Peacekeeping Force in Sri Lanka* (Dehradun: Natraj Publishers, 2001), p. 85.
17. Gautam Das and M.K. Gupta-Ray, *Sri Lanka Misadventure: India's Military Peace-Keeping Campaign, 1987-1990* (New Delhi: Har-Anand, 2008), p. 109.
18. Warren Chin, 'Examining the Application of British Counterinsurgency Doctrine by the American Army in Iraq', *Small Wars and Insurgencies*, vol. 18, no. 1 (2007), p. 1.
19. *The Kautilya Arthashastra*, pt. III, *A Study*, by R.P. Kangle (1965, rpt., New Delhi: Motilal Banarsidass, 2000), p. 11.

20. Ashok S. Chousalkar, *A Comparative Study of Theory of Rebellion in Kautilya and Aristotle* (Delhi: Indological Book House, 1990), p. 65.
21. P.V. Kane, *History of Dharmasastra (Ancient and Medieval Religious and Civil Law in India)*, vol. 1, pt. 1 (Pune: Bhandarkar Oriental Research Institute, 1968), pp. 152-3.
22. Surendra Nath Mittal, *Kautilya Arthashastra Revisited* (2000, rpt., Centre for Studies in Civilizations, distributed by Munshiram Manoharlal: New Delhi, 2004), p. 65.
23. *KA*, pt. III, by Kangle, pp. 19-20.
24. Kane, *History of Dharmasastra*, vol. 1, part 1, p. 197.
25. *KA*, pt. III, by Kangle, p. 63.
26. C. Dale Walton, 'The Strategist in Context: Culture, the Development of Strategic Thought, and the Pursuit of Timeless Truth', *Comparative Strategy*, vol. 23, no. 1 (2004), p. 95.
27. Torkel Brekke, 'The Ethics of War and the Concept of War in India and Europe', *NUMEN*, vol. 52 (2005), p. 80; 'Wielding the Rod of Punishment: War and Violence in the Political Science of Kautilya', *Journal of Military Ethics*, vol. 3, no. 1 (2004), p. 46.
28. Michael I. Handel, *Masters of War: Classical Strategic Thought* (1992, rpt., London: Frank Cass, 1996), p. 31.
29. Victoria Tin-Bor Hui, *War and State Formation in Ancient China and Early Modern Europe* (New York: Cambridge University Press, 2005), pp. 2, 15.
30. *The Kautilya Arthashastra*, pt. II. An English Translation with Critical and Explanatory Notes, by R.P. Kangle (1972, rpt., Delhi: Motilal Banarsidass, 1992), p. 314.
31. Hew Strachan, 'Clausewitz and the Dialectics of War', in Strachan and Andreas Herberg-Rother (eds.), *Clausewitz in the Twenty-First Century* (Oxford: Oxford University Press, 2007), p. 43.
32. Chousalkar, *Theory of Rebellion in Kautilya and Aristotle*, p. 77.
33. *KA*, pt. II, by Kangle, p. 515.
34. Singh, *Doda*, p. 141.
35. James D. Fearon and David D. Laitin, 'Ethnicity, Insurgency, and Civil War', *American Political Science Review*, vol. 97, no. 1 (2003), p. 75.
36. Fearon and Laitin, 'Ethnicity, Insurgency, and Civil War', p. 80.
37. Azeem Ibrahim, 'Conceptualisation of Guerrilla Warfare', *Small Wars and Insurgencies*, vol. 15, no. 3 (2004), pp. 120-1.
38. Lieutenant-Colonel Vivek Chadha, *Low Intensity Conflicts in India: An Analysis* (New Delhi: Sage, 2005), p. 403.
39. Thomas A. Marks, 'Ideology of Insurgency: New Ethnic Focus or Old Cold War Distortions?', *Small Wars and Insurgencies*, vol. 15, no. 1 (2004), pp. 107-12.
40. Halvard Buhaug and Scott Gates, 'The Geography of Civil War', *Journal of Peace Research*, vol. 39, no. 4 (2002), pp. 418-19.

41. Havard Hegre, Tanja Ellingsen, Scott Gates and Nils Petter Gleditsch, 'Towards a Democratic Civil Peace? Democracy, Political Change, and Civil War, 1816-1992', *American Political Science Review*, vol. 95, no. 1 (2001), p. 37.
42. Satish Kumar, 'Sources of Democracy and Pluralism in India', in Vice-Admiral K.K. Nayyar and Jorg Schultz (eds.), *South Asia Post 9/11: Searching for Stability* (New Delhi: Rupa, 2003), p. 74.
43. Vivek Chadha, 'India's Counterinsurgency Campaign in Mizoram',. in Ganguly and Fidler (eds.), *India and Counterinsurgency*, pp. 32-3.
44. Chadha, *Low Intensity Conflicts in India*, pp. 405-6, 419.
45. Manoj Kumar Sinha, 'Hinduism and International Humanitarian Law', *International Review of the Red Cross*, vol. 87, no. 858 (2005), p. 289.
46. K. Santhanam, 'Sources of Terror: India', in Nayyar and Schultz (eds.), *South Asia Post 9/11*, p. 34.
47. Lawrence E. Cline, 'The Insurgency Environment in Northeast India', *Small Wars and Insurgencies*, vol. 17, no. 2 (2006), p. 141.
48. Jaideep Saikia, 'Changing Contours of Separatism', *Aakrosh*, vol. 6, no. 18 (2003), pp. 72-3.
49. Das and Gupta-Ray, *Sri Lanka Misadventure*, p. 19.
50. Singh, *Indian Peacekeeping Force in Sri Lanka*, p. 193.
51. Herfried Munkler, 'The Wars of the 21st Century', *International Review of the Red Cross*, vol. 85, no. 849 (2003), pp. 7-9.
52. Walter C. Ladwig III, 'Insights from the Northeast: Counterinsurgency in Nagaland and Mizoram', in Ganguly and Fidler (eds.), *India and Counterinsurgency*, p. 50.
53. *KA*, pt. II, by Kangle, p. 493.
54. Ibid., p. 491.
55. *KA*, pt. III, by Kangle, p. 261.
56. Ibid., pp. 236-9.
57. Ibid., p. 234.
58. *MODAR: 1999-2000*, pp. 110, 113.
59. Rajagopalan, *Fighting like a Guerrilla*, p. 107.
60. Singh, *Doda*, p. 141.
61. Singh, *Indian Peacekeeping Force in Sri Lanka*, p. 107.
62. Chadha, *Low Intensity Conflicts in India*, p. 412.
63. Ladwig III, 'Insights from the Northeast', in Ganguly and Fidler (eds.), *India and Counterinsurgency*, pp. 46, 48.
64. Singh, *Indian Peacekeeping Force in Sri Lanka*, p. 191.
65. Ibid., p. 107.
66. Thomas M. Kane, *Ancient China on Postmodern War: Enduring Ideas from the Chinese Strategic Tradition* (London/New York: Routledge, 2007), p. 25.
67. *KA*, pt. III, by Kangle, pp. 242-3.
68. Das and Gupta-Ray, *Sri Lanka Misadventure*, p. 85.

69. R.K. Nanavatty, 'Use of the Armed Forces in Aid to Civil Authority: Problems & Prospects', *National Defence College Journal*, vol. 15 (1993), p. 14.
70. Singh, *Doda*, Appendix D.
71. Ibid.
72. *KA*, pt. III, by Kangle, p. 242.
73. Stathis N. Kalyvas, 'The Paradox of Terrorism in Civil War', *Journal of Ethics*, vol. 8 (2004), p. 101.
74. Gavin Bulloch, 'Military Doctrine and Counterinsurgency: A British Perspective', *Parameters* (Summer 1996), p. 2, http://www.carlisle.army.mil/usawc/Parameters/96 summer/bulloch.htm
75. Rod Thornton, 'The British Army and the Origins of its Minimum Force Philosophy', *Small Wars and Insurgencies*, vol. 15, no. 1 (2004), p. 83.
76. Niccolo Machiavelli, *The Prince*, tr. with an Introduction by George Bull (1961, rpt., Harmondsworth, Middlesex: Penguin, 1981), p. 91.
77. P.C. Chakravarti, *The Art of War in Ancient India* (1941, rpt., Delhi: Low Price Publications, 1989), p. vii.
78. *KA*, pt. II, by Kangle, p. 4.
79. Ibid., p. 490.
80. Ibid.
81. Ibid., p. 489.
82. Athale, 'Insurgency and Counter-Insurgency in Modern India', in Prasad (ed.), *Historical Perspectives of Warfare in India*, p. 403.
83. Kalyvas, 'The Paradox of Terrorism in Civil War', p. 99.
84. Austin Long, *On "Other War": Lessons from Five Decades of RAND Counterinsurgency Research* (Santa Monica: RAND Corporation, 2006), p. 45.
85. Chadha, 'India's Counterinsurgency Campaign in Mizoram', in Ganguly and Fidler (eds.), *India and Counterinsurgency*, pp. 34, 38.
86. Mittal, *Kautilya Arthashastra Revisited*, pp. 40-1.
87. *KA*, pt. II, by Kangle, p. 22.
88. A.K. Srivastava, *Ancient Indian Army: Its Administration and Organization* (New Delhi: Ajanta Publications, 1985), p. 101.
89. *KA*, pt. II, by Kangle, p. 25.
90. S.K. Sinha, 'Counter Insurgency Operations', *JUSII*, vol. 100, no. 420 (1970), p. 267.
91. Rajagopalan, *Fighting Like a Guerrilla*, p. 109.
92. Singh, *Indian Peacekeeping Force in Sri Lanka*, pp. 191-2.
93. Das and Gupta-Ray, *Sri Lanka Misadventure*, p. 60.
94. Thomas R. Mockaitis, 'Winning Hearts and Minds in the "War on Terrorism"', *Small Wars and Insurgencies*, vol. 14, no. 1 (2003), p. 21.
95. Thomas X. Hammes, *The Sling and the Stone: On War in the 21st Century* (St Paul, MN: Zenith Press, 2006), p. 208.

96. Ibid., p. 274.
97. Walter C. Ladwig III, 'A Cold Start for Hot Wars? The Indian Army's New Limited War Doctrine', *International Security*, vol. 32, no. 3 (Winter 2007/8), pp. 158, 160.
98. Herfried Munkler, 'Clausewitz and the Privatization of War', in Strachan and Herberg-Rothe (eds.), *Clausewitz in the Twenty-First Century*, pp. 220-1; Mary Kaldor, *New and Old Wars: Organized Violence in a Global Era* (1999, rpt., Dehradun: Natraj, 2005).
99. Ibid., pp. 226-7.
100. Kanchan L., 'Negotiating Insurgencies: The Naga Imbroglio', in K.P.S. Gill and Ajai Sahni (eds.), *FAULTLINES: Writings on Conflict and Resolution*, vol. 11 (New Delhi: Institute for Conflict Management and Bulwark Books, 2002), pp. 127-36.
101. Herfried Munkler, 'New Wars: Characteristics and Commonalities', in John Andreas Olsen (ed.), *On New Wars* (Oslo: Norwegian Institute of Defence Studies, 2007), pp. 69, 82.
102. David Lonsdale, 'Clausewitz and Information Warfare', in Strachan and Herberg-Rothe (eds.), *Clausewitz in the Twenty-First Century*, p. 243.
103. Mockaitis, 'Winning Hearts and Minds in the "War on Terrorism"', p. 22.
104. Julian Reid, 'Foucault on Clausewitz: Conceptualizing the Relationship between War and Power', *Alternatives*, vol. 28, no. 1 (2003), p. 18.
105. Kane, *Ancient China on Postmodern War*, pp. 4-5, 8, 103.
106. Chousalkar, *Theory of Rebellion in Kautilya and Aristotle*, p. 75.
107. Alan Beyerchen, 'Clausewitz and the Non-Linear Nature of Warfare: Systems of Organized Complexity', in Strachan and Herberg-Rothe (eds.), *Clausewitz in the Twenty-First Century*, pp. 45-56.
108. Christopher Bassford, 'The Primacy of Policy and the "Trinity" in Clausewitz's Mature Thought', in Strachan and Herberg-Rothe (eds.), *Clausewitz in the Twenty-First Century*, p. 77.
109. *KA*, pt. II, by Kangle, p. 10.
110. Ibid., p. 5.
111. Ulrike Kleemeier, 'Moral Forces in War', in Strachan and Herberg-Rothe (eds.), *Clausewitz in the Twenty-First Century*, p. 111.
112. Jose Fernandez Vega, 'War as "Art": Aesthetics and Politics in Clausewitz's Social Thinking', in Strachan and Herberg-Rothe (eds.), *Clausewitz in the Twenty-First Century*, pp. 128-9.
113. Kleemeier, 'Moral Forces in War', in Strachan and Herberg-Rothe (eds.), *Clausewitz in the Twenty-First Century*, pp. 116-9.
114. Rupert Smith, 'Thinking about the Utility of Force in War Amongst the People', in Olsen (ed.), *On New Wars*, p. 33.
115. Singh, *Indian Peacekeeping Force in Sri Lanka*, p. 3.
116. Das and Gupta-Ray, *Sri Lanka Misadventure*, pp. 98, 104.
117. *KA*, pt. II, by Kangle, p. 23.

118. Ibid., p. 492.
119. Srivastava, *Ancient Indian Army*, p. 108.
120. *KA*, pt. III, by Kangle, p. 243.
121. Ibid., pp. 250-1.
122. Handel, *Masters of War*, p. 45.
123. Chadha, *Low Intensity Conflicts in India*, p. 19.
124. Beatrice Heuser, 'Clausewitz's Ideas of Strategy and Victory', in Strachan and Herberg-Rothe (eds.), *Clausewitz in the Twenty-First Century*, p. 161.
125. *KA*, pt. II, pp. 495-7, 511.
126. Ibid., p. 494.
127. Ibid., pp. 26-7.
128. Caleb M. Bartley, 'The Art of Terrorism: What Sun Tzu can Teach Us about International Terrorism', *Comparative Strategy*, vol. 24 (2005), p. 245.
129. Rajagopalan, *Fighting Like a Guerrilla*, pp. 56, 108.
130. Long, *On "Other War"*, p. 44.
131. *KA*, pt. III, by Kangle, p. 257.
132. Ibid., p. 259.
133. Ibid., pp. 258-9.
134. *MODAR: 1999-2000*, p. 94.
135. Bulloch, 'Military Doctrine and Counterinsurgency: A British Perspective', p. 1.
136. Alice Hills, 'Basra and the Referent Points of Twofold War', *Small Wars and Insurgencies*, vol. 14, no. 3 (2003), pp. 24-5.
137. Antulio J. Echevarria II, 'Clausewitz and the Nature of the War on Terror', in Strachan and Herberg-Rothe (eds.), *Clausewitz in the Twenty-First Century*, p. 215.
138. Christopher Dasse, 'Clausewitz and Small Wars', in Strachan and Herberg-Rothe (eds.), *Clausewitz in the Twenty-First Century*, p. 194.
139. Andreas Herberg-Rothe, 'Clausewitz and a New Containment: The Limitation of War and Violence', in Strachan and Herberg-Rothe (eds.), *Clausewitz in the Twenty-First Century*, pp. 301, 304.
140. *KA*, pt. III, by Kangle, p. 232.
141. Ibid., pp. 255-6.
142. Singh, *Indian Peacekeeping Force in Sri Lanka*, p. 112.
143. E.N. Rammohan, 'Manipur: A Degenerated Insurgency', in Gill and Sahni (eds.), *Faultlines*, vol. 11, p. 15.

CHAPTER 3

Warfare in Pre-British South Asia: A New Interpretation

KAUSHIK ROY

INTRODUCTION

Most of the Western scholars argue that warfare in pre-British India involved skirmishes by the armed rabbles. Intrigues and treacheries resulting in mass desertions and political negotiations rather than decisive set piece battles and sieges decided the outcomes. This was because of the weak state structure and the peculiar social fabric and cultural ethos of India. In accordance with this line of argument, pre-British India lacked the concept of a strong state. The Indian political culture did not have the concept of a ruler enjoying a monopoly over armed forces in his realm. Frontiers fluctuated frequently and the reach of the state was limited. This was partly because of the Hindu concept of *dharma* and the Islamic concept of *fitna*. *Dharma* which encouraged the divisive political culture of *bheda* (divide and rule) and the fourfold caste system of Hindu society caused further fragmentation of political authority. The ecology of the subcontinent functioned as another break as regards the emergence of a strong centralized state. Abundance of war horses and the numerous armed populace of the countryside backed by rural fortifications prevented any central government before 1800 to demilitarize the society. The net result was continuous private warfare among the large number of small war bands led by the powerbrokers within the subcontinent. All these resulted in stagnation in the techniques and technology of warfare in the subcontinent. It was only in the eighteenth century the British established a strong polity which was able to demilitarize society and deploy an armed force capable of conducting decisive battles and sieges. Many historians have harped on this decisive British breakthrough as a Military Revolution in South Asia which in turn gave birth to a military-fiscal state.[1]

Most books dealing with the military history of India devote only a few pages on pre-British military developments in South Asia.[2] Did India experience any other Military Revolution before the British intervention? And how far could the concept of Military Revolution be used as a heuristic device for explaining military changes, especially when the very concept is coming under challenge by many historians dealing with West European warfare?

Geoffrey Parker's use of the concept of Military Revolution for explaining 'rapid and radical' innovations in early modern West European warfare, which gave the West European states an edge over the extra-European states, have been challenged by Bert S. Hall and Kelly R. DeVries. Both assert that Parker's concept of Military Revolution smacks of technological determinism and further unpacking of the concept, especially in relation to Renaissance technology is required.[3] To be fair to Parker, the latter in my view gives adequate space to managerial changes to explain the Military Revolution of early modern West Europe.[4] However, the timing and rate of military changes emphasized by Parker could be challenged.

Quite a few other scholars also challenge the concept of Military Revolution. They argue that a series of incremental military innovations occurred from 1400 onwards and the slow development of military techniques continued even after 1800. Hence, Jeremy Black argues for using the concept of Military Evolution. Black points out the dynamic growth of non-Western warfare till the eighteenth century.[5] Nicholas Hooper and Matthew Bennett assert that the process of Military Evolution in West Europe started in the 'Dark Age'.[6]

The proponents of Military Revolution have tried to salvage their theory by two means. One group argues that only West Europe (including Greece and Rome) experienced a series of Military Revolutions from Salamis to Vietnam. Clifford J. Rogers has come up with the concept of successive Military Revolutions. The Infantry Revolution occurred between 1420-40 and then came the Artillery Revolution, which in turn generated a Fortress

Revolution, etc.[7] These successive revolutions gave rise to the Western Way of Warfare. A few among them claims that the Western Way of Warfare was characterized by a penchant for decisive battles, supremacy of drilled and disciplined heavy infantry and emphasis on technological superiority.[8] The implication is that the non-Europeans were incapable of conducting decisive battles due to their cultural peculiarities. Interestingly, a British historian of colonial India asserted that due to racial defects, the Indians were incapable of conducting decisive battles.[9] In recent times, it seems that instead of racial superiority of the West Europeans, the far more politically innocent term of cultural uniqueness of the West Europeans is used.

Another group argues that West Europe experienced both Military Revolutions and Revolutions in Military Affairs (RMA). A RMA could be defined as a military organization adopting new technology and techniques of warfare very quickly which revolutionizes the nature of warfare within a short span. But RMA had no lasting effect on broader society. RMA is somewhat equivalent to the concept of Military Technical Revolution but the latter term is mainly used for contemporary military innovations. Military Revolution in this paradigm means comparatively slower changes which have larger ramifications on society. In addition, social and cultural dialectics also influence the configuration of military forces.[10] In such a schema, spatially and temporarily the concept of Military Revolution is so broadened that it seems better to coin the term Military Transformation. John Stone's article somewhat points in this way as he argues that the so-called Infantry Revolution of the fourteenth century was the product of larger socio-economic changes and military technology and techniques improved very slowly in halting stages throughout the medieval era.[11] In the context of Asia, Jos Gommans claims that a Horse Warrior Revolution occurred between CE 1000 and 1800. This did not involve any radical technological breakthrough but introduction of horse warrior economic institutions in the settled agrarian societies of India and China. And horse traders became rulers of these polities which experienced agricultural expansion.[12] If out of 2,500 years of recorded history of India, 800 years is given to a particular revolution and the chief causative factors behind the revolution are non-military, then we need to change the analytical concept from Military Revolution to Military Transformation.

When the concept of Military Revolution appears to be standing on shaky ground, can we fall back on the late Victorian concept of using the schema of decisive battles? Yuval Noah Harari writes that very few battles could be regarded as decisive in the sense of them changing the course of world history. Rightly, he says that a battle-centric military history is far too simplistic.[13] This is not to say that decisive battles never occurred in world history in general and in Indian history in particular. The battles of Adrianopole, Hastings and three Panipats[14] were definitely important though the quantum of their significance could be challenged. Further, mere concentration on battles would leave the broader social, cultural and political aspects of warfare outside our purview.

A new scheme which attempts to explain the nature of warfare in India is the 'Military Synthesis' model. Military Synthesis stands for a complex amalgam of foreign revolutionary military inputs with indigenous military elements. The Indian context made imported military institutions inadequate for the military domination of the subcontinent. Hence, a balance between innovative foreign technologies, managerial techniques and primordial elements in the Indian society was necessary.[15] However, the Military Synthesis model cannot really account for all the military changes that occurred at different speed, at different levels of the society and at different times in South Asian history.

In this article, the author argues that at some moments of time, RMAs had occurred in history. And some spheres of military activities had witnessed Military Evolution. Military Evolution means slow changes which occurred over a large period of time but did not have long-term structural relationship with society. Nor societal dynamics in case of Military Evolution did substantially effect incremental changes in the styles of warfare. And for most of the cases Military Transformation had occurred. We will take South Asia as a case study. However, South Asia's military history will not be assessed in isolation but by putting it on a global perspective. This will help us to reconstruct how the various parts of Eurasia interacted with each other, especially as regards military changes. Peter A. Lorge's recent work is a credible attempt to get military history of Eurasia out of the Eurocentric perspective. He claims that the paradigm of early modern warfare (the use of gunpowder and the evolution of a bureaucracy for administering the military organization) first emerged in China during the eleventh and twelfth centuries. And then from China it spread to other parts of Asia.[16]

Textbooks have often analysed South Asian history on the basis of rise and fall of empires. Many have divided Indian history into three neat categories—ancient, medieval and modern.[17] Some historians have introduced the era of early medieval period which extended from CE 600 to 1300.[18] Instead of going for periodization, this essay will analyse South Asia's military history through the three concepts of RMA, Military Evolution and Military Transformation and follow a chronological perspective whenever possible. Before that a cross-cultural comparative analysis of the South Asian military organizations throughout history is necessary in order to see whether the pre-British indigenous armies were armed rabbles or cohesive disciplined entities.

SIZE AND COHESION OF THE ARMIES OF SOUTH ASIA IN A EURASIAN PERSPECTIVE

Like Parker, John Lynn accepts that the size of the West European armies registered exponential growth during the seventeenth and the eighteenth centuries. Lynn says that the size of the French Army jumped from 55,000 in 1610 to 400,000 in 1689. The growth of the army was sustained by international finances and expansion of the government bureaucracy.[19] Mahinder S. Kingra in an article challenges Parker's linkage between the spread of *trace italienne* and the rise in the size of the armies.[20]

In pre-British India, the indigenous powerbrokers due to the demographic resources of India and high agricultural productivity mobilized huge armies which can only be compared with the pre-modern Chinese armies. To give an example, in CE 546, the Eastern Wei army of Kau Huan numbered 200,000 men.[21]

When Alexander invaded the subcontinent, Porus who was the ruler of central Punjab had, in Diodoros Siculus' estimate 50,000 foot and 3,000 cavalry.[22] In Pliny's estimate, the Nanda Army at that time comprised 600,000 foot and 30,000 cavalry.[23]

One can challenge the estimates of these authors. Similar doubts have also been raised by the historians as regards the size of the Roman Army at different moments of time.[24] At the Third Battle of Panipat (14 January 1761), Ahmad Shah Abdali, the ruler of Afghanistan, and his allies (Muslim rulers of Rohilkhand and Awadh) deployed 41,800 cavalry, 38,000 infantry and about 80 pieces of cannon.[25] Abdali's army was quite big by the standards of eighteenth century. In fact, after the British intervention, during the late eighteenth century, the size of the armies declined in South Asia.[26]

Unlike V.D. Hanson, Lynn argues that the early modern West European Military Revolution could not be traced back to Classical Greece. But early modern West Europe witnessed the emergence of the 'battle culture of forbearance', drill and the regimental system which in turn gave rise to the centralized bureaucratic state.[27] In general, historians of Western armies agree that the regimental ethos along with uniforms and units' standards generated bonds of cohesion among the personnel of the West European armies during the early modern era.[28]

The Roman Army of the Early and Middle Republic comprised citizens under arms led into warfare by the elected magistrates. During the Late Republic, military service was for a minimum of six years which could be extended by the magistrates. However, many individuals volunteered to serve longer. Under the Roman Empire, the individual legions and the auxiliary regiments remained permanently in commission with particular names, numerals and titles. Many soldiers served throughout their lives. In CE 13, Emperor Augustus ordered that military service in the legions should be fixed for 16 years followed by a four-year service in reserve. After retirement, the soldiers were rewarded with a grant of gratuity.[29] After Augustus, the Roman Empire's army was a standing professional force whose soldiers normally served for 25 years.[30] All this generated legionary cohesion and gave the army a strong formal institutional ethos and infrastructure.

However, the standing professional Roman Army started to dissolve under the later Roman Empire. As the structure of the armies became loose, political treacheries often decided the outcome of campaigns in West European history. Castinus' Roman force was crippled in CE 422 against the Vandals in Spain, when the former's Gothic federates deserted.[31] During the fifth and the sixth centuries CE, the military officers became landlords and turned the peasants into armed retainers. The retainers were bound by personal ties to their leaders. The net result was the growth of semi-private armies which engaged in brigandage. Many barbarian chieftains also became landlords and pursued the above-mentioned policy. The landlords' power increased as they expanded their estates which in turn brought more dependents under their folds. The estates of the landlords were further fortified by the construction of castles.[32]

The pre-British South Asian armies were not collections of disorganized mobs but comprised veteran soldiers organized in well-defined units and the armies were able to manoeuvre in the battlefields. Tiruvalluvar, the author of the Tamil Classic *Kural* which was composed probably during the sixth century CE emphasizes the importance of administration for retaining loyalty of the personnel in order to maintain the combat effectiveness of the army.[33] The *Kural* points out: 'An army will win when there is no desertion.'[34] J. Sundaram asserts that the Chola infantry was organized into regiments. One such unit was named as Rajaraja *terinda vil.* Rajaraja refers to the king who raised the unit during the mid-eleventh century. *Terinda* was the title given to the unit due to its exemplary field record and *vil* means that the unit comprised archers. The *valavans* were in charge of training the infantry units and the *nayakans* were commandants of the infantry units.[35] The army of the Delhi Sultanate (CE 1206-1526) had a hierarchical officer corps: *naqib* was in charge of 100 troops, *sarhang* commanded 500 soldiers, and so on. The core of the sultanate army comprised regular troops who had uniform.[36] The Afghan cavalry of Ahmad Shah Abdali which fought the Marathas at the Third Battle of Panipat comprised 24 *dustas* (equivalent of regiments). Each *dusta* had 1,200 horsemen.[37]

During the Battle of Hydaspes (June 326 BCE), Porus ordered the cavalry on his right wing to come in support of the cavalry which was deployed on his left wing.[38] Zahir-ud-din Babur just before the First Battle of Panipat (21 April 1526) describes his order of battle in the following words:

> Our right was Humayun, Khwaja Kalan . . . our left was Muhammad Sultan Mirza, Mahdi Khwaja, Adil Sultan. . . . The right hand of the centre was Chin Timur Sultan, Sulaiman Mirza. . . . The left of the centre was

> Khalifa, Khwaja, Mir-i-miran. . . . The advance was Khusrau Kukuldash. . . . Abdul-aziz the Master of Horse was posted as the reserve. For the turning party (*tulghuma*) at the point of the right wing, we fixed on Red Wali . . . with their Mughals; for the turning party at the point of the left wing, we arrayed Qara-quzi . . . these two parties . . . were to turn his rear, one from the right, the other from the left.[39]

The proponents of the Western Way of Warfare assert that drilled and disciplined heavy infantry emerged in Classical Greece. The civic sense of small Greek farmers encouraged decisive close quarter combat.[40] By about 650 BCE, the phalanx formation emerged in Classical Greece. The tactics of phalanx depended on *osthismos* (shoving and pushing).[41]

Combat-effective infantry was not totally absent in the Asian armies. Though Achaemenid Persia depended on cavalry as the premier arm, they, at least in the opinion of one French historian, did not neglect heavy infantry totally. In addition to dependence on Greek mercenary heavy infantry, the Persians also introduced heavy infantry equipped somewhat like the hoplites under the direct control of the Great King.[42]

Arrian spoke highly of Porus' infantry:

> The foot-soldiers carry a bow made of equal length with the man who bears it. This they rest upon the ground, pressing against it with their left foot thus discharge the arrow, having drawn the string far backwards: for the shaft they use is little short of being three yards long, and there is nothing which can resist an Indian archer's shot—neither shield nor breastplate.[43]

Infantry known as *paiks* in the medieval age played an important though a secondary role in the armies of the Delhi Sultanate. The *paiks* recruited from Bengal were equipped with bows and arrows as well as swords and shields and protected by armour. Most of the *paiks* were Hindus. Ali Athar writes that they were trained with their weapons during the hunting expeditions launched by the sultans and they also underwent drill. The *paiks* were deployed in battle either at the front or in the flanks. At times, they undertook offensive operation in the battle by attacking the cavalry and preparing the way for a charge by the elephants. The *paiks* were especially important in sieges. They scaled the walls with ladders and also made breaches in the walls of the forts.[44] Even before the British intervention, the use of firearms spread among the infantry. In the Third Battle of

Panipat, the Maratha Army had 9,000 infantry equipped with firelocks.[45]

The *acharyas* (Hindu classical theorists) argued that only the Kshatriyas should join military service. The reality is different. A story of Somadeva's *Kathasaritsagara* (composed between sixth and eighth centuries CE) tells us that many Brahmins acquired training in weapons and developed their physique through wrestling.[46] Obviously such Brahmins joined the army. The *chaturvarga* system was the ideal for the orthodox Brahmin thinkers. In reality, at the beginning of the Common Era, there were 60 mixed castes.[47]

Decisive battles were not unique to the Western military culture. Battle-centric strategy, which was a crucial component of the so-called Western Way of Warfare, was not favoured by the medieval West European commanders. Medieval West European warfare for most of the time comprised small-scale skirmishes based on regional network of fortifications.[48] The *acharyas* also preferred siege warfare rather than battles. Manu's *Manava-Dharmashastra* composed at the beginning of the Common Era notes: 'When a king launches a military expedition against the realm of an enemy, he should advance at a measured pace towards the enemy's fort.'[49] However, when occasionally the South Asian armies engaged in decisive battles, their military effectiveness cannot be neglected. Plutarch wrote about the Battle of Hydaspes: 'The contest, which began early in the morning, was so obstinately maintained that it was fully the eighth hour of the day before the Indians renounced all attempts at further resistance.'[50] At Hydaspes, against the army of a regional monarch of South Asia, Alexander lost 1,000 soldiers and several thousands were wounded. In contrast, at Gaugemela (331 BCE) against the Persian army of Darius, Alexander lost 500 men and many were wounded.[51]

The state was not always predominant in West European history and the polity was not always marginal throughout South Asian history. According to a modern historian, even the Romans did not have the concept of possessing a military border.[52] In post-Roman West Europe, rather than the state, the chieftains and the captains took over the task of feeding, maintaining, clothing and equipping the soldiers. And these units were then hired out to the feudal magnates.[53]

Strabo tells us that the Mauryan Empire had a monopoly over horses and elephants and the soldiers were equipped by the government.[54] Each elephant consumed about 600 pounds of green fodder per day.[55] The presence of large number of elephants in the South Asian armies during ancient and medieval eras and the requirements of feeding them necessitated the emergence of a bureaucracy. The Delhi Sultanate had a centralized machinery (*diwan-i-ariz*) for administering and supplying the army.[56] The army in pre-British India maintained the state and was also one of the motors behind its administrative expansion and penetration into the society.[57] One of the indices of John Brewer's concept of military fiscal state, i.e. high taxation for maintaining large armies, if applied to the British-Indian period will not shed much light on the contours of the Raj, because the trend of extracting a huge surplus from the soil and utilizing it for military purpose was also present in the Mughal polity.[58]

RMAs IN SOUTH ASIAN HISTORY

RMA IN THE GUPTA ERA

The Inner Asian steppe nomads skilled in mounted archery from the dawn of history till the supremacy of the gunpowder weapons (*c.* 1800) threatened all the agrarian bureaucratic empires located along the rim of Eurasia. The techniques of cavalry warfare registered innovations with time. Between the fifth and seventh centuries BCE, the bit, bridle, headstall and reins together with the form of S-shaped cheek bars and saddles were developed by the steppe nomads.[59]

In order to combat the steppe nomadic cavalry, the Assyrians, writes John W. Eadie, came up with unarmoured mounted archers in the ninth century BCE. In order to protect the unarmoured mounted archers, the Assyrians introduced partially armoured cavalrymen. The horsemen were covered with mail shirts (metal plates sewn to the tunics) and equipped with pikes.[60] The principal threat to the Han Empire came from the Hsiung-nu nomads. Around 100 BCE, the Han Emperor Wu-Ti was aware of the importance of Central Asian steppe horses and acquired them by sending military expeditions into Ferghana. The Han military establishment was able to deploy between

30,000 to 100,000 horses for a campaign.[61] The horsemen of Classical Greece were equipped with swords and shields. Classical Greece encountered Scythian mounted archers equipped with compound bows and dressed in breeches, high shoes and riding caftans with long sleeves. The Sarmatians in the age of Roman Emperor Trajan were equipped with the compound bows used by the nomads.[62]

The earliest historical evidence we have of the use of horse archers in battles fought in the subcontinent was the Battle of Hydaspes. Alexander used the horse archers to throw Porus' left wing into confusion.[63] Large-scale attacks by mounted archers occurred under the Gupta Empire (which lasted from CE 320 to 527). Around CE 350, the Huns established themselves near Herat in Afghanistan and started attacking Persia and India.[64] Skandagupta (Gupta emperor from CE 455-67) was able to defeat the Huns who were called Hunas by the *acharyas*.[65]

How does one explain the success of the Guptas against the Huns? The Gupta Army recruited from the traditional four sources: *maula* (soldiers from hereditary military families who constituted the regular army), *bhrata* (mercenaries), *mitra* (soldiers provided by the vassal rulers and allies) and *srenis* (soldiers provided by various commercial guilds).[66] Ashvini Agrawal on the basis of literary and numismatic evidence asserts that under Chandragupta Vikramaditya (emperor from CE 375-414), the Gupta Army became cavalry centric.[67] Thanks to the impact of the Parthians who attacked the subcontinent during the first century CE, the Guptas took to mounted horse archery. Several Gupta emperors were depicted in their coins as equipped with the nomadic compound bows on horseback.[68] The coins of Samudragupta (emperor from CE 335-80) show the presence of a close fitting cap on the horsemen.[69] The Gupta cavaliers adopted riding trousers, belts and boots introduced by the Kushanas (Yueh-Chi tribes of Central Asia who invaded India). King Wu Ling of Chao who reigned from 325 to 299 BCE despite facing cultural obstruction from the Chinese nobility imitated the nomadic practice of riding and introduced trousers. The nomadic riders wore short jackets but the Chinese considered wearing long gowns as obligatory for a man of status.[70]

Kalidasa's *Raghuvamsam* composed between fourth and fifth

centuries CE gives a picture of force structures and combat between Raghu (probably Skandagupta) and the *Yavanas* (Huns) in north-west India. Kalidasa says that the *Yavanas* mainly depended on mounted archers. The mobile cavalry of the *Yavanas/Turangas* created terror among the *samantarajas* (warden of the marches). Hence, Raghu also deployed mounted archers. When the mounted archers launched their barrage of arrows, the sky became dark. Though the *Yavanas* were encased in armour still they fell victim to the barrage of arrows. When the mounted archers dislocated the *Yavana* formation, then Raghu's heavy cavalry equipped with *bhalas* (heavy spears, somewhat equivalent to lances) charged the *Yavanas*. Kalidasa implies that the Indian rulers were aware of the importance of Central Asian mounts. Raghu with his victorious cavalry army acquired many good horses as tributes from Kamboja (Gandhara).[71]

According to Denis Sinor, the earliest known representations of the stirrup come from Korea and Japan and can be dated between the fourth and fifth century CE. He argues that the Huns (CE 370-450) did not use stirrups but its use was known to the Avars (Juan-juan of the Inner Asian Steppes) during the sixth century CE. From the Avars, the Byzantines learnt the use of stirrups. And even in the early seventh century CE, stirrups were not used in Iran. Sinor continues that the use of iron horseshoe was known in Europe as early as fifth century CE but was used in Inner Asia only by the Mongols.[72]

Gurcharn Sandhu asserts that the superiority of the Gupta heavy cavalry over the Huns was due to the introduction of loop stirrups by the former. The reliefs and sculptures of Barhut and Sanchi prove the use of loop stirrup by the Hindus. The Huns rode without stirrups. Hence, the rider had an unsteady seat and the mount could function only as a precarious weapon platform. Due to the use of loop stirrup, the Gupta cavaliers were able to press a charge and engage in close-quarter combat. In contrast, the Hun cavaliers without stirrup were easily pulled down from their horses during close quarter combat. For implementing a cavalry charge and close-quarter combat, the Guptas introduced heavy cavalry which meant that the riders wore knee-length chain mail and the horses were also covered with plate armour. The Gupta heavy cavalry equipped with lances was able to deliver a compact close order charge. In addition, the Guptas also had light cavalry. The riders belonging to light cavalry wore only a padded jacket and were equipped with bows and arrows.[73]

Mounted archery of the Guptas and ecological constraints of steppe nomadic warfare resulted in failure of the Huns to conquer India. Lack of adequate amount of grasslands and absence of arid zone explains the Huns' and the Mongols' failure to conquer West Europe[74] and also the warm and moist fertile plains of north India. However, mounted archery vanished from the Hindu armies after the demise of the Guptas. This was because the steppe nomadic threat vanished in the sixth century CE and India lacked arid climate and adequate amount of grasslands for breeding high quality horses. Finally, the Gupta emperors failed to establish any linkage between the corps of mounted archers they introduced as a military measure with contemporary political economy and the social fabric.

A Naval RMA

In ancient and medieval South Asia, the hull of a ship was constructed by means of planks joined horizontally. The planks were joined by means of stitches of coconut fibre. However, the twine broke under the impact of violent wind and waves. To prevent leakage, the small holes were covered with wool and wax. The ships had two to three masts and square type sails were used. The speed of the ships was about 6 miles per hour. The anchors were made of rocks. In early medieval India, some of the ships were quite big and were used for transporting horses and elephants by the monarchs from peninsular India to Sri Lanka. For direction finding, the ancient Hindus used a device comprising an iron fish that floated in a vessel of oil and pointed to the north. The use of magnetic needle as mariner's compass came to South Asia from China around the end of the eleventh century.[75] During the first half of the eleventh century, the Chola Navy made an expedition to Malaysia and in the course of campaign captured the Nicobar Islands in the Bay of Bengal.[76] This was the first and last case of naval power projection by a South Asian polity in the pre-British era. Unfortunately, we lack details regarding the Chola Blue Water Navy.

The Delhi Sultanate had no navy. The Mughal Empire made

sporadic attempts to construct a navy. Towards the end of the seventeenth century, when Aurangzeb decided to construct a navy for protecting foreign trade, his *Wazir* (Prime Minister) Jafar Khan responded by saying that there was no lack of money or timber for building ships but trained men were lacking to direct the navy. The Siddis sponsored by Aurangzeb maintained a coastal navy along the shores of west India. Each ship was of 300 to 400 tons armed with heavy ordnance, and carried matchlock men and spearmen. The largest Mughal ship had 800 guns and 400 musketeers on board. The big Mughal ships were preceded by pilot vessels known as *khelna* boats which sounded the depth of the water and guided the heavy ships across the channels of the rivers during high and low tides.[77]

During the sixteenth and the seventeenth centuries, the West European navies intruded into the Indian Ocean and established maritime supremacy. The Portuguese were superior to the Indians in scientific knowledge, ship-borne firearms and naval technology.[78] The Portuguese vessels which the South Asian maritime states encountered during the sixteenth century were more strongly built and had more effective cannons.[79] Anirudh Deshpande after making a case study of naval warfare along the west Indian coast during the seventeenth and eighteenth centuries claims that as far as coastal warfare was concerned, the so-called West European Naval Revolution did not play any significant part. The British East India Company (EIC) was able to establish coastal supremacy along India's west coast through huge financial investment and by exerting diplomatic pressure. The EIC's policy of divide and rule against the indigenous powers rather than superior naval hardware played a more important role in establishing British maritime supremacy.[80]

As regards coastal warfare, inadequate knowledge about shallow water, winding creeks, sandbanks and changing tides definitely proved to be a handicap for the British mariners. However, there is no doubt that in the high sea, the West European Naval RMA proved to be supreme. During the second half of the eighteenth century, Haidar Ali, the ruler of Mysore and his son Tipu Sultan attempted to construct an ocean-going navy with the aid of Dutch and British mercenaries. Haidar and Tipu's attempt was unsuccessful because of the treachery of the mercenaries, inadequate funding and lack of skilled shipwrights at their disposal.[81]

MILITARY EVOLUTION: SIEGE WARFARE IN SOUTH ASIA

Ancient India developed a theory of siege warfare. Kautilya in the *Arthashastra*, composed roughly around 300 BCE, notes: 'Among them a river fort and a mountain fort are places for the protectors of the country, a desert fort and a jungle fort are places . . . of retreat in times of calamity.'[82] Tiruvalluvar writes that a fort should be surrounded by a high and massive wall. It should have large space inside but the approaches from the outside should be narrow. He further emphasizes that foodstuffs and other supplies should be stored within the forts.[83]

Towards the end of the second century BCE, the Andhras in the Deccan had towns protected by walls.[84] Several types of fortifications are depicted in bas-reliefs: rectangular towers, towers topped with hemispherical domes and towers with loopholes above the curtain walls in Amaravati during the second century CE. The Kushanas introduced Central Asian traditions of fortifications in the subcontinent. The chief features were curtain walls pierced with long loopholes and hollow semi-circular towers. This type of architectural remains are found in Mathura and depicted in Gandhara arts. Mathura was one of the capitals of the Kushanas during the beginning of the Common Era. During second century CE, in the types of fortifications depicted at the bas-reliefs about Mathura, we find quadrangular towers and rectangular merlons, as well as hollow spherical towers. However, the indigenous tradition was construction of solid and not hollow towers. Except north-west India, fortifications in the subcontinent during the Common Era were characterized by the presence of massive ramparts flanked by solid quadrangular towers.[85]

We are not sure whether ancient and early medieval India, like the Greeks and the Romans, possessed artillery machines like catapult for shooting arrows, ballista for throwing stones and scorpio for discharging javelins. The motive power of these machines was based on twisted cords or thongs. Dionysius the Elder, the tyrant of Syracuse around 400 BCE, constructed catapults with the help of the engineers brought from all over the Mediterranean world. According to one view, towers with battlements for shooting arrows and stones were invented around 750 BCE in the Near East. In Leigh Alexander's view, the Phoenicians possessed such machines and through them, the siege engines passed into the hands of the Carthaginians and then through them to Sicily. Alexander continues that the knowledge of ballista came to Near East from Assyria around 800 BCE. It is to be noted that other siege engines like battering rams and moving towers on wheels were also used by the Assyrians.[86] The Hittites used the technique of building an earthen ramp at a low spot in the wall and rolling large covered battering rams into place. The Assyrians constructed wooden siege towers taller than the walls of the cities under siege and used archers to provide covering fire for the battering ram crews working below.[87] The Hindu *acharya*s spoke about the *yantra*s (machines) but unfortunately have given no details.

The mangonels/*manjaniq*s (stone-throwing catapults) were first introduced in Sind during the first half of eighth century CE by the Arab commander Muhammad Kasim. The *Chach-Nama* describes: 'Muhammad Kasim ordered the mangonels to be placed on boats, and went towards the fort of Nirun.'[88] The Arabs got the knowledge of stone-throwing machines either from China or from the Byzantine Empire or from both. From Sind, the mangonels were introduced in north and central India by Alauddin Khalji (Sultan of Delhi from 1296 till 1315). During the fifteenth century, for siege warfare mangonels and crossbows were used.[89]

The use of gunpowder technology in siege warfare gradually spread to Deccan and south India. In 1472, the military engineers of the Bahmani Army used explosive mines while besieging the Belgaum Fort which belonged to an ally of the Vijayanagara Empire. In 1502, the Portuguese naval squadron was bombarded from the hilltop overlooking the Bhatkal Port. In 1504, the ruler of Ahmadnagar maintained artillery in the Chaul Port. In 1510, when the Portuguese occupied Goa, the Bijapuri garrison used artillery.[90] Due to the introduction of artillery, from the late fifteenth century onwards, the fortress architecture registered slow changes. The enclosed space within the fort was expanded in order to make it difficult for the besieging force to aim its cannons at the built-up areas of the fort.[91] Many Mughal forts had wet ditch and inner rampart walls on which heavy guns were mounted.[92]

Trace italienne reached the subcontinent through the European mercenaries during the late eighteenth century. But this practice was not widely followed. In the last decade of the eighteenth century, Tipu Sultan of Mysore with the help of the French engineers refortified Seringapatnam (his capital) in accordance with the *trace italienne* design. Though Seringapatnam was redesigned in accordance with *trace italienne* design it fell quite easily to the EIC in 1799. But the Bharatpur Fort which was protected in the traditional Indian style with earthen embankments and ditch proved to be a hard nut to crack for the EIC during the two sieges of 1805 and 1826.[93]

MILITARY TRANSFORMATIONS IN SOUTH ASIA

Chariots and Transformation of South Asian Warfare: 1200-300 BCE

A Military Transformation occurred in the Near East about 3,000 years ago. It involved the use of horse-drawn spoked-wheeled war chariots.[94] P.R.S. Moorey asserts that the technique of constructing spoked wheels spread from south Russia to Anatolia around 2000 BCE. The introduction of light war chariots into Western Asia by the Hurrian-speaking people in the middle of the second millennium BCE also witnessed the use of the composite bows by the charioteers.[95] While the Egyptians focused on small, light two-wheeled chariots geared for speed and manoeuvrability, the armies of the Levant adopted the four-wheeled chariots with a crew of three to four. These chariots emphasized better protection, firepower and shock potential.[96] In China, horse-driven chariots operated in the Shang (1765-1123 BCE) and Chou (1122-256 BCE) periods.[97] The chariots reached both China and India from the same source, i.e. Central Asia, around roughly the same time, i.e. 1300-1200 BCE.[98]

The war chariots enabled the Aryans from Central Asia to invade the subcontinent in about 1200 BCE. The Indian epic *Mahabharata* (though composed around 500 BCE) depicting the scenario between 1000 and 800 BCE speaks about the aristocratic warriors fighting with bows and arrows from their chariots.[99] The Indo-Aryans used simple bow and drew the bowstring up to the ear before discharging the arrow. The arrow was made of wood, reed or bamboo. The rear end of the shaft was fletched with three or four feathers in order to steady the flight path of the arrow.[100]

The Indians used both heavy and light chariots. Initially, the

Vedic chariots were light two-wheeled vehicles with four to six spokes in each wheel. The wheels were mounted on a wooden axle and the body of the chariot was made up of a wooden frame covered with animal hide or wicker work.[101] The *Rigveda* speaks of encounters between hostile *rathas* (chariots). Ajatashatru, the ruler of Magadha (493-462 BCE) during his war with Vaishali introduced *rathamusala*. It was a chariot to which a mace was attached.[102] This was similar to the scythed chariots which Darius used against Alexander at the Battle of Gaugemela. Alexander K. Nefiodkin asserts that scythed chariots emerged in Persia between 467-458 BCE in response to the threat posed by the Greek hoplites.[103] It is to be noted that sources concerned with ancient India refer to scythed chariots under Ajatashatru but no mention is made of such chariots in Porus' army or in the Mauryan times. We know that Darius I's admiral Scylax annexed Punjab and many men from Punjab joined Xerxes' army which invaded Greece.[104] Either scythed chariots emerged in north India and then passed into Persia through the Indian mercenaries or it emerged in Persia and then through the same medium came to north India, or the weapon system emerged separately and almost simultaneously in two different locations.

The Hindu charioteer controlled the horse by means of a bit in the horse's mouth. It comprised of a metal spike mounted in the middle of a leather collar which went round the muzzle of the horse.[105] The bit was present in Shang times and the bit with cheek bars appeared in China, Inner Asia and in West Asia. Whether there was a common source of origin or not is still debated.[106] Probably, India acquired bits from the Aryans who came from Central Asia. The Indo-Aryans integrated the chariots with indigenous military techniques—*paiks*, cavalry and elephants. Thus, emerged the *chaturanga bala* (four-limbed army) of the Mauryas (324-187 BCE).[107]

Elephants and Transformation of South Asian Warfare

The post-Mauryan era witnessed the abolition of the chariot corps. After the collapse of the Guptas, the South Asian armies depended mainly on elephants supported by the *paiks*. War elephants of South Asia impressed the foreigners. Q. Curtius Rufus in his *History of Alexander the Great* notes: 'The elephants are more powerful than those tamed in Africa, and their size corresponds to their strength.'[108] From the fragments of *Indika* preserved in Diodoros' account, we know that Megasthenes, the Greek ambassador in Chandragupta Maurya's court, was also impressed by the elephants. He states:

> It is prolific, besides, in elephants, which are of monstrous bulk, as its soil supplies food in unsparing profusion, making these animals far to exceed in strength those are bred in Libya. It results also that, since they are caught in great numbers by the Indians and trained for war, they are of great moment in turning the scale of victory.[109]

Elephants proved to be adequately effective in warfare outside South Asia. Seleucus took 500 elephants from Chandragupta and used some of these beasts with decisive effect during the wars of Diadochi. Seleucus's successor Antiochus III of Syria invaded north-west India around 206 BCE. He received 150 elephants from an Indian ruler named Subhagsena and went back.[110] Probably in response to the threat posed by the Seleucid Empire, Ptolemid Egypt also acquired elephants from the upper reaches of Nile. From the Ptolemids, the tradition of using war elephants passed on to the Carthaginians who used Libyan bush elephants. Hannibal (247-183 BCE) from Spain marched towards Italy in 218 BCE with 80 elephants.[111] The Carthaginian elephants during the passage of Rhone River and at the Battle of Metarus during the Second Punic War had Indian drivers. The *mahout* (Indian elephant driver) controlled the beast with the help of an *ankush* (iron goad).[112] The use of war elephants by the Carthaginians probably encouraged the Romans to use them. In 197 BCE, at Cynoscephalae, the Roman commander Flamininus used elephants to smash the left wing of the Macedonian Army of Philip.[113]

Elephants were also used for warfare in South China. In 506 BCE, the Ch'u used elephants against the Wu. In December CE 554, at the Battle of Chiang-ling, armoured elephants carrying towers and guided by Malayan trainers were sent against the army of Western Wei by the Liang defenders. The elephants were turned back by the flight of arrows. The Southern Han used elephants with towers (called *howdah* by the Indians). Each tower carried 10 men. The elephant battalion was effective during the Han invasion of Ch'u in CE 948. However, in CE 971, the Sung soldiers equipped with crossbows decimated the elephant corps of the Southern Han.[114]

Harsa's (ruler of north India from CE 606-48) army mainly depended on elephants. Bana in *Harsacharita* (composed between seventh and eighth centuries CE) says that the forest-covered Vindhya Mountains in central India were full of elephants.[115] The elephants constituted the strike corps of the Rashtrakuta Army during the seventh century CE. The *Nitisara* of Kamandaka (composed between sixth and eighth centuries CE) tells us: 'The consolidated units of the enemy should be liquidated by units of furious elephants.'[116] The Hindu armies of early medieval period continued to depend on elephants because of their availability in large numbers and problems of raising good-quality horses. Elephants became extinct in South China and also in Libya. Further, the elephant like the main battle tank functioned as a battering ram. It was also useful as a command vehicle as the *senapati* (commander) from his *howdah* could survey the battlefield from a higher position. However, the commander could also easily be targeted by the enemy bowmen. To a great extent, the elephant is somewhat like a helicopter. It provides the commander a top-down view of the battlefield but the command vehicle is also vulnerable to enemy fire from the ground.

Military Transformation under the Turks

According to Charles Oman, the Battle of Adrianopole witnessed the advent of the age of horse which continued till the advent of gunpowder infantry armies during the early modern age.[117] Bernard S. Bachrach challenges the view of cavalry-centric warfare in medieval West Europe. In his view, small-scale siege warfare by the infantry equipped with longbows and pikes constituted the principal feature of the medieval West European military landscape.[118] However, the importance of cavalry in medieval Eurasia could not be discounted. R.J. Barendse rightly says that between the tenth and thirteenth centuries, Eurasia witnessed the weakening of the central government and rise of the horse warrior aristocracy. They usurped the power of the monarchs and exploited the peasantry.[119] In the view of at least one scholar, Charles Martel the Carolingian ruler developed heavily armed cavalry in response to the mounted warriors of

Islam.[120] Heavy cavalry predominated in the Umayyad army. Iron stirrups replaced wooden ones by the 680s.[121]

K.S. Lal opines that the Turks were able to overwhelm the Hindus due to their possession of better horses, the institution of *ghulams* and the ideology of *jihad*.[122] *Ghulam* was an institution used by the Islamic polities from Egypt in the west till India in the east, for integrating the Turks into the Islamic military culture as slave soldiers. The *ghulams* were equivalent to the *mamluks*.[123] Between ninth and eleventh centuries CE, the elephant-*paik* centric Hindu armies had no successful counter to the mounted archery of the Turks. Even the Byzantine Army in the eleventh century collapsed against the Turkish mounted archers in Anatolia.[124] Ranabir Chakravarti asserts that both the Palas (CE 750-1175) and the Senas (CE 1096-1225) of Bengal built-up cavalry-centric armies. The powerful cavalry at the disposal of the Palas enabled them to project power into north India and to fight the other two premier powers: the Rashtrakutas of Deccan and the Pratiharas of west India, respectively. The horses were brought at the fairs organized by the traders who came from the lands beyond north-west India through the Karakoram Highway. When the supply of the Central Asian horses ceased due to the invasion of the Turks across north-west India, the rulers of Bengal acquired horses through Kamrup (Assam). These horses came either from Tibet and Bhutan or even Yunnan province of south China. The rulers of Bengal even exported some of these horses by ships to the Hindu kingdoms located in southern India.[125] Besides better horses, the steppe nomads possessed better bows. The composite bows of the nomads had greater range and penetrating power compared to the simple bows (which were made of single piece of wood) used by the Hindus of north and south India. The Hoysala (of western Deccan) cavalry during the twelfth and thirteenth centuries reached quite a high level of technical perfection. The horses were shoed and saddle with high pommel and cantle became common. However, being armed with lances and swords, the Hoysala cavalry failed against the mounted archery of the Turkish cavalry of the Delhi Sultanate.[126]

The horse archers threatened all the agrarian civilizations of the world. The men on horseback were able to dislocate the foot slogging *paik*-elephant-centric armies of the agrarian states of South Asia by increasing the 'speed of battle'. The element of surprise and mobility provided by horse archery enabled the Inner Asian nomads to dominate the north Indian plain. But, India did not possess the huge grasslands necessary for grazing the Central Asian steppe horses. And the nomads' composite bows were vulnerable to moisture.[127] India's climate is characterized by heavy rainfall and high temperature. Moreover, the nomadic warriors gradually merged with the settled agrarian populace. The Turkish rulers used the institution of *iqta* (during the Delhi Sultanate) and *jagirs* granted as part of the *mansabdari* system (under the Mughals) in order to integrate the horse warriors and their cavalry retainers with the subcontinent's rural society. The *iqtas* and the *mansabs* were somewhat equivalent to the fiefs of medieval West Europe. However, the *iqtas* and the *mansabs* were not hereditary like the medieval West European fiefs. Moreover, the *mansabdars* were periodically transferred from one *jagir* to another. Gradually, Indian ecology and the existence of numerous forts forced the Delhi Sultans and the Mughal *badshahs* to integrate their horse archers with *paiks*, siege cannons and elephantry. Thus, the Sultanate and the Mughal period witnessed a transition from an Empire on Horseback towards a rudimentary Gunpowder Empire.[128]

Gunpowder Warfare in Land and Military Transformation in British-India: 1740-1850

Long before the advent of the British, the South Asian monarchs were tinkering with gunpowder technology. Iqtidar Alam Khan claims that the skill of manufacture and use of gunpowder were first introduced in north-west India during the thirteenth century by the Mongol invaders. The Mongols got it from the Chinese. The Delhi Sultanate used gunpowder for constructing pyrotechnical devices. For instance, the Sultanate military establishment had at its disposal *bans* (rockets). At the Battle of Ghagra (1529), the Afghans of Bengal used *bans* against Babur. Initially, the knowledge of *bans* came to India from China. By the last decade of the sixteenth century, the *bans* were packed with sharpnels. Towards the second half of the eighteenth century, Tipu Sultan further developed the *bans* and then the British tried to develop it as Congreve rockets.[129]

In the fifteenth century, the artillery pieces of India were made with brass or bronze. In the sixteenth century, heavy mortars (*kazans*) were introduced. Khan asserts that this was the result of improved designs and casting methods learnt from the West. This knowledge was brought into India both by Babur and the Portuguese. The *kazans* were used both in siege warfare and in the battlefield. However, the *kazans* were not battle winning weapons. Despite the presence of 21 *kazans* at the Battle of Kanauj (1540), Sher Shah Suri's cavalry charge was successful in destroying Humayun's (Babur's son) firepower-heavy Mughal Army. Besides the low rate of fire, the weight of these pieces seriously impeded the mobility of the army. Hence, Babur during his campaigns in India (1526-30) used *zarb-zans* (light cannons), which were miniature replicas of the *kazans*.[130] The argument that the Asian rulers regarded cannons as sacro-magical replicas[131] cannot be sustained. Despite the use of gunpowder weapons, unlike in West Europe, handguns-equipped infantry supported by mobile field artillery did not become the dominant elements of warfare in Persia, China and India. This was because the bows of mobile steppe nomadic-mounted archers remained more effective than muskets at least till the 1750s. And due to the beginning of industrialization, the West European states unlike the agrarian bureaucratic polities of Asia mastered cast iron technology.[132]

William R. Thompson asserts that advanced West Eurasian political economy and military superiority accelerated political disintegration of the Asian polities.[133] The elements of war making which gave the West a crucial superiority if not supremacy from the eighteenth century onwards needs to be analysed. Between 1600 and 1800, north-west Europe experienced a military transformation which gave birth to disciplined infantry equipped with hand-held firearms, supported by mobile field artillery. Every significant power seemed to imitate the West European military transformation. During Selim III's reign (1789-1807), the Ottoman government acquired muskets and bayonets from Britain. Later, rifles were obtained from France and Sweden. German and Russian renegades were brought in as drill masters.[134]

The EIC imported this lethal combination into South Asia in mid-eighteenth century. However, the South Asian demography, economy and geography resulted in several changes. Instead of the underemployed urban proletariat on which the European infantry forces mainly depended,[135] the British-led Sepoy Army had to depend on the peasants. The EIC maintained four armies (Bombay Army, Madras Army, Punjab Frontier Force and the Bengal Army), which recruited Indians. The Bengal Army was the largest among these four. It recruited mainly *Purbiyas* (Brahmins and Rajputs from Awadh [Eastern U.P.] and Bihar), who were generally younger sons of the small peasants.[136] The EIC was forced to tune its military machine to the demands of the agrarian scenario: for example, *furlough* was given to the sepoys at harvest time. Faced with the enormous manpower reserve of South Asia and its cultural, linguistic and religious heterogeneity, the army had to chalk out a strategy for selecting groups for military recruitment. Neither the quantum of coercion nor the amount of perquisites delivered had to be overwhelming, because there were no other potential employers left in the subcontinent to whom the sepoys and the sowars could turn for employment. The Indian economy was backward. So, there were no competing civilian sectors which could attract the recruits away from a long-term volunteer military force. The British techniques which transformed the rural recruits into disciplined and obedient soldiers were the court martial mechanism, the regimental fabric and the welfare bureaucracy. The end result was the construction of a bureaucratic standing army.[137]

In early modern West European warfare, light cavalry became useless due to the rise of the volley firing technique by the infantry. But in India the lack of light cavalry initially hampered the EIC's military capability. The heavy British cavalry of the EIC was geared for shock actions. But for reconnaissance, foraging and screening, hit and run attacks, the British needed light cavalry.[138] Further, the vast theatre of military operations in India made it necessary for the combatants to maintain light cavalry.[139] So the EIC was forced to raise sowars on the Indian model through the *siladari* system. It is to be noted that light cavalry played an important role in Austria and Poland at least till the seventeenth century.[140] In the final analysis, the British were able to dominate South Asia militarily not merely through the Western modelled infantry, but because they were able to construct a hybrid military machine by integrating institutions from the East and the West.

This is not to suggest that the construction of a hybrid military machine is an example of European exceptionalism. With the help of the French military mercenaries, both the Marathas and the Sikhs were able to modernize their military machines. Ranjit Singh effectively integrated the field artillery of the European armies with the light irregular cavalry of the Sikhs. However, Ranjit Singh's successors failed to build a stable political infrastructure. This proved to be the Achilles heel of the *Dal Khalsa* during the Anglo-Sikh Wars. Overall, both the Marathas and the Sikhs failed to build a professional indigenous officer corps capable of commanding their Westernized armies in the battlefields. The European mercenary officers on whom both these two powers depended proved to be disloyal.[141]

CONCLUSION

The presence of centralized bureaucratic states and standing professional armies have been overemphasized in West European history and underemphasized in South Asian historical studies. RMAs, Military Evolutions and Military Transformations occurred due to diffusion of technologies and techniques of warfare through the various parts of Eurasia. Global flow of technologies and military mercenaries were common even before the advent of globalization. Till 1700, military techniques and technology from Central Asia rather than West Europe shaped the South Asian military landscape. Both in the case of RMAs and Military Transformations, military mercenaries who migrated over long distances played a crucial role in shaping the flow of technologies and techniques of warfare. External threat and dynamic leadership from the men at the top are definitely important factors in ushering a successful RMA. For the Gupta Age RMA, the threat posed by the Huns and the charismatic leadership provided by Chandragupta Vikramaditya and Skandagupta were crucial factors. Since RMA has no integral relationship with wider societal aspects, it dies out quickly. The Gupta RMA is such a case.

However, in case of a Military Transformation, a sizeable segment of society has a vested interest in the shaping of the military organization. Hence, the paradigm of warfare established by a Military Transformation remains in the historical canvas for a longer period. Chariots and elephants influenced civilian values so much that in ancient India the chariot became the symbol of sovereignty and the elephant the sign of royalty from the medieval era till the advent of the British. The long-lasting paradigms of Turkish and EIC's military machines were due to the fact that both the Turkish warlords and the British were able to integrate the crucial systems of warfare with the indigenous social fabric.

ACKNOWLEDGEMENTS

My thanks to Suhrita for her comments on an earlier draft. I am grateful to Pratyay Nath for supplying me with some of the sources.

NOTES

1. For the concept of segmentary state in pre-British India see Dirk H.A. Kolff, *Naukar, Rajput and Sepoy: The Ethnohistory of the Military Labour Market in Hindustan, 1450-1850* (Cambridge: Cambridge University Press, 1990), 'A Millennium of Stateless Indian History?', in Rajat Datta (ed.), *Rethinking a Millennium: Perspectives on Indian History from the Eighth to the Eighteenth Century: Essays for Harbans Mukhia* (Delhi: Aakar Books, 2008), pp. 51-67 and Burton Stein, *A History of India* (1998, rpt., New Delhi: Oxford University Press, 2004). Andre Wink in 'Sovereignty and Universal Dominion in South Asia', in Jos J.L. Gommans and Dirk H.A. Kolff (eds.), *Warfare and Weaponry in South Asia: 1000-1800* (New Delhi: Oxford University Press, 2001), pp. 99-130 speaks of the effects of *bheda* and *fitna* on politics. For the stagnation of pre-British India's military heritage, see T.A. Heathcote, *The Military in British India: The Development of British Land Forces in South Asia, 1600-1947* (Manchester: Manchester University Press, 1995), pp. 1-20; G.J. Bryant, 'Asymmetric Warfare: The British Experience in Eighteenth-Century India', *Journal of Military History* (*JMH*), vol. 68, no. 2 (2004), pp. 431-69. Stephen Peter Rosen in *Societies and Military Power: India and its Armies* (New Delhi: Oxford University Press, 1996) asserts that due to the divisive caste system, the Indian armies remained incoherent. For the concept of Mughal warfare being a political theatre backed by coercive force see Lorne Adamson, 'The Mughal Armies: A Re-appraisal', *Journal of the Bihar Research Society*, vol. LIX, pts. I-IV (1973), pp. 138-44 and Jos Gommans, *Mughal Warfare: Indian Frontiers and High Roads to Empire, 1500-1700* (London/New York: Routledge, 2002). John

F. Richards, in 'Warriors and the State in Early Modern India', *Journal of the Economic and Social History of the Orient* (*JESHO*), vol. 47, pt. 3 (2004), pp. 390-400 speaks of a weak Mughal state always busy in struggling against the armed peasantry. One author asserts that warfare in medieval South Asia was not about battles but about sacrifice, feasting and laughing, etc., all of which constituted crucial components of sovereignty. Daud Ali, 'Violence, Gastronomy and the Meaning of War in Medieval South India', *Medieval History Journal*, vol. 3, no. 2 (2000), pp. 261-89. C. Bayly, influenced by John Brewer's categorization of the early modern British state as a Fiscal-Military State (*The Sinews of Power: War, Money and the English State, 1688-1783*, London: Unwin Hyman, 1989) asserts that Britain's Indian Empire was a Fiscal-Military State. C.A. Bayly, 'The British Military-Fiscal State and the Indigenous Resistance: India 1750-1820', in idem, *Origins of Nationalism in South Asia: Patriotism and Ethical Government in the Making of Modern India* (New Delhi: Oxford University Press, 1998), pp. 238-75.

2. Pradeep Barua's *The State at War in South Asia* (Lincoln/London: University of Nebraska Press, 2005) devotes only 45 pages out of 306 pages of text for warfare in pre-British India.
3. Bert S. Hall and Kelly R. DeVries, 'Essay Review: The "Military Revolution" Revisited', *Technology and Culture*, vol. 31, no. 3 (1990), pp. 500-7.
4. Geoffrey Parker, *The Military Revolution: Military Innovation and the Rise of the West, 1500-1800* (Cambridge: Cambridge University Press, 1988).
5. Jeremy Black, *A Military Revolution? Military Change and European Society, 1500-1800* (Basingstoke/London: Macmillan, 1981), *Cambridge Illustrated Atlas of Warfare: Renaissance to Revolution, 1492-1792* (Cambridge: Cambridge University Press, 1996).
6. Nicholas Hooper and Matthew Bennett, *Cambridge Illustrated Atlas of Warfare: The Middle Ages, 768-1487* (Cambridge: Cambridge University Press, 1996) p. 153.
7. Clifford J. Rogers, 'The Military Revolutions of the Hundred Years War', and Geoffrey Parker, 'In Defence of *The Military Revolution*', in Rogers (ed.), *The Military Revolution Debate: Readings on the Military Transformation of Early Modern Europe* (Colorado/Boulder: Westview Press, 1995), pp. 55-93, 337-65. For a Military Revolution and the emergence of modern warfare in Classical Greece, refer to Arthur Ferrill, *The Origins of War from the Stone Age to Alexander the Great* (1985, rpt., London: Thames and Hudson, 1986), pp. 149-223.
8. Geoffrey Parker (ed.), *The Cambridge Illustrated History of Warfare: The Triumph of the West* (Cambridge: Cambridge University Press, 1995). See especially the Introduction titled 'The Western Way of War', by Parker, pp. 2-9. Victor Davis Hanson, *Carnage and Culture: Landmark Battles in the Rise of the West* (New York: Random House, 2001).

9. William Irvine, *The Army of the Indian Moghuls: Its Organization and Administration* (rpt., Delhi: Low Price Publications, 1994).
10. Williamson Murray and MacGregor Knox, 'Thinking about Revolutions in Warfare', in Knox and Murray (eds.), *The Dynamics of Military Revolution: 1300-2050* (2001, rpt., Cambridge: Cambridge University Press, 2003), pp. 1-14.
11. John Stone, 'Technology, Society, and the Infantry Revolution of the Fourteenth Century', *JMH*, vol. 68, no. 2 (2004), pp. 361-80.
12. Jos Gommans, 'Warhorse and Post-nomadic Empire in Asia, *c.* 1000-1800', *Journal of Global History*, vol. 2 (2007), pp. 1-21.
13. Yuval Noah Harari, 'The Concept of "Decisive Battles" in World History', *Journal of World History* (*JWH*), vol. 18, no. 3 (2007), pp. 251-66. Harari is wrong in arguing that nobody wrote about decisive sieges. G.B. Malleson in *The Decisive Battles of India* (rpt., Jaipur: Aavishkar Publishers, 1986) wrote about the decisive siege of Bharatpur.
14. For decisive battles in Indian history see Kaushik Roy, *India's Historic Battles: From Alexander the Great to Kargil* (New Delhi: Permanent Black, 2004).
15. For Military Synthesis in Indian military history see Kaushik Roy, *From Hydaspes to Kargil: A History of Warfare in India, 326 BC to AD 1999* (New Delhi: Manohar, 2004).
16. Peter A. Lorge, *The Asian Military Revolution: From Gunpowder to the Bomb* (Cambridge: Cambridge University Press, 2008).
17. One representative is R.C. Majumdar et al., *An Advanced History of India* (1946, rpt., Madras: Macmillan, 1991).
18. Ranabir Chakravarti, 'Early Medieval Bengal and the Trade in Horses: A Note', *JESHO*, vol. 42, no. 2 (1999), p. 195.
19. John A. Lynn, 'The Growth of the French Army during the Seventeenth Century', *Armed Forces and Society*, vol. 6, no. 4 (1980), pp. 568-85.
20. Mahinder S. Kingra, 'The *Trace Italienne* and the Military Revolution during the Eighty Years' War, 1567-1648', *JMH*, vol. 57, no. 3 (1993), pp. 431-46.
21. Benjamin J. Wallacker, 'Studies in Medieval Chinese Siegecraft: The Siege of Yu-pi, AD 546', *Journal of Asian Studies*, vol. 28 (1969), p. 792.
22. *The Invasion of India by Alexander the Great* as described by Arrian, Q. Curtius, Diodorous, Plutarch and Justin, tr. with an Introduction by J.W. McCrindle (1896, rpt., New Delhi: Cosmo Publications, 1983), p. 274.
23. *Ancient India as described by Megasthenes and Arrian, being a translation of the fragments of the Indika of Megasthenes collected by Dr. Schwanbeck, and of the first part of the Indika of Arrian,* with Introduction by John W. McCrindle (1926, rpt., New Delhi: Munshiram Manoharlal, 2000), p. 141.

24. For the problems regarding the chroniclers' estimate of the size of the Byzantine Army see Warren Treadgold, *Byzantium and its Army: 284-1081* (Stanford, California: Stanford University Press, 1995), pp. 43-86.
25. *An Account of the Last Battle of Panipat and of the Events Leading to it*, by Casi Raj Pandit, tr. from Persian into English by James Browne, ed. with an Introduction by H.G. Rawlinson (London: Oxford University Press, 1926), p. 18.
26. Kaushik Roy, *The Oxford Companion to Modern Warfare in India from the Eighteenth Century to Present Times* (New Delhi: Oxford University Press, 2009), pp. 11-107.
27. John A. Lynn, 'Forging the Western Army in Seventeenth Century France', in Knox and Murray (eds.), *Dynamics of Military Revolution*, pp. 35-56.
28. Stephen Wilson, 'For a Socio-Historical Approach to the Study of Western Military Culture', *Armed Forces and Society*, vol. 6, no. 4 (1980), pp. 527-52.
29. Lawrence Keppie, *The Making of the Roman Army: From Republic to Empire* (1984, rpt., London: Routledge, 1998), pp. 55, 76, 146-7.
30. Brian Campbell, 'The Roman Empire', in Kurt Raaflaub and Nathan Rosenstein (eds.), *War and Society in the Ancient and Medieval Worlds: Asia, the Mediterranean, Europe, and Mesoamerica* (Centre for Hellenic Studies, Washington DC, Cambridge, Massachusetts, distributed by Harvard University Press, 1999), p. 219.
31. Wolfgang Liebeschuetz, 'The End of the Roman Army in the Western Empire', in John Rich and Graham Shipley (eds.), *War and Society in the Roman World* (1993, rpt., London/New York: Routledge, 1995), p. 267.
32. Dick Whittaker, 'Landlords and Warlords in the Later Roman Empire', in Rich and Shipley (eds.), *War and Society in the Roman World*, pp. 281-3, 292-3.
33. P. Sensarma, *Military Thoughts of Tiru Valluvar* (Calcutta: Noya Prakash, 1981), p. 17.
34. *Tiruvalluvar, The Kural*, tr. from the Tamil with an Introduction by P.S. Sundaram (New Delhi: Penguin, 1990), p. 97.
35. J. Sundaram, 'Chola and other Armies—Organization', in S.N. Prasad (ed.), *Historical Perspectives of Warfare in India: Some Morale and Material Determinants* (Delhi: Centre for Studies in Civilizations in association with Motilal Banarsidass, 2002), pp. 189, 197.
36. Ali Athar, 'Military Hierarchy and Designations under the Sultans of Delhi (13th-14th Centuries)', *Journal of Asiatic Society*, vol. XLII, nos. 1-2 (2000), pp. 2-3.
37. *An Account of the Last Battle of Panipat and of the Events Leading to it*, p. 17.

38. J.R. Hamilton, 'The Cavalry Battle at the Hydaspes', *Journal of Hellenic Studies*, vol. 76 (1956), p. 30.
39. *Babur-Nama*, tr. from the original Turki text of Zahir-ud-din Muhammad Babur by A.S. Beveridge, vol. 2 (New Delhi: Saeed International, 1989), pp. 472-3.
40. Victor Davis Hanson, 'Genesis of the Infantry 600-350 BC', 'From Phalanx to Legion 350-250 BC', and 'The Roman Way of War 250 BC-AD 300', in Parker (ed.), *Cambridge Illustrated History of Warfare*, pp. 12-61; Hanson, *Warfare and Agriculture in Classical Greece* (Berkeley: University of California Press, 1998).
41. Kurt Raaflaub, 'Archaic and Classical Greece', in Raaflaub and Rosenstein (eds.), *War and Society in the Ancient and Medieval Worlds*, p. 133.
42. Pierre Briant, 'The Achaemenid Empire', in Raaflaub and Rosenstein (eds.), *War and Society in the Ancient and Medieval Worlds*, pp. 114-25.
43. *Ancient India as Described by Megasthenes and Arrian*, p. 225.
44. Ali Athar, 'Military Exercise and Training in the Sultanate of Delhi during the 13th and 14th Centuries', and 'The Role of the *Piyadah* in the Sultanate of Delhi', *Journal of the Asiatic Society*, vol. 37, no. 1 (1995), pp. 19-25, no. 2 (1995), pp. 24-7.
45. *An Account of the Last Battle of Panipat and of the Events Leading to it*, p. 19.
46. Somadeva Bhatta, *Kathasaritsagara*, tr. from Sanskrit into Bengali by Hirendralal Biswas, 5 vols. (1975, rpt., Kolkata: Academic Publishers, 1983), vol. 1, p. 58. All English translations are by the author.
47. K.M. Shrimali, 'Religions in Complex Societies: The Myth of the "Dark Age"', in Irfan Habib (ed.), *Religion in Indian History* (New Delhi: Tulika, 2007), p. 49.
48. Dennis E. Showalter, 'Caste, Skill, and Training: The Evolution of Cohesion in European Armies from the Middle Ages to the Sixteenth Century', *JMH*, vol. 57, no. 3 (1993), pp. 407-8.
49. *Manu's Code of Law: A Critical Edition and Translation of the Manava-Dharmasastra* by Patrick Olivelle, with the editorial assistance of Suman Olivelle (2005, rpt., New Delhi: Oxford University Press, 2006), p. 164.
50. *The Invasion of India by Alexander the Great*, p. 308.
51. Hans Delbruck, *Warfare in Antiquity: History of the Art of War*, vol. 1, tr. from the German by Walter J. Renfroe, Jr (1975, rpt., Lincoln/London: University of Nebraska Press, 1990), pp. 214, 224.
52. John France, 'Recent Writing on Medieval Warfare: From the Fall of Rome to *c.* 1300', *JMH*, vol. 65, no. 2 (2001), p. 442.
53. Showalter, 'Caste, Skill, and Training: The Evolution of Cohesion in European Armies from the Middle Ages to the Sixteenth Century', p. 419.

54. *Ancient India as Described by Megasthenes and Arrian*, pp. 88, 90.
55. Richard Glover, 'The Elephant in Ancient War', *Classical Journal*, vol. 39, no. 5 (1944), p. 257.
56. Ali Athar, 'The Ministry of War in the Delhi Sultanate', *Journal of the Asiatic Society*, vol. 37, no. 3 (1995), pp. 1-6.
57. The linkage between the expansion of the administrative tentacles of the indigenous states in south India and armies undergoing military innovations is pointed out by Sanjay Subrahmanyam in 'Warfare and State Finance in Woodeyar Mysore, 1724-25: A Missionary Perspective', *Indian Economic and Social History Review* (*IESHR*), vol. 26, no. 2 (1989), pp. 203-33.
58. K.K. Trivedi, 'The Share of Mansabdars in State Revenue Resources: A Study of the Maintenance of Animals', *IESHR*, vol. 24, no. 4 (1987), pp. 411-22.
59. Chauncey S. Goodrich, 'Riding Astride and the Saddle in Ancient China', *Harvard Journal of Asiatic Studies*, vol. 44, no. 2 (1984), pp. 294-5.
60. John W. Eadie, 'The Development of Roman Mailed Cavalry', *Journal of Roman Studies*, vol. 57, nos. 1-2 (1967), p. 161.
61. Stanley J. Olsen, 'The Horse in Ancient China and its Cultural Influence in Some Other Areas', *Proceedings of the Academy of Natural Sciences of Philadelphia*, vol. 140, no. 2 (1988), pp. 173-5.
62. M. Rostovtzeff, 'The Parthian Shot', *American Journal of Archaeology*, vol. 47, no. 2 (April-June 1943), pp. 176-7.
63. Hamilton, 'Cavalry Battle at the Hydaspes', p. 30.
64. Ashvini Agrawal, *The Rise and Fall of the Imperial Guptas* (Delhi: Motilal Banarsidass, 1989), pp. 165, 212.
65. Radhakumud Mookerji, *The Gupta Empire* (1973, rpt., New Delhi: Motilal Banarsidass, 1997), p. 92.
66. Ibid., p. 100.
67. Agrawal, *Rise and Fall of the Imperial Guptas*, p. 166.
68. Bimal Kanti Majumdar, *The Military System in Ancient India* (Calcutta: Firma KLM, 1960), p. 85; Murray B. Emeneau, 'The Composite Bow in India', *Proceedings of the American Philosophical Society*, vol. 97, no. 1 (1953), p. 86.
69. Agrawal, *Rise and Fall of the Imperial Guptas*, p. 91.
70. H.G. Creel, 'The Role of the Horse in Chinese History', *American Historical Review*, vol. LXX, no. 3 (1965), pp. 650-1.
71. *Mahakabi Kalidas Birachitam Raghuvamsam, Pratham Sarga to Chaturdash Sarga*, Mallinath krita Tikapetam, Gurunath Vidhyanithi Bhattacharyamanudithancha, ed. Ashok Kumar Bandopadhyay (Kolkata: Sanskrit Pustak Bhandar, 1411) [in Bengali, all English translation by the author], Chaturtha Sarga, 62, 64, 67, 70-1, Sastha Sarga, 33, Nabam Sarga, 56, pp. 79-81, 107, 163.

72. Denis Sinor, 'The Inner Asian Warriors', *Journal of the American Oriental Society*, vol. 101, no. 2 (1981), pp. 137-8.
73. Gurcharn Singh Sandhu, *A Military History of Ancient India* (New Delhi: Vision Books, 2000), p. 372.
74. R.P. Lindner, 'Nomadism, Horses and Huns', *Past and Present*, no. 92 (1981), pp. 3-19; Greg S. Rogers, 'An Examination of Historians' Explanations for the Mongol Withdrawal from East Central Europe', *East European Quarterly*, vol. 30, no. 1 (1996), pp. 3-26.
75. Lallanji Gopal, 'Art of Shipbuilding and Navigation in Ancient India', *Journal of Indian History* [hereafter *JIH*] (1962), pp. 313-27.
76. K.G. Krishnan, 'Cola Rajendra's Expedition to South-East Asia', *JIH*, Golden Jubilee Volume (1973), pp. 109-16.
77. Syed Hasan Askari, 'Mughal Naval Weakness and Aurangzeb's Attitude Towards the Traders and Pirates on the Western Coast', and Atulchandra Roy, 'Naval Strategy of the Mughals in Bengal', *Proceedings of the 24th Session of Indian History Congress* (Delhi, 1961), pp. 162-70, 174.
78. Naim R. Farooqi, 'Moguls, Ottomans and Pilgrims: Protecting the Routes to Mecca in the Sixteenth and Seventeenth Centuries', *International History Review*, vol. 10, no. 2 (1988), p. 200.
79. John Law, 'On the Social Explanation of Technical Change: The Case of Portuguese Maritime Expansion', *Technology and Culture*, vol. 28, no. 2 (1987), pp. 246-7.
80. Anirudh Deshpande, 'Limitations of Military Technology: Naval Warfare on the West Coast, 1650-1800'. *Economic and Political Weekly*, vol. 27, no. 17 (1992), pp. 900-4.
81. Surendranath Sen, 'Haidar Ali's Fleet', in Sen, *Studies in Indian History: Historical Records at Goa* (1930, rpt., New Delhi: Asian Educational Services, 1993), pp. 146-54.
82. *The Kautilya Arthashastra*, pt. II, An English Translation with Critical and Explanatory Notes by R.P. Kangle (1972, rpt., Delhi: Motilal Banarsidass, 1992), p. 61.
83. Sensarma, *Military Thoughts of Tiru Valluvar*, pp. 40-1.
84. Majumdar, *Military System in Ancient India*, p. 78.
85. Jean Deloche, *Studies on Fortification in India* (Pondicherry: Institut Francais de Pondichery and Ecole Francaise d'Extreme Orient, 2007), pp. 26, 32-3, 37.
86. Leigh Alexander, 'The Origin of Greek and Roman Artillery', *Classical Journal*, vol. 41, no. 5 (1946), pp. 208-12.
87. Richard A. Gabriel, *The Great Armies of Antiquity* (Westport, Connecticut, London: Praeger, 2002), p. 15.
88. *The History of India as Told by its Own Historians*, ed. from the posthumous papers of H.M. Elliot, by John Dawson, 8 vols., vol. 1 (1867-77, rpt., New Delhi: Low Price Publications, 2001), p. 157.
89. Iqtidar Alam Khan, 'Early Use of Cannon and Musket in India: AD 1442-1526', *JESHO*, vol. 24, pt. II (1981), pp. 147, 152.

90. Richard M. Eaton, '"Kiss My Foot", said the King: Firearms, Diplomacy, and the Battle for Raichur, 1520', *Modern Asian Studies*, vol. 43, no. 1 (2009), pp. 296-7.
91. Iqtidar Alam Khan, 'Gunpowder and Empire: Indian Case', *Social Scientist*, vol. 33, nos. 3-4 (2005), p. 55.
92. J. Burton-Page, 'A Study of Fortification in the Indian Subcontinent from the Thirteenth to the Eighteenth Century AD.', *Bulletin of the School of Oriental and African Studies* (*BOAS*), vol. 23 (1960), p. 522.
93. Kaushik Roy, 'A Military Revolution in Mysore? Technological Changes, Social Transformation and Military Modernization under Haidar Ali and Tipu Sultan between 1752-99', *Contemporary India*, vol. 3, no. 3 (2004), p. 29; Roy, *Oxford Companion to Modern Warfare*, pp. 62-107.
94. John Keegan, *War and Our World: The Reith Lectures 1998* (1998, rpt., London: Pimlico, 1999), p. 29; Doyne Dawson, *The First Armies* (London: Cassell, 2001), pp. 210-11. According to Stanley J. Olsen, the horse replaced the onager as draught animal for pulling the chariot in 2300 BCE in the Middle East. Olsen, 'The Horse in Ancient China and its Cultural Influence in Some Other Areas', p. 185.
95. P.R.S. Moorey, 'The Emergence of the Light, Horse-drawn Chariot in the Near-East *c.* 2000-1500 BC', *World Archaeology*, vol. 18, no. 2 (1986), pp. 202-3, 208.
96. Gabriel, *Great Armies of Antiquity*, p. xv.
97. Creel, 'The Role of the Horse in Chinese History', p. 649.
98. For China see Sun Tzu, *Art of War*, tr. with a Historical Introduction by Ralph D. Sawyer (Boulder, San Francisco: Westview Press, 1994), p. 36.
99. G.N. Pant, 'The Saga of Indian Arms', *JIH*, Golden Jubilee Volume (1973), p. 248.
100. Sandhu, *Military History of Ancient India*, pp. 105, 127.
101. Ibid., p. 97.
102. B.P. Sinha, 'Art of War in Ancient India (600 BC-300 AD)', in Guy S. Metraux and Francois Crouzet (eds.), *Studies in the Cultural History of India* (Agra: Shiva Lal Agarwal & Co., 1965), pp. 121, 157.
103. Alexander K. Nefiodkin, 'On the Origin of the Scythed Chariot', in Everett L. Wheeler (ed.), *The Armies of Classical Greece* (Aldershot, Hampshire: Ashgate, 2007), pp. 493-502.
104. Murray B. Emeneau, 'Composite Bow in India', p. 85.
105. Sandhu, *Military History of Ancient India*, p. 99.
106. Goodrich, 'Riding Astride and the Saddle in Ancient China', p. 286.
107. P.C. Chakravarty, *The Art of War in Ancient India* (1941, rpt., Delhi: Low Price Publications, 1989), pp. 15-57, 76-125, 181-97; V.R.R. Dikshitar, *War in Ancient India* (1944, rpt., Delhi: Motilal Banarsidass, 1987), pp. 157-66.

108. *The Invasion of India by Alexander the Great*, pp. 186-7.
109. *Ancient India as Described by Megasthenes and Arrian,* p. 30.
110. Majumdar, *Military System in Ancient India*, p. 77; B. Bar-Kochva, *The Seleucid Army: Organization and Tactics in the Great Campaigns* (1976, rpt., Cambridge: Cambridge University Press, 1979), p. 76.
111. Daneta Billau and Donald A. Graczyk, 'Hannibal: The Father of Strategy Reconsidered', *Comparative Strategy*, vol. 22, no. 4 (2003), pp. 336, 351.
112. Glover, 'Elephant in Ancient War', p. 269.
113. Keppie, *Making of the Roman Army*, pp. 41-3.
114. Edward H. Schafer, 'War Elephants in Ancient and Medieval China', *Oriens*, vol. 10, no. 2 (December 1957), pp. 290-1.
115. *The Harsacharita by Banabhatta*, tr. E.P. Cowell and P.W. Thomas, ed. R.P. Shastri (Delhi: Global Vision, 2004), pp. 130, 284-5, 304.
116. *The Nitisara by Kamandaki*, ed. Rajendra Lala Mitra, revised with English translation by Sisir Kumar Mitra (1849, rpt., Calcutta: The Asiatic Society, 1982), p. 442.
117. C.W.C. Oman, *The Art of War in the Middle Ages:* AD *378-1515* (London: Oxford University Press, 1885).
118. Bernard S. Bachrach, 'Charles Martel, Mounted Shock Combat, the Stirrup, and Feudalism', in Bachrach, *Armies and Politics in the Early Medieval West* (Aldershot, Hampshire: Variorum, 1993), Essay XII, pp. 49-75; Bachrach and Rutherford Aris, 'Military Technology and Garrison Organization: Some Observations on Anglo-Saxon Military Thinking in light of the Burghal Hidage', *Technology and Culture*, vol. 31, no. 1 (1990), pp. 1-17.
119. R.J. Barendse, 'The Feudal Mutation: Military and Economic Transformations of the Ethnosphere in the Tenth to Thirteenth Centuries', *JWH*, vol. 14, no. 4 (2003), pp. 503-29.
120. France, 'Recent Writings on Medieval Warfare', p. 445.
121. Patricia Crone, 'The Early Islamic World', in Raaflaub and Rosenstein (eds.), *War and Society in the Ancient and Medieval Worlds*, p. 321.
122. K.S. Lal, 'The Striking Power of the Army of the Sultanate', *JIH*, vol. LV, pt. II (1977), pp. 85-110.
123. David Ayalon, 'The Mamluks of the Seljuks: Islam's Military Might at the Crossroads', *Journal of the Royal Asiatic Society*, series 3, vol. 6, no. 3 (1996), pp. 305-33.
124. Walter Emil Kaegi, Jr., 'The Contribution of Archery to the Turkish Conquest of Anatolia', in John Haldon (ed.), *Byzantine Warfare* (Aldershot, Hamshire: Ashgate, 2007), pp. 237-49.
125. Chakravarti, 'Early Medieval Bengal and the Trade in Horses', pp. 196-207.
126. Jean Deloche, *Military Technology in Hoysala Sculpture (Twelfth and*

Thirteenth Century) (New Delhi: Sitaram Bharti Institute of Scientific Research, 1989), pp. 12, 30-1, 34.

127. Gabriel, *Armies of Antiquity*, p. 28.
128. David Nicolle, 'Medieval Warfare: The Unfriendly Interface', *JMH*, vol. 63, no. 3 (1999), p. 581; J.D. Latham, 'Notes on Mamluk Horse Archers', *BOAS*, vol. 32 (1969), pp. 257-67; A. Jan Qaisar, 'Horsehoeing in Mughal India', *Indian Journal of History of Science* (*IJHS*), vol. 27, no. 2 (1992), pp. 133-44; Jagadish Narayan Sarkar, *The Art of War in Medieval India* (New Delhi: Munshiram Manoharlal, 1984), pp. 33-89, 93-142, 273-331; M. Athar Ali, 'Organization of the Nobility: Mansab, Pay, Conditions of Service', in Commans and Kolff (eds.), *Warfare and Weaponry in South Asia*, pp. 232-74.
129. Iqtidar Alam Khan, 'Origin and Development of Gunpowder Technology in India: AD 1250-1500', *Indian Historical Review*, vol. 4, no. 1 (1977), pp. 24-9; Kaushik Roy, 'Rockets under Haidar Ali and Tipu Sultan', *IJHS*, vol. 40, no. 4 (2005), pp. 635-55.
130. Khan, 'Gunpowder and Empire: Indian Case', pp. 55-7.
131. C.R. Boxer, 'Asian Potentates and European Artillery in the 16th-18th Centuries: A Footnote to Gibson-Hill', *Journal of the Malaysian Branch of the Royal Asiatic Society*, vol. 38, no. 208 (1966), pp. 156-72.
132. Iqtidar Alam Khan in *Gunpowder and Firearms: Warfare in Medieval India* (New Delhi: Oxford University Press, 2004), p. 196 asserts that the Mughal gunsmiths failed to manufacture reliable medium weight wrought iron cannons. For the absence of steppe nomadic threat and rise of gunpowder infantry in West Europe see Kenneth Chase, *Firearms: A Global History to 1700* (Cambridge: Cambridge University Press, 2003).
133. William R. Thompson, 'The Military Superiority Thesis and the Ascendancy of Western Eurasia in the World System', *JWH*, vol. 10, no. 1 (1999), pp. 143-78.
134. Stanford J. Shaw, 'The Origins of Ottoman Military Reform: The Nizam I Cedid Army of Sultan Selim III', *Journal of Modern History*, vol. 37, no. 3 (1965), pp. 291-300.
135. David Fitzpatrick, 'Militarism in Ireland, 1900-22', in Thomas Bartlett and Keith Jeffrey (eds.), *A Military History of Ireland* (Cambridge: Cambridge University Press, 1996), pp. 380-1; V.G. Kiernan, *European Empires from Conquest to Collapse* (Bungay, Suffolk: Fontana, 1982), p. 15.
136. James W. Hoover, 'The Recruitment of the Bengal Army: Beyond the Myth of the Zemindar's Son', *Indo-British Review*, vol. 21, no. 2 (1996), pp. 144-56.
137. Kaushik Roy, 'Military Synthesis in South Asia: Armies, Warfare, and Indian Society, *c.* 1740-1849', *JMH*, vol. 69, no. 3 (2005), pp. 651-90,

Brown Warriors of the Raj: Recruitment and the Mechanics of Command in the Sepoy Army, 1859-1913 (New Delhi: Manohar, 2008).

138. Pradeep Barua, 'Military Developments in India, 1750-1850', *JMH*, vol. 58, no. 4 (1994), pp. 599-616; G.J. Bryant, 'The Cavalry Problem in the Early British Indian Army, 1750-85', *War in History*, vol. 2, no. 1 (1995), pp. 1-21; idem, 'The Military Imperative in Early British Expansion in India, 1750-85', *Indo-British Review*, vol. 21, no. 2 (1996), pp. 18-35.
139. For the idea that the vast distances resulted in stretching the Asian war machines to the maximum, which in turn made the employment of cavalry necessary, see Jeremy Black, 'War and the World, 1450-2000', *JMH*, vol. 63, no. 3 (1999), p. 677.
140. Jeremy Black, 'Military Organisations and Military Change in Historical Perspective', *JMH*, vol. 62, no. 4 (1998), p. 876.
141. Colonel Mouton, 'The First Anglo-Sikh War (1845-46)', *Panjab Past and Present*, vol. 15 (1981), pp. 116-27; Kaushik Roy, 'Firepower-Centric Warfare in India and the Military Modernization of the Marathas: 1740-1818', *IJHS*, vol. 40, no. 4 (2005), pp. 597-634, 'Technology and Transformation of Sikh Warfare: *Dal Khalsa* Against the *Lal Paltans*, 1800-1849', *IJHS*, vol. 41, no. 4 (2006), pp. 383-410.

CHAPTER 4

Siege Warfare in Mughal India: 1519-1538

PRATYAY NATH

The sixteenth century was a time when warfare was undergoing rapid structural changes in West Europe and in the Ottoman-dominated Middle East. Several historians argue that the rise of firepower heavy armies from this period onwards set the stage for the beginning of modern warfare. Both land and sea warfare was transformed. As regards land warfare, set piece battles and sieges witnessed decisive changes. This essay attempts to show how far sixteenth-century siege warfare in South Asia exhibited changes and continuities, especially in comparison to sieges conducted in other parts of Eurasia.

THE MILITARY REVOLUTION AND SIEGE WARFARE

Military history of early modern era remains one of the hotbeds of historiographical debates. In 1955, Michael Roberts formulated the hypothesis of a Military Revolution, which he saw taking place in western Europe between 1560 and 1660 due to rapid changes in tactical aspects of warfare brought about by Maurice of Nassau and Gustavus Adolphus. Roberts' Military Revolution was characterized by the rise of the firearms-equipped disciplined infantry over the feudal heavy cavalry.[1] In 1976, Geoffrey Parker modified Roberts' thesis substantially. Parker dated the Military Revolution as unfolding between 1530 and 1710. He argued that faced by artillery bombardment, fort architects in Italy came up with innovative designs resulting in the *trace italienne* between 1450 and 1520s, which once again restored the strategic balance in favour of defensive warfare. With the initial advantage neutralized, the besieging armies once again had to go for long-drawn sieges with the objective of surrounding the expanding fortress architecture and hence it became imperative on their part to increase the size of their armies.[2] A detailed demonstration of this transformation of fortress warfare in different parts of the world, particularly in West Europe, under the impact of artillery bombardment, came from Christopher Duffy in 1979.[3] Duffy showed how the sophisticated design of the *trace italienne* gradually evolved over the entire first half of the sixteenth century in the Italian peninsula under the impact of constant military campaigns in this region by the French and the Spanish (Austro-Spanish after 1519) forces and their ample, and on occasions effective, deployment of siege artillery. The basic features of this new design of forts involved spacious and low-lying ramparts to provide a good platform for deploying heavy guns and to resist the blows of enemy shots; a formidable ditch and a wall to resist escalade, and finally, a star-shaped ground plan that left no dead ground that might allow an effective breaching and successful escalade by the besieging force.[4]

In West Europe where emerging nation states, empowered by fast-improving firepower technology, were fighting it out among themselves in the west on the one hand and a viciously offensive Ottoman Empire was encroaching into mainland Europe as well as into the Mediterranean basin from the east, this design of fort architecture spread like an epidemic via the medium of professional engineers and architects. While the fifteenth century saw more of the victorious march of the cannon, with Ottoman cannons blasting open the walls of Constantinople in 1453 and the united Catholic monarchy of Spain literally blasting away the forts of the Moorish province of Granada between 1483 and 1492,[5] the development of the *trace italienne* made the fortress a secure place once again.

The Ottomans since 1453 continued to celebrate the victory of the siege artillery over fortifications in the Middle East. During the first half of the sixteenth century, they brought up a formidable navy that was particularly designed to transport their siege artillery. In 1522, artillery fire won for the Ottomans the island fortress of Rhodes in eastern Mediterranean. Around the same time, cannonade, mining, sapping and assault by Ottoman forces subverted Belgrade. During the Siege of Vienna in 1529, Ottoman miners succeeded in breaching the walls of the city before they had to retreat due to difficulties of maintaining overstretched supply lines. In Persia, cannonade became a regular part of siege warfare by the first decade of the sixteenth century. Shah Ismail used siege artillery and musket fire against the Fort of Yazd in 1504 and Jizra in 1506. By 1516, the Safavid arsenal had around 2,000 muskets and 40 guns copied from a cannon left behind by the Ottomans.[6]

As in West Europe, the advent of gunpowder warfare triggered off a transformation of castle architecture in Japan, though on completely different, and essentially indigenous, lines. The new castles that started coming up during the later part of the sixteenth century combined the fortress, palace and the barrack in one single unit. Such buildings had strong granite walls and housed gun ports and stone-throwing ports inside these walls.[7] In the Muscovite empire and the Korean peninsula, artillery became common in the second half of the sixteenth century.[8] In South-East Asia, several polities were acquainted with gunpowder artillery by the agency of the Portuguese around the same time, while in China, firearms were not at all widely used until well into the seventeenth century.[9]

THE HISTORIOGRAPHY OF SIEGE WARFARE IN EARLY MODERN SOUTH ASIA

In the context of the Indian subcontinent, the use of the cannon and musket spread gradually from the mid-fifteenth century onwards. In 1456, Mahmud Khalji's cannonade blew up a reservoir inside the fort of Mandalgarh. Cannons were also used by Sultan Mahmud Bighara during his Siege of Champanir in 1484-5, by Sikandar Lodi, the Sultan of Delhi (1489-1517) as well as by the rulers of Kashmir, Gujarat and the Deccan.[10] These, however, were more of isolated instances of deployment of the artillery than symptoms of any systemic transformation of the besieging armies. In the 1520s, the Timurid prince, and subsequently the founder of the Mughal dynasty, Zahiruddin Muhammad Babur revolutionized the structure of pitched battles in medieval north India by deploying artillery in the centre and mounted archery on the flanks to effect an enveloping manoeuvre (*taulqama*) on the enemy. This was the same tactics that the Ottomans successfully used at Chaldiran against the Safavids in 1514 and at Mohacs against the Hungarians in 1526, and the Safavids against the Uzbegs near Herat in 1528. Babur further displayed the adaptability of his artillery-cavalry force in his later campaigns. As he campaigned further eastwards along the Ganga basin, ecological and geographical factors prevented him from deploying his cavalry during military engagements on a similar scale and his victories in the two riverine engagements on the Ganga and Ghaghra owed hardly anything to the warhorse. It was the masterful use of artillery and boats that saw Babur victorious there.[11]

These battles of Babur, particularly the first two at Panipat and Khanua, along with the sieges undertaken by Babur's grandson Akbar in the 1560s, prompted many scholars to view the Mughal Empire as a Gunpowder Empire. The Gunpowder Empire hypothesis views the artillery essentially as a means of empowerment of the political centre at the expense of the region and hence a tool for the creation as well as sustenance of centralized empires like those of the Mughals, Ottomans and the Safavids. Within this framework, gunpowder was considered as the most important factor in shaping the political and sociocultural landscape all over the world.[12]

William McNeill came out in support of the Gunpowder Empire thesis in his world history of warfare. Working in a similar framework, he views gunpowder artillery as the chief pivot on which early-modern empires were built and consolidated. McNeill argued that in case of all these polities, political centralization and stability were direct corollaries of near monopolistic use of gunpowder technology. As for the Mughals, he pointed out that in the interior parts of the Indian subcontinent, imperial consolidation remained precarious precisely because it was not possible for them to convey their cannons and mortars to these parts in the absence of much scope for transportation by water.[13]

In recent years, interest in the role of artillery in Mughal warfare and state-formation has been stoked by two works. Iqtidar Alam Khan, in the Introduction of his book, more or less accepts this argument about the centrality of gunpowder technology in early modern state-formation in general and the Mughal case in particular.[14] Another important work that has lent support to this framework is the world history of firearms by

Kenneth Chase. His work examines the crucial impact of firearms on warfare across the vast landmass of Eurasia.[15]

Such an understanding of Mughal warfare with gunpowder artillery at its centre has been seriously challenged from various quarters. Duffy in his *magnum opus* on siege warfare pointed out that unlike the case in sixteenth-century Europe, the use of gunpowder artillery in the Indian subcontinent did not bring any major change in the art of fortification.[16] This observation implies that unlike in Europe, gunpowder artillery could not threaten the security that the forts of this region offered; i.e. artillery was not much successful against existing fortifications, and hence fort architects did not feel it necessary to improve upon existing structures. Douglas Streusand claims that gunpowder artillery played only a marginal role even in the most famous Mughal sieges under Akbar, namely Chittor, Ranthambhor and Kalinjar, and that the Mughals continued to rely on traditional means of conducting siege, like escalade, mining and storming.[17]

A collateral dimension of the marginality of firepower in Mughal warfare was the rise of the importance of the warhorse in Mughal historiography. From the 1950s, the 'Aligarh School' focused on the Mughal *mansabdars* and the agrarian system that sustained them. All the works by the Aligarh School downplayed the role of artillery in Mughal warfare. The central assumption of their work was the idea that not artillery, but cavalry was the principal fighting force of the empire, as reflected by its centrality in the military organization as well as it being the chief consumer of the surplus production of the empire.[18] Irfan Habib spells out this assertion clearly: 'Artillery was not the decisive arm of their army and they were never able to employ it successfully against really strong forts. Their real strength lay in cavalry and it was in battle in the open field, in rapid movements, that they remained invincible.'[19] Several works studying the institutional aspects of Mughal military and administrative apparatus also give greater importance to cavalry compared to artillery.[20]

Recent works by Jos Gommans, from a completely different methodological position and historical understanding of the Mughal state, also hold the warhorse at the centrestage of Mughal warfare. Gommans looks at the Mughal state as a post-nomadic state, just like the contemporary empires of the Manchus, the Safavids and the Ottomans. He defines these entities as primarily horse-economies, as it was the continual production and deployment of the warhorse that sustained their empires. While this gave the Mughals superiority over their opponents inside the subcontinent, the warhorse also functioned in another way—allowing them to have strategic mobility across the vast landmass of the subcontinent.[21]

With such a diverse historiography to draw upon, the present paper will look at the period between the triumphal march of the Mughals into the subcontinent riding on a supposedly decisive advantage given to them by their possession of gun-powder artillery on the one hand and subsequent reverses in the hands of Sher Shah and their overthrow from the position of political and military hegemony over large parts of north India. While Babur's pitched battles at Panipat (1526) and Khanua (1527) have received much limelight,[22] and even the encounters between Humayun and Sher Shah at Chausa (1539) and Kanauj (1540) have not gone unattended,[23] Mughal sieges of this period still await serious attention. This neglect of siege warfare, in fact, is a chronic weakness that plagues Indian military historiography. While works on military technology do not direct much attention to the study of forts and fortifications and siege appliances other than gunpowder artillery, a second approach focuses only on pitched battles and do not study sieges at all. Indian military history has paid surprisingly low attention to the study of fortress warfare and logistics in general, and the role of siege warfare in state-formation has remained entirely unexplored.[24] This paper looks at five important, and incidentally successful, sieges conducted by the Mughals under Babur and Humayun between 1519 and 1537 in the Indian subcontinent—Bajaur (1519), Chanderi (1528), Mandu (1535), Champanir (1536) and Chunar (1537)—and assess the role of gunpowder artillery, or for that matter the Mughal warhorse, in these sieges. The observations made about these sieges will be utilized to reflect on the nature of Mughal warfare in this period and the position of siege warfare within it.

BABUR'S SIEGES

SIEGE OF BAJAUR, CE 1519

On Tuesday, 4 January CE 1519, Babur camped near Bajaur and an envoy was sent into the city demanding submission. This denied, a siege was decided upon. The army stayed at the same site the next day and siege appliances like mantelets, ladders, and so on were made ready for commencing the operation.[25] On Thursday, the soldiers put on their mail armour and proceeded towards the fort. The left wing moved to the north of the fort, the centre dismounted on the plains to the north-west and the right wing to the west of the lower gate of the fort. The left wing got engaged in a skirmish with some Bajauris, who came out of the fort on foot, shooting arrows. Babur's men shot back from horseback and drove them right inside.[26]

It was here that Babur used matchlocks for the first time. This weapon was still unknown in these quarters and Babur mentions how the Bajauris in fact made fun of it in the beginning. But, the matchlocks proved to be very effective. On that day, Ustad Ali-Quli shot five Bajauris, while two were brought down by Wali Khizanchi and still more by other matchlock men. Their bullets pierced the cuirasses and shields of the defenders. Enemy casualty to matchlock fire figured nearly ten that day. At dawn, Babur ordered that the whole army was to advance towards the fort without breaking their formation and try to escalade the walls. The left wing and centre advanced from their position with mantelets and after reaching the ramparts, they tried to escalade. The left wing, reinforced by the left of the centre, advanced to the north-eastern bastion of the fort and concentrated their efforts in trying to pull the wall down. Ustad Ali Quli fired his *firingi* twice and Wali also shot a Bajauri. Babur's men climbed up with ladders at various points and engaged the Bajauris in close-range fighting with swords and spears, while heavy firing of arrows by the besiegers standing below hardly allowed the defenders to put their head out. The defenders still tried to ward off those of Babur's men engaged in undermining the fort's ramparts by directing stones and arrows towards them, but apparently the latter showed great courage and perseverance in being persistent in their efforts and as a result, these soon bore fruits. By breakfast-time, the left wing under Dost Beg had breached the north-eastern tower, stormed into the fort and put the enemy to flight. Almost at the same time, soldiers of the centre also escaladed the walls successfully and climbed their way in. A large part of the adult male population, numbering almost 3,000 was massacred while many women and children were taken captives.[27]

Babur did not have at his disposal any cannon at Bajaur. The only variety of firearms available to him was matchlocks, which were used to provide covering fire while his men engaged themselves in undermining the walls of the fort manually and trying to scale its walls. Gunpowder weapons had hardly played an important role in capturing the fort. Matchlocks were used in conjunction to archery to provide cover to the escaladers, and since the number of matchlocks at Babur's disposal must have been low at this point, they must have had a secondary role to archery as far as causing casualties were concerned.

SIEGE OF CHANDERI, CE 1528

After the victory at Khanua, Babur subjugated the powerful chief Medini Rao and occupied the latter's fort, Chanderi. The citadel of Chanderi, which Medini Rao was holding with 4,000-5,000 men, rests on a low flat-topped hill on the edge of the table-land overlooking the valley of the Betwa. The bed of the river opposite Chanderi is 1,050 ft above the sea level; the city of Chanderi, on the table-land, is 1,300 ft above the sea and the fort 230 ft above the city. The fort is one and quarter mile in length from north to south and three-fourths a mile in breadth. It is irregular in shape on the north and east faces, with the circuit of the walls more than 4 miles. This includes the *Bala-kila*, or citadel, which stands in the north-west quarter of the fort occupying somewhat less than one-fourth of the whole area. The outer ramparts of the fort exceed 5 miles in length.[28] In 1438, this fort had been captured by Alauddin Mahmud of Malwa by surprise escalade during the night after a prolonged siege of eight months.

On 22 January 1528, Babur assigned the centre, right and left wings of his army their respective positions around the outer ramparts of the fort. Ustad Ali Quli chose a flat table-land probably to the north of the fort. Overseers and men with spades were instructed to raise a place for firing his mortar. This was probably one of Babur's *kazans*, which were very heavy, though short, mortars mounted on four-wheeled carts, cast in two pieces and drawn by 400-500 persons or two to three elephants.[29] The rest of the army was ordered to get the mantelets, ladders and other siege appliances ready. The outer fort was not heavily manned, and as a result it fell to escalade by Babur's men on the night of 28 January. The guards fled to the inner citadel. The next day Babur ordered every man to arm, go to their posts, provoke to fight, and attack the enemy from his place. Ustad Ali Quli fired his mortar repeatedly, much to the amusement of Babur, but the strong stone walls of the fort did not give way. Babur indicates that the mortar's effectiveness was also reduced by the fact that the ground on which the mortar had been placed, was not high enough, at least not higher than the walls of the citadel.[30]

The greatest concentration of the besiegers was on a double-walled road (*du-tahi*) that ran between the citadel and a nearby source of water. This 'nearby source of water' described by Babur must have been the Kirat Sagar, mentioned by Alexander Cunningham in the following words: 'The fort is badly supplied with water, the citadel being dependant on a single tank inside, which frequently dries up, and on a large tank at the foot of the hill outside, which is connected with the upper works by a covered way. This tank is called the *Kirat Sagar*.'[31] The royal troops at the centre attacked this covered way, which Babur says to have been 'the one place for attack'. This was also logical because if the garrison's access to this tank could be cut-off, they would never be able to sustain defence of the fort for a long time. The garrison also fought hard to repel the assault and threw stones and 'flung flaming fire' on the besiegers. But, very soon, a commander named Shahin escaladed the walls and other men followed his example. The *du-tahi* was thus taken by escalade.[32]

How the citadel proper was captured after this is unclear. Even so, from the impression that the *Baburnama* gives, it does not seem that artillery had much to do with it. The main modality of assault seems to have been escalade. Babur says, '. . . when a number of our men swarmed up, they [i.e. the defendants] fled in haste.'[33] Since, Babur had a lot of matchlocks in action in the pitched battles of Panipat and Khanua before this siege, we may assume that matchlocks and arrows provided an effective covering fire for the besiegers, as at Bajaur. But there is no evidence of artillery causing any decisive damage to the ramparts of the fort. In fact, other than the mortar under the command of Ustad Ali Quli, there is no mention of the use of any other cannon or mortar during the siege.

Driven back from the walls, the Rajputs retreated briefly, only to come back as a force of forlorn hope. Firishta says that in the fighting that ensued, 6,000 of the defendants perished.[34] As the garrison was losing the battle, the Rajput elite performed the ceremony of *jauhar*, whereby they put their women and children to death to save them from dishonour and slavery. Then, they performed acts of self-sacrifice in the house of Medini Rao, in which the latter also died. The fort was thus captured by the Mughals without much difficulty in less than a month. A pillar of heads was built on a hill north-west of Chanderi.[35]

HUMAYUN'S SIEGES

SIEGE OF MANDU, CE 1535

In November 1534, Humayun marched against the Afghan Sultanate of Gujarat. Humayun decided to confront Bahadur Shah, the Afghan ruler of Gujarat, after the latter captured the fort of Chittor. The two armies encamped opposite each other in Mandasor and remained in this position for several days. After Bahadur Shah evaded a direct confrontation and retreated to the Fort of Mandu, Humayun decided to lay siege to the fort and in May 1535, encamped at Nalca, around three miles to the north of the Delhi gate of the fort.

The city of Mandu stands on a hill, some 2,000 ft above the sea level. The outer ramparts are around 25 miles in length. The whole hill juts out from the main Vindhyan range, to which it is connected by a narrow neck of land only, and its sides slope precipitously to the plains of the Narmada Valley, the height of the southern scarp being some 1,200 ft above ground level. The slopes and the top of the hill were mostly covered with thick jungle. The fort had nine gates at the time of Humayun's siege, each being strongly fortified, and architecture giving the garrison an advantageous position in relation to the assailants. The gates were provided with guard rooms, barbicans and a steep slope. Escalade of the walls must have been easier in the northern parts than southern ones, because of the abruptly high walls on other sides. Water was pretty abundant inside the fort, and it was drawn from numerous wells, brooks, pools and large reservoirs.[36]

One night, taking advantage of the slackness of the guards, a contingent of Humayun's soldiers, 200-300 in number, escaladed the back walls using ladders and ropes and got into the outer fort. Then they opened the gates of the fort and brought in their horses and other soldiers by that way under the cover of night.[37] Soon the officer-in-charge of the garrison, Mallu Khan was informed of the infiltration. He immediately rushed off to Sultan Bahadur. Awakened from his sleep, Bahadur had to hasten on his horseback towards the gate of the fort with 25 horsemen. They were intercepted by a group of around 200 Mughal cavalrymen, but Bahadur Shah successfully overcame them. Reaching the inner citadel of Mandu (named Sungad), he lowered down several horses by ropes and then came down himself and fled towards Gujarat with five or six associates.[38]

Humayun was informed of this night assault in the morning and when he entered the fort through the Delhi Gate, a part of the garrison under Sadr Khan was still resisting the besiegers near the entrance of his house. Ultimately, they took refuge in Sungad along with Alam Lodi, the nephew of Ibrahim Lodi. After much deliberation, Sadr Khan and Alam Khan were made to surrender and ultimately forgiven, although probably Alam Khan's foot was cut-off owing to Humayun's ill-disposition towards him. The fort was plundered for three days by the victorious Mughal Army.[39]

Siege of Champanir, CE 1536

After Humayun's forces had taken the Mandu Fort by a night assault, Sultan Bahadur Shah, who had taken refuge there, fled to the fort of Champanir in Gujarat. In pursuit, the Mughal troops arrived at Champanir and encamped near the Pipli Gate of the fort near the enormous tank of Imad ul-Mulk which Abul Fazl says to have been 3 *kos* in circumference.[40] On receiving this information, Bahadur Shah strengthened the defences and provisions of the fort and himself slipped out of the fort by way of the gate near the Shukr tank and fled towards Cambay, and eventually to Diu. A garrison under the command of Ikhtiyar Khan was placed at the Champanir Fort.

The citadel of the Champanir Fort stood on a rectangular plan and enclosed about 80 acres. The walls were strongly constructed of a light brown sandstone in large blocks, strengthened at intervals by bastions and rising to the height about 30 ft. There were two large gateways and each of these gateways was provided with an elaborate defense mechanism to intercept any infiltration.[41]

Even four months after the siege had begun, the Mughals failed to make any headway. One day, while going round the fort, Humayun discovered that the garrison in the fort was regularly supplied with provisions by the local *Kolis*. This transaction was done at a point of the fort where the walls were about 60-70 *gazs* (yards) in height at the back of the fort. This area was approached through a forested tract. The garrison lowered money by ropes and pulled up the goods in the same way. The walls were vertical and smooth, which made the possibility of a successful escalade almost impossible. Humayun guessed that for this very reason, this was a spot that the garrison felt very secure about and hence did not care to guard properly. Humayun decided to take advantage of this over-confidence of the garrison and take the fort by escalading this section. Hence, he had his workmen make about 70 to 80 steel spikes to facilitate the escalade. Then, he returned to the spot at night with a select and highly specialized contingent of soldiers.[42] The iron nails were planted into the mortar joints of the wall, left and right, at a distance of a yard from each other. By daybreak around 300 soldiers had infiltrated using this iron ladder, including Bairam Khan and Humayun himself. Then, early in the morning, by means of a prearranged signal, a combined attack was launched on the fort from all sides. The main army assaulted from outside. This attack comprised attempts at escalade from outside, while Humayun and his men attacked the garrison from inside the fort walls and showered arrows upon them. The latter party was quick to get hold of one of the gates of the fort and let the assailants from outside enter. The garrison was overpowered while the fort-commandant Ikhtiyar Khan took refuge in the citadel. Since he was a man with great learning and sagacity, he was pardoned and inducted into Mughal service.[43]

Jauhar gives a slightly different version of the siege. He says that after the siege had continued for some time, a person informed Humayun that if trusted, he could lead the Mughal troops across the summit of a mountain to a high position which completely commanded the fort. Humayun accompanied him with a select body of soldiers and two drummers and a trumpeter. After a difficult ascent, they got into the fort. At Humayun's signal, the drummers and trumpeter started playing aloud. Hearing this sound, the different detachments of Humayun's army made assaults at different points of the fort and its bastions. Discovering that the besiegers were carrying out a mighty night assault, Bahadur Shah escaped towards Cambay. The fort and the treasure inside it fell into the hands of Humayun.[44]

Jauhar's version differs with the other accounts with respect to the agency of discovery of the weak point in the fort's defences. But overall, the key points of this account match with Abul Fazl's version—knowledge about a difficult but weak point in fort defenses, secret operation under the cover of night by a specialized detachment accompanying Humayun and a difficult escalade followed by a concerted attack at the sounding of a signal to take the garrison by surprise.

Siege of Chunar, CE 1538

When Humayun led an offensive against Sher Shah in the Bengal-Bihar region during the latter half of the 1530s, the siege of Chunar was one of the few occasions when his military operation proved to be successful. But, how the Mughals actually subverted this impressive fort is difficult to decipher from the available sources. There is a general consensus among the contemporary authorities that the main credit lay with the military genius of Rumi Khan, the *Mir Atish* of Humayun's army and a deserter from the Gujarat Sultanate after the Mughal victory at Mandasor. But what he actually did in this campaign is hazy.

An early nineteenth century English traveller described the Chunar Fort in the following words:

> The view of Chunar is, from the river, very striking. Its fortress . . . covers the crests and sides of a large and high rock, with several successive enclosures of walls and towers, the lowest of which have their base washed by the Ganges. . . . The site and outline are very noble; the rock on which it stands is perfectly insulated, and either naturally or by art, bordered on every side by a very beautiful precipice, flanked, wherever it has been possible to obtain a salient angle, with towers . . . and bastions of various forms and sizes. . . . Colonel Robertson, however, told me, that the ammunition on which he would

> most depend for the defence of Chunar, are stone cylinders, rudely made, and pretty much like garden-rollers, which are piled up in great numbers throughout the interior of the fort, and for which the rock on which the fort stands affords an inexhaustible quarry, these, which are called 'mutwallas', (drunkards) from the staggering motion, are rolled over the parapet down the steep face of the hill, to impede the advances and overwhelm the ranks of an assaulting army, when a place has not been regularly breached, or where, as at Chunar, the scraped and sloping rock itself serves as a rampart, few troops will so much as face them.[45]

In absence of any evidence to prove the contrary, it would perhaps not be very odd to assume that such defensive mechanisms were also used by the defendants against Humayun's possible efforts at escalade.

The most detailed account of the siege comes from Jauhar. He narrates how in order to gain knowledge about the strong and weak points of the fort's defence, Rumi Khan flogged a faithful black slave of his, Khilafat, and asked him to go over to the defending garrison and act like a deserter to them, only to get an entry into the fort and return with useful information about the fort's defences. Khilafat did his job pretty well, and after possessing adequate knowledge about the fortification, returned. He advised Rumi Khan to attack the bastion on the side of the Ganga River, to run a sap on the land side and to cut-off the supply line of the fort by encircling it. Accordingly Rumi Khan directed his cannons towards the river-side bastions and set-up batteries at other places round the ramparts too. However, finding it hard to make any impression on the walls facing the river from his current position, he decided to strike from the river itself. Accordingly, a high wooden tower was built over a few months, resting on three large boats. Then, under the cover of the night, this mechanism was drawn near the ramparts of the fort and anchored there. This was followed by a general attack. However, before the tower could make any impression on the fort walls, Humayun's army suffered heavily due to a counter-attack by the garrison. The floating battery was damaged and 700 Mughal soldiers were killed. Next morning, while the tower was being repaired, the garrison, threatened by the determination of the Mughals to take the fort, surrendered. Under the instruction of Rumi Khan, the hands of 300 artillerymen of the defending army were severed.[46]

Abul Fazl's account varies in the modality of the actual siege. He says that Rumi Khan constructed a *sabat* (a covered trench-like approach) and provided it with a roof made out of wooden planks, so that the miners and sappers could be protected against the missiles thrown by the garrison. This facilitated setting up of mines under the ramparts, which when blown off shook up the whole earth. The commandant of the fort, Qutb Khan fled and Humayun's men occupied the fort, probably through the breach made by the explosion of the mine.[47]

Badauni's account reveals no more than that it was principally Rumi Khan's efforts that were responsible for the success of the siege.[48] The accounts of Firishta and Nizamuddin both confirm that the main assault was made from the side of the river. Nizamuddin also mentions that the tower that was constructed at Rumi Khan's order was placed on several large boats. But as to how exactly this tower was used is unclear from Firishta and Nizamuddin's narratives. Firishta mentions the tower to have been launched, floated towards the walls of the fort and that walls being low here, were surmounted easily.[49] While Abul Fazl records Rumi Khan having constructed a covered way on the boats (*sabat*) to get his soldiers to the walls to facilitate an escalade from the river side, the translators of Nizamuddin seem to be convinced that the walls were battered down and hence, they conclude that it was a battering ram that was built on the boats.[50]

Now, the use of siege towers was known since the days of Alexander and the Mauryas. The siege tower was used to counter the advantage that the defending garrison commanded by virtue of their position on top of the rampart of the fort as opposed to the besiegers standing below. Jagadish Narayan Sarkar points out that such a tower had two purposes. Sometimes archers or some siege engines would be placed on top of the tower to enable the besiegers to command the inside of the fort and provide covering fire for an advancing party of assailants. Also, in many cases, such towers would have a drawbridge which would be lowered so as to provide a way for the assailants to advance. The Mughals called this *siba*. Such devices were erected by Akbar in the siege of Ranthambhor, by Dara during the siege of Qandahar and by Shaista Khan for taking Chakan. Aurangzeb used massive siege towers in Golconda and Satara.[51] The tower at Chunar can be seen as such a siege tower, its novelty lying solely in the fact that it was constructed and placed on the boats and operated on water. But even so, we are in darkness regarding exactly how this tower was used in Chunar.

According to Erskine, Rumi Khan had set-up a cannon on top of the platform he had built on the boats, which being mobile due to its location on water could be brought close to the ramparts and then fired, so as to effect a breach in the walls of the fort. He says that this breach was used by the assailants to attempt a storming of the fortress, but the garrison inflicted heavy casualties on them. The tower was also half-destroyed by the shots, probably from the recoil of the cannon set on it.[52]

Now, none of the sources says clearly that it was a cannon that was mounted at the top of the tower and that this cannon effected a breach in the fortifications of Chunar. In fact, Jauhar states clearly that little impression was made on the fort walls by the machine devised by Rumi Khan. While Erskine's words regarding the breach of walls by cannonade are difficult to accept, the damage of the tower due to the recoil of a cannon mounted on it is a more plausible argument. In any case, it is certain that Humayun did possess several cannons during the siege, as in his battles with Sher Shah a little later he used a number of them. But, at the same time, it is also certain that it was not bombardment by these cannons that earned him the Chunar Fort. The account of Firishta gives the impression that the fort fell due to successful scaling of the walls from the side of the river. However, the translator of Nizamuddin sees the machine built by Rumi Khan as a battering ram, while Abul Fazl gives the main credit to the blasting of gunpowder in mines near the fort walls which terrified the garrison. Even if we accept the fact that it was a battering ram that was built on the boats, it would still be doubtful how much damage such a floating ram could cause, in the face of resistance of water on which the platform carrying it was floating. Besides, no source mentions any breach being effected by Rumi Khan's machine. It was probably a combination of escalade and mining that led to the surrender of the garrison. But, in any case, it seems more or less certain that Chunar was not a case of cannon-conquest.

MUGHAL SIEGES: A COMPARATIVE OVERVIEW

From the above discussion, it becomes quite evident that artillery played only a negligible role in Mughal sieges in the subcontinent right up to 1538. Forts, with all the political and financial significance that they symbolized, were surely taken by both Babur and Humayun, but the modality of these operations comprised more of traditional methods like escalade. Even mining is not mentioned by any of the sources. Although by this time the advent of gunpowder had brought drastic changes in the structure of pitched battles in north India at the organizational as well as operational levels, the mechanism of siege warfare seems to have remained relatively unaffected by it, at least at this stage. The sieges of Bajaur and Chanderi were hardly different from previous sieges that Babur had conducted in Transoxiana prior to his advent in the subcontinent (and prior to his acquaintance with firearms) like that of Samarkand in 1500, where the fort was taken by escalade.[53] The only difference at Bajaur and Chanderi was that in addition to bows and arrows, here Babur deployed matchlocks to provide covering fire. The mortar he used at Chanderi clearly did not have much impact on the fortifications. All the sieges described here owed their success to escalade. Firepower played only a limited and indeed, a secondary, role. The location of the forts of Chanderi, Mandu and Champanir on top of steep hills must have made deployment of heavy siege artillery of the time against them extremely difficult.

On the other hand, the celebrated warhorse had little role to play in these sieges. The Mughal post-nomadic heritage of mounted archery, whatever edge it might have given them in pitched battles, was hardly of any use in helping them take these forts. What counted here was concerted action and prosaic labour. It was more of these apparently mundane aspects of militarism, than the artillery or warhorse, which aided the early Mughals to take the strongholds in the subcontinent.

Interestingly enough, evidence suggests that at this time, the

Afghan sultanates of Gujarat and Bengal were probably at par with, if not ahead of, the Mughals in terms of gunpowder technology. B.P. Ambashthya points out that the Bengal Sultanate's army had a formidable artillery park in the Battle of Surajgarha against Sher Shah, which fell in the latter's hand following his victory. Prior to this, Sher Shah's army was mainly a cavalry army, augmented by a few elephants and some matchlockmen and a few artillery pieces. It was probably the presence of this formidable artillery park in the Bengal Army and the weakness of this arm in his own army that dictated Sher Shah's battle tactics at Surajgarha. He decided not to go near the enemy formation himself and lure the Bengal cavalry away from the support of its firepower and engage cavalry with cavalry without facing enemy artillery. Ambashthya makes it clear that it was chiefly owing to his capture of the Bengal artillery park after Surajgarh battle that Sher Shah's firepower increased drastically.[54] In all probability, the technological knowhow for maintaining such an artillery park had come to the Bengal sultans from the Portuguese trading community settled in the coastal areas of the region.

Humayun faced a similar military difficulty against the Afghans in Gujarat in 1535, but proved to be less wise than Sher Shah. When the two armies remained encamped opposite each other for quite some time at Mandasor, the Gujaratis had a better artillery park than the Mughals. It was owing to this that instead of engaging in cavalry warfare, the former took up an entrenched position, arranging guns on carts behind a moat. It was clearly owing to such a fortified position of the Afghans of Gujarat, guarded by field pieces and matchlockmen that Humayun refrained from storming the enemy camp. On 4 April 1535, however, 500-600 of Humayun's horsemen under Muhammad Zaman Mirza were lured by a group of Bahadur Shah's horsemen within the reach of the latter's artillery. The Afghan horsemen first discharged two or three volleys of arrows, and then dispersed. The Mughal detachment pursuing them, suddenly found themselves within the range of the enemy's cannons, which they had been so careful to avoid so long. The Afghan artillery opened fire, causing considerable Mughal casualties.[55] Till the end, the Mughals never dared to come to confront the Afghans of Gujarat face to face, clearly owing to the latter's superior cannons, safeguarded in strongly entrenched positions. The case is similar to that narrated by Babur about one of his engagements against the Uzbegs prior to Panipat. The Mughals had already used fortified artillery position in pitched battles to their own advantage and clearly knew the dangers implicit in leading a charge against such a formation.

The *Mirat-i Ahmadi* gives some other instances as well which show the advanced state that Gujarati gunpowder technology was at this period:

He [Bahadur Shah] then occupied Bhilsa and encamped, by continual marches, on a river-bank near the fort of Raisin which was held by Lakhmi Sen Silhadi. He appointed Amirs to entrenchments for conquest of the fort. Rumi Khan, who was matchless of the age, in pyrotechnics, demolished one turret within twinkling of an eye, by a cannon-ball. Twelve thousand Dekhani soldiers who were attendants of the Sultan mined a fort wall and flew one turret of it to a distance of the remotest arrow.[56]

The *Mirat-i Ahmadi* notes '. . . The Sultan went to the port of Div [Diu], he sent two very big cannons along with one hundred [smaller] others to Mahmudabad with an intention to conquer Chitor.'[57]

In a article published in 2009, Richard Eaton has shown that during this period, the Deccani polities also were quite advanced in terms of gunpowder technology. The Bahamani state was in possession of a large number of cannons which they deployed in the Battle of Raichur in 1520 and also to defend the fort of Raichur against the Vijayanagara Army. The fact that this large arsenal could not win the day for the Bahamanis is a different question. The source of such an elaborate arsenal seems to have been Ottoman professionals, who, after their arrival on the Indian coast to counter the Portuguese around 1510, stayed back and took up service under the Bahamanis. The Vijayanagara state in turn took in service a group of Portuguese matchlockmen from Goa during their siege of Raichur, who proved to be extremely effective.[58]

These examples show the validity of Iqtidar Alam Khan's comment that during this period, it was the coastal polities that were the first to receive latest technological know how about gunpowder weaponry from Europe. This knowledge then proliferated among landlocked polities with relative slowness.[59] This might be the reason why Humayun and Sher Shah were still behind Bahadur Shah and Nusrat Shah in firearms technology in the 1530s. Babur brought with him a tradition of gunpowder warfare that owed its origin to the arid zone of Eurasia. But, in case of Gujarat, Bengal or the Deccani polities, firearms technology came via two sources through maritime channels—the Portuguese and the Ottomans. By the first decades of the sixteenth century, European gunners had started making single-piece brass/bronze cannons. Unlike Europe, in the larger parts of the Islamic world, heavy mortars still used to be cast in two pieces—the powder-chamber and the stone-chamber separately.[60] Since the two gunners of Babur came from Ottoman realm and Safavid Persia respectively,[61] it can be assumed that this was the method by which Babur's cannons were also cast. Thus, if by the time of Babur's invasion of Hindustan, technological know how regarding the casting of single-piece cannons had proliferated into coastal polities of the Deccan via Portuguese deserters, the large guns that these polities possessed at this juncture were technologically more sound than those which Babur possessed.

Clearly, right into the mid-1530s, the Mughals were not the ultimate military hegemon in the Indian subcontinent. Probably, Bahadur Shah had more advanced, and more effective artillery than Humayun. Even though the former could not defend his forts against Mughal sieges with these guns, they surely had caused decisive damage to the fortifications at Raisin, and probably at Chittor. Bahadur Shah also tactically neutralized the celebrated Mughal warhorse in the field by means of superior firepower. Mughal artillery failed to breach the walls of Chunar. And finally, Sher Shah demonstrated at Chausa and Kanauj that the Mughals could actually be beaten at the game they were supposed to be the kings—in pitched battles. Thus, even though Mughals had established themselves in northern India by means of decisive military superiority demonstrated repeatedly between 1526 and 1529 in four battles, the course of military interface with other powers inside the subcontinent was clearly more complicated and multilayered than is usually imagined. The picture becomes more nuanced when the focus is shifted beyond the conventional study of pitched battles to include other aspects of war-making, the principal of them being siege warfare. More detailed studies of the latter are hence essential to complicate existing notions of warfare in the Indian subcontinent and unravel its multiple layers.

NOTES

1. Michael Roberts, 'The Military Revolution, 1560-1660', cited in Mahindar S. Kingra, 'The *Trace Italienne* and the Military Revolution during the Eighty Years' War, 1567-1648', *Journal of Military History*, vol. 57, no. 3 (1993), pp. 431-46.
2. Geoffrey Parker, 'The "Military Revolution", 1560-1660–a Myth?', *Journal of Modern History*, vol. 48, no. 2 (1976), pp. 195-214.
3. Christopher Duffy, *Siege Warfare: The Fortress in the Early Modern World, 1494-1660* (1979, rpt., London and New York: Routledge, 1997).
4. Duffy, *Siege Warfare*, p. 2.
5. Weston F. Cook Jr., 'The Cannon Conquest of Nasirid Spain and the End of the Reconquista', *Journal of Military History*, vol. 57, no. 1 (1993), pp. 43-70.
6. Kenneth Chase, *Firearms: A Global History to 1700* (Cambridge: Cambridge University Press, 2003), p. 117.
7. Duffy, *Siege Warfare*, pp. 237-46; Chase, *Firearms*, pp. 181-8.
8. Duffy, *Siege Warfare*, pp. 166-74, 238-9.
9. Chase, *Firearms*, pp. 136-40, 141-71; Duffy, *Siege Warfare*, pp. 232-7.
10. Iqtidar Alam Khan, 'Early Use of Cannon and Musket in India: AD. 1442-1526', *Journal of the Economic and Social History of the Orient*, vol. 24, no. 2 (1981), pp. 146-64.
11. Pratyay Nath, 'Rethinking Early Mughal Warfare: Babur's Pitched Battles, 1499-1529', in Raziuddin Aquil and Kaushik Roy (eds.), *Warfare, Religion and Society in South Asia* (forthcoming).
12. Marshall G.S. Hodgson, *The Venture of Islam: Conscience and History in a World Civilization*, vol. 3, *The Gunpowder Empires and Modern Times* (Chicago: Chicago University Press, 1974), pp. 17-18.
13. William H. McNeill, *The Pursuit of Power: Technology, Armed Forces, and Society Since AD 1000* (Chicago: University of Chicago Press, 1982), p. 95.
14. Iqtidar Alam Khan, *Gunpowder and Firearms: Warfare in Medieval India* (New Delhi: Oxford University Press, 2004), pp. 1-8.
15. Chase, *Firearms*.
16. Duffy, *Siege Warfare*, p. 229.
17. Douglas Streusand, *The Formation of the Mughal Empire* (New Delhi: Oxford University Press, 1999), pp. 10-13, 51-81.
18. Ahsan Jan Qaisar's research shows that 445 *mansabdars* belonging to the ranks of 500 *zat* and above, amounting to a mere 5.6 per cent of the total *mansabdari* corps consumed 61.5 per cent of the total assessed revenue income (*jama*) of the empire. Ahsan Jan Qaisar, 'Distribution of the Revenue Resources of the Mughal Empire Among the Nobility', in Muzaffar Alam and Sanjay Subrahmanyam (eds.), *The Mughal State, 1526-1750* (New Delhi: Oxford University Press, 2006), pp. 252-8.

19. Irfan Habib, *Agrarian System of Mughal India* (rpt., New Delhi: Oxford University Press, 1999), p. 364.
20. William Irvine, *The Army of the Great Moghuls: Its Organization and Administration* (rpt., New Delhi: Low Price Publications, 2004), p. 57; U.N. Day, *The Mughal Government, AD 1556-1707* (New Delhi: Munshiram Manoharlal, 1970), pp. 145-6; I.H. Qureshi, *The Administration of the Mughal Empire* (Patna: N.V. Publications, n.d.), p. 126; Abdul Aziz, *The Mansabdari System and the Mughal Army* (Delhi: Idarah-i Adabiyat-i Dilli, 1972), p. 181.
21. Jos Gommans, 'Warhorse and Post-Nomadic Empire in Asia, *c.* 1000-1800', *Journal of Global History*, vol. 2 (2007), pp. 1-21; Gommans, *Mughal Warfare: Indian Frontiers and High Roads to Empire, 1500-1700* (London: Routledge, 2002), pp. 99-111.
22. See relevant chapters of the following books: Jadunath Sarkar, *Military History of India* (New Delhi: Orient Longman, 1970); Kaushik Roy, *From Hydaspes to Kargil: A History of Warfare in India from 326 BC to AD 1999* (New Delhi: Manohar, 2004); Roy, *India's Historic Battles: From Alexander the Great to Kargil* (New Delhi: Permanent Black, 2005).
23. B.P. Ambashthya, *Decisive Battles of Sher Shah* (Patna: Janaki Prakashan, 1977).
24. Existing works on fortification inside the subcontinent seldom emphasize military underpinnings of these structures and even more rarely relate their observations to the larger debates of military history. A major exception to this literature is a recent welcome publication by Jean Deloche which deals with military fortifications of a large part of the subcontinent. Jean Deloche, *Studies on Fortification in India* (Pondicherry: Institut Francais de Pondichery, 2007). Also see Kaushik Roy, 'Science and Secularization of Warfare: Transition in Siege Warfare in South Asia from Medieval to Modern Times', in Raziuddin Aquil and Kaushik Roy (eds.), *Warfare, Religion and Society in South Asia* (forthcoming).
25. Zahiruddin Muhammad Babur, *Baburnama*, tr. A.S. Beveridge (rpt., New Delhi: Munshiram Manoharlal, 1998), pp. 367-8.
26. *Baburnama*, tr. Beveridge, p. 368.
27. Ibid., pp. 369-70.
28. Alexander Cunningham, *Archaeological Survey of India: Four Reports Made During the Years 1862-63-64-65*, vol. 2 (New Delhi: Archaeological Survey of India, 2000), p. 404.
29. Khan, *Gunpowder and Firearms*, pp. 64-8.
30. *Baburnama*, tr. Beveridge, pp. 592-5.
31. Cunningham, *Four Reports*, vol. 2, p. 405.
32. *Baburnama*, tr. Beveridge, p. 595.
33. Ibid.
34. Mahomed Kasim Firishta, *History of the Rise of Mahomedan Power in*

India, tr. John Briggs, vol. 2 (rpt., Delhi: Low Price Publications, 1990), p. 38.

35. *Baburnama*, tr. Beveridge, pp. 595-6.
36. Sidney Toy, *The Fortified Cities of India* (London: Heinemann, 1965), pp. 67-74.
37. Abul Fazl puts the number at 200 while Firishta estimates the force to be 300 in number. Abul Fazl, *The Akbarnama*, tr. H. Beveridge, vol. 1 (rpt., Delhi: Low Price Publications, 1989), p. 305; *History of the Rise of Mahomedan Power*, tr. John Briggs, p. 48.
38. Firishta says that Bahadur escaped with 5,000 cavalrymen, but this is highly improbable. Abul Fazl says that on his way to Champanir, he was joined by 1,500 men. *History of the Rise of Mahomedan Power*, tr. John Briggs, p. 48; *Akbarnama*, tr. H. Beveridge, p. 306.
39. *Akbarnama*, tr. H. Beveridge, pp. 304-6; *History of the Rise of Mahomedan Power*, tr. John Briggs, pp. 48-9; Khwajah Nizamuddin Ahmad, *Tabaqat-i-Akbari*, tr. B.N. De and Beni Prashad, vol. 2 (rpt., Delhi: Low Price Publications, 1992), pp. 51-2.
40. *Akbarnama*, tr. H. Beveridge, p. 307.
41. Toy, *Fortified Cities*, pp. 62-3.
42. Although Nizamuddin says that the number of special troopers initially selected was 600, he mentions that the number of soldiers who actually entered the fort with Humayun by daybreak was 300. Firishta says that Humayun selected 300 men for the task. Abul Fazl also mentions the number of soldiers to have entered the fort with Humayun to have been 300. We thus may conclude that even if the number of troopers initially selected might as well have exceeded 300, the latter represented the approximate number of soldiers who ultimately got into the fort with Humayun that night. *Akbarnama*, tr. H Beveridge, p. 312; *Tabaqat*, tr. B.N. De and Beni Prashad, p. 55; *History of the Rise of Mahomedan Power*, tr. John Briggs, p. 49.
43. *Akbarnama*, tr. H. Beveridge, pp. 310-12; *Tabaqat*, tr. B.N. De and Beni Prashad, pp. 54-6; *History of the Rise of Mahomedan Power*, tr. John Briggs, pp. 49-50.
44. Jouher, *Tezkereh al Vakiat or Private Memoirs of the Moghul Emperor Humayun*, tr. Major Charles Stewart (London: Oriental Translation Fund of Great Britain and Ireland, 1832), p. 5.
45. Reginald Heber, *Narrative of a Journey Through Upper Provinces of India from Calcutta to Bombay, 1824-1825* (rpt., Delhi: Low Price Publications, 1993), pp. 401-4.
46. *Tezkereh*, tr. Charles Stewart, pp. 9-10.
47. *Akbarnama*, tr. H. Beveridge, pp. 331-2.
48. Badaoni, *Muntakhabu-t-Tawarikh*, tr. B.P. Ambashthya and George S.A. Ranking, vol. 1 (rpt., Delhi: Renaissance Publishing House, 1986), pp. 456-7.

49. *History of the Rise of Mahomedan Power,* tr. John Briggs, p. 53.
50. *Tabaqat,* tr. B.N. De and Beni Prashad, pp. 62-3.
51. Jagadish Narayan Sarkar, *The Art of War in Medieval India* (New Delhi: Munshiram Manoharlal, 1984), pp. 167-8.
52. William Erskine, *History of India under Humayun* (rpt., New Delhi: Atlantic Publishers and Distributors, 1995), pp. 140-2.
53. *Baburnama,* tr. Beveridge, pp. 109, 132-3.
54. Ambashthya, *Decisive Battles,* pp. 18-24.
55. *Akbarnama,* tr. H. Beveridge, pp. 301-3.
56. Ali Muhammad Khan, *Mirat-i Ahmadi: A Persian History of Gujarat,* tr. M.F. Lokhandwala (Baroda: Maharaja Sayajirao University, 1965), p. 60.
57. *Mirat,* tr. Lokhandwala, p. 61.
58. Richard M. Eaton, '"Kiss My Foot", Said the King: Firearms, Diplomacy and the Battle for Raichur, 1520', *Modern Asian Studies,* vol. 43, no. 1 (2009), pp. 289-313.
59. Khan, *Gunpowder and Firearms,* p. 63.
60. Ibid., pp. 60-2.
61. Ibid., p. 69.

CHAPTER 5

Cannons of Chittagong University Museum: New Studies on the Artillery of Shah Jahan

ZIAUDDIN CHOWDHURY
PRANAB K. CHATTOPADHYAY

INTRODUCTION

Cannons played an important role in medieval warfare. Sometimes in the tenth century CE, the Chinese discovered gunpowder and invented 'primitive' cannons. From China, the cannon technology spread through different routes to South-East Asia and South Asia. The dominant view is that Zahir-ud-din Muhammad Babur introduced cannons in north India. And in India, cannons were first used during the First Battle of Panipat by Babur in 1526. The theory of the introduction of cannons in India by Babur cannot be accepted fully and the statement has to be reassessed thoroughly.

GUNS AND GUNPOWDER IN BENGAL

In Bengal, the earliest reference to the cannon is found in the *Rajamala*, which states that the Bengal Sultan Hussein Shah used cannons while fighting against Dhanyamanikya in the early part of the sixteenth century CE (perhaps in 1513-14). Bhattasali has referred to the autobiography of Babur (*The Baburnama*) translated by A.S. Beveridge, where it has been mentioned that on 4-5 May 1529, Babur had an encounter at Kharid with Nusrat Shah, the Sultan of Bengal. Babur commented on the use of cannons by the Bengalis noting that their aiming was not proper and they had resorted to random shots. Some scholars argue that the Afghans of Bengal used *bans* (rockets, which were pyrotechnic device) against Babur at the battle of Ghagra. Babur noted that a person came from usta (east) saying the stone is ready (perhaps cannon balls), and fired the stone. Babur had written that he saw one large stone was fired along with several *firingi* (probably cannons copied from the Western models. The word *firingi* was derived from Franks, meaning French, but it was applied for all the West Europeans including the Portuguese) ones and the Bengalis had a reputation for making *bans* (probably rockets). Both the facts related to Hussein Shah and Nusrat Shah clearly indicate that cannons were used by the Bengal sultans during the invasion of the Mughals.[1]

The discovery of cannon was related with the discovery of gunpowder. The credit goes to the Chinese, who discovered that combination of saltpetre (potassium nitrate or other nitrate salts), sulphur and charcoal when burnt gives an excellent propellant property. The mixture turns into an explosive if it was fired within a container. The container explodes with a heavy sound—this was the basic principle of a detonator. If the former is made into an optimal proportion of saltpetre, sulphur and charcoal as 75:13:12 in a metal barrel of cannon made of adequate strength, then on firing the mixture produces suddenly 3,000 times of bulk gas—nitrogen, oxides of carbon and sulphur. The temperature suddenly reaches around 3880 °C.[2] This explosion is conducted in a place known as chamber of the cannon.

It is surprising that Mahau, a Chinese traveller, had referred to the presence of 'guns' in Bengal around CE 1406, but he does not describe whether they were made of metal or they were bamboos filled with explosives, which were known as *bans*. Probably, the transfer of gunpowder technology from China to Bengal had occurred prior to the arrival of Babur in eastern India. The diffusion of knowledge on saltpetre in Bengal occurred directly from the Chinese.[3] The *Rumis* (Ottoman Turks) are believed to initiate bronze casting in manufacturing cannons in the subcontinent. The Turks were employed in the service of Indian monarchs for manufacturing cannons. Cannon technology though initially originated outside South Asia, further development took place in late-medieval India based on local/regional needs.

Three types of cannon were referred by Babur. These were *kazan, zarb zan* and *firingi*. *Kazan* was a type of large cannon.

The *Ain-i-Akbari* refers to the invention of different cannons by Akbar. The names of cannons appearing in the text indicate *gajnals*, which could be easily carried by a single elephant; *narnals* was a device, which could be carried by a single man; *shatrunal* was a type of small cannon which was mounted on camels. We are aware that in late medieval Bengal the small cannons were also mounted on boats. In a temple plaque in Hooghly district country boats were seen with mounted cannons. Several naval cannons are preserved at the Tai Museum in Sibsagar, Assam.

In India and Bangladesh a number of scholars have made intensive studies on the cannons of the subcontinent. Mohammad Abul Hashem Miah (1991) studied the cannons of this part of the subcontinent on the whole. The cannons of Dhaka have been studied by Stapleton[4] and in a Bengali book titled *Dhakar Itihas* by J.M. Roy (1913). These works analyse the cannons which are at present located at the National Museum of Bangladesh. For specific studies on the cannons of Bengal we may mention the works of P. Neogi (1914). Originally this book was published from Rajshahi.[5] In Varendra Research Museum's *Annual Report*, N.B. Sanyal has thrown considerable information on Sher Shah's cannons,[6] while S. Sharf-ud-din has studied the Persian inscriptions on them.[7] This collection includes two cannons of Sher Shah. From the analysis of Sher Shah's cannon of Varendra Research Museum, we know that they were made of bronze, though the tin content very low, and we may call it gunmetal. Shamsul Hussain (1985, 1988) first threw light on the cannons of the Chittagong University Museum.[8] Our objective in this paper is to further elaborate the studies he initiated.

ANALYSIS OF THE GUNS OF BENGAL

Basically we find two types of cannons in Bengal. These are cast brass or bronze and forged welded iron cannons. We have also found composite cannons where both the alloys and iron were used. First we shall discuss cast bronze cannons. Babur described the casting of brass or bronze cannons. His chief gun-founder, a Rumi, was Ustad Ali Quli Khan. Mustafa Rumi was another gun-founder of Babur.[9] The description of casting techniques in *Baburnama*, however, lacks technical details. By the time of

Akbar, the process of casting of fully complete bronze cannons was well established in India.

We have three brass/bronze cannons of Sher Shah in the National Museum of Bangladesh and two cannons in the Varendra Research Museum. The cannons of Shah Jahan, now located in the Chittagong University Museum, were manufactured by Saiyad Ahmad Rumi. The brass cannons of Sher Shah, presently located in Varendra Research Museum, were made by another Rumi namely Saiyad Ahmad. These two cannons are known as the *Sherdahan* type.

The casting methods of the cannons are basically of two types. Small cannons like those in the museum which we are studying were made in single cast, whereas large cannons were manufactured as two pieces—the barrel and the gunpowder chamber were manufactured separately and then joined together. The mould and casting technique of the cannons located in the Chittagong University Museum is explained in Figure 5.1. The main part of the cannon mould was made hollow to fit over the

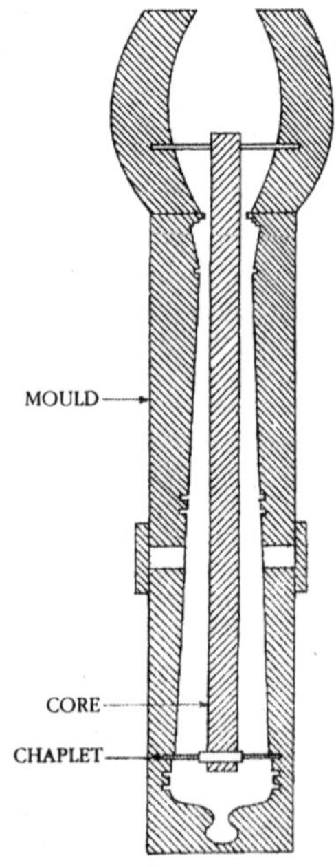

FIGURE 5.1: CANNON MOULD.

core. The cannon was cast around the core that was lowered into the main sleeve of the mould. The core was maintained in position and centred by means of a iron cross or web known as chaplets and they held the core in place during the casting process.[10]

The chemical and metallographic studies on the bronze-brass cannons of Bengal are too meagre. The two cannons of the Chittagong University Museum have been identified as bronze, though chemical analysis has not been obtained. The brass or bronze cannons in this subcontinent have been hardly analysed. However, we are aware with the chemical analysis of the bronze cannons of Sher Shah of the Varendra Research Museum. The composition indicates 84.72 per cent of copper, 13.32 per cent of zinc and iron and 1.83 per cent tin. The analysis of a Mughal cannon indicated a composition of 87.72 per cent of copper, 10.32 per cent of zinc, 1.83 per cent of tin and 0.13 per cent of iron.[11] The cannons of Sher Shah and Isa Khan of National Museum of Bangladesh are of brass. It may be mentioned that XRD (X-ray diffraction, a technique used to identify the several elements and their phases in an object) of the two cannons of the State Archaeological Museum, West Bengal, indicated mostly zinc alloying, that is brass. Unless the tin content of the Chittagong University cannons are identified, we cannot accept them as bronze cannon.

We have massive forged welded iron cannons in Dhaka, Murshidabad and Bishnupur. The manufacturing technique of forged welded iron cannons of Bengal was as follows. Iron rings were joined together by forging at red hot temperature. The blacksmiths were well conversant with the physical properties of iron and steel. They knew the critical assessment of thermal expansion and contraction of iron rings. The rings were placed over the iron staves and over each other by shrink fitting. In forging barrels often two or three layers of ring forged together.

THE CANNONS OF CHITTAGONG UNIVERSITY MUSEUM

The Chittagong University Museum was established on 14 June 1973. In brief, the following are the descriptions of four cannons of the Chittagong University Museum. In 1966, the then National Bank of Pakistan had purchased the four cannons from Karimunnesa Begum of Chittagong who donated them to the Museum. It is important to note that Karimunnesa Begum was a lineal descendant of Adhu Khan Hazari, a Mughal officer posted at Chittagong. It was the reason for the possession of those cannons by Karimunnesa Begum.

The most important brass/bronze cannon of Bangladesh are the *Sar Jang* (head of the war) preserved at the Chittagong University Museum. This cannon's muzzle face bears an excellent engraved design with befitting ornamentation. In Table 5.1 gives the measurements of various components. From the three inscriptions on the first noted cannon, one may conclude that it belonged to Shah Jahan and was built by Muhammad Hussain Izzat in AH 1066, under the supervision of Barqandaz Khan.

Description of Cannon no. 1 (Accession no. 124), The cannon of Shah Jahan (Figure 5.2): This cannon contains three embossed *Nastaliq* inscriptions. There are two hand-written inscriptions also which we could not decipher correctly. The measurements of the cannons are shown in Figure 5.3. The first inscription tells to whom that cannon belonged to. Shamsul Hossain had correctly identified the title *Shah Buland Iqbal*—it was the most favourite title of Emperor Shah Jahan. The transliterated first inscription is: *Tope Shah Buland Iqbal Ast* (This is the cannon of Shah Buland Iqbal). The second inscription indicates that it was built under the supervision of a particular person. The transliterated second

TABLE 5.1: DIMENSIONS OF THE FOUR CANNONS

	Cannon no. 1 Accession 124	Cannon no. 2 Accession 125	Cannon no. 3 Accession 126	Cannon no. 4 Accession 127
Length				
Total length	197 cm	197 cm	216 cm	171 cm
Barrel	165 cm	171 cm	175 cm	110 cm
Backplate to trunnion	78 cm	71 cm	82 cm	48 cm
Trunnion	11 cm	5.5 cm	7 cm	5 cm
Trunnion to muzzle	106 cm	108 cm	112 cm	79 cm
Diameters				
Muzzle face	22 cm	18 cm	22 cm	14 cm
Bore	10 cm	7.5 cm	10 cm	6.5 cm
Backplate	28 cm	18 cm	30 cm	14 cm
Trunnion	10 cm	6 cm	8 cm	3 cm
Vent	4 cm	1 cm	2 cm	2 cm

FIGURE 5.2: THE CANNON *SAR JANG* OF SHAH JAHAN.

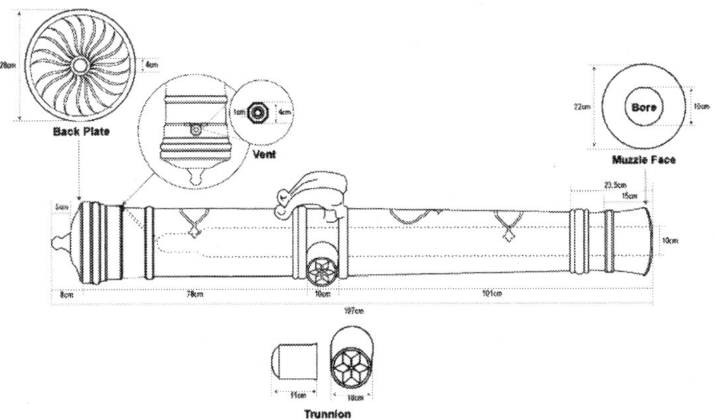

FIGURE 5.3: DIMENSIONS OF THE CANNON *SAR JANG*, #1.

inscription is: *Tope bad darogha Barqandaz Khan tyar shud* (This cannon was built under the supervision [Daroghaship] of Baraqandaz Khan). Hussain commented further that this title of Baraqandaz Khan was awarded to one Baha-ud-din *Topchi*—an officer in the Mughal artillery. The third inscription indicates the name of the cannon, the name of the builder and the year of manufacture. The transliterated third inscription is *Tope 1066 Sar Jang Amal Muhammad Hussain Izzat* (This cannon, *Sar Jang*, is the work of Muhammad Hussain Izzat, dated (AH 1066). Figure 5.4 indicates the details of the inscriptions on this cannon and the ornamental designs engraved on it.

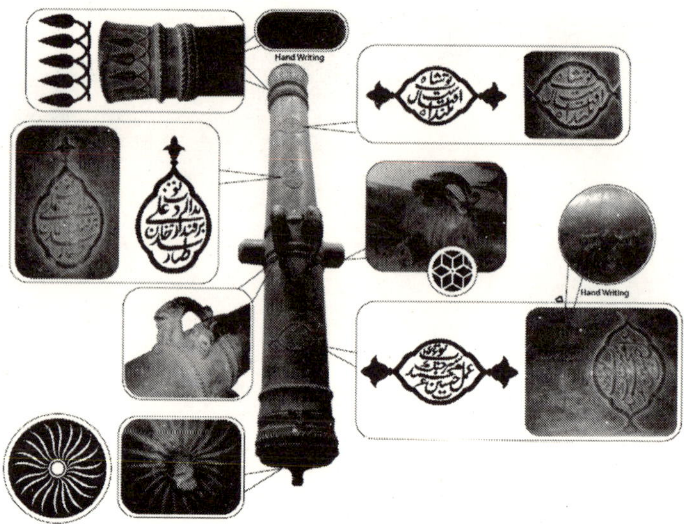

FIGURE 5.4: PERSIAN INSCRIPTIONS AND ORNAMENTAL DESIGN IN CANNON #1.

Cannon no. 2 (Accession no 125) bears Arabic and Bengali inscriptions with two fish motifs. The inscriptions are not so prominent as in Cannon no. 1. Only a few Persian letters are prominent—those are seen as *Mauzij (?)—zurrat (?) 33 seer—4 seer*, as shown in Figure 5.6. We conclude that the weight of each shot was 33 *seers* (27.63 kg) and firing each shot required 4 *seers* (3.35 kg) of gunpowder. Figures 5.5 and 5.5A indicate the details of this cannon.

Cannon no. 3 (Accession no. 126) is a forge welded iron cannon. This does not bear any inscription. Figure 5.7 indicates the details of this cannon. It bears no inscription.

Cannon no. 4 (Accession no. 127) was originally mounted on a naval ship, probably used in the battle with the Magh troops in CE 1666. This cannon was possibly the *firingi* one as described in the *Baburnama*. This cannon is attached with a swivel connecting the trunnions. Figure 5.8 indicates the details of it. The tail portion with a long handle helped the gunners to aim the target easily.

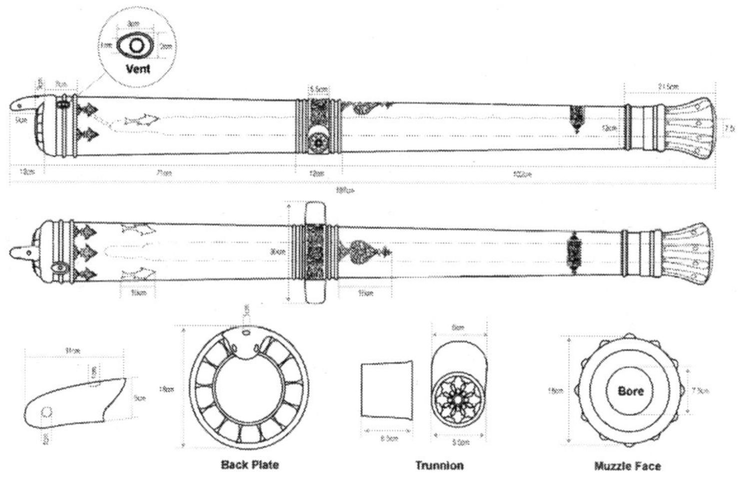

FIGURE 5.5: DIMENSIONS OF CANNON #2.

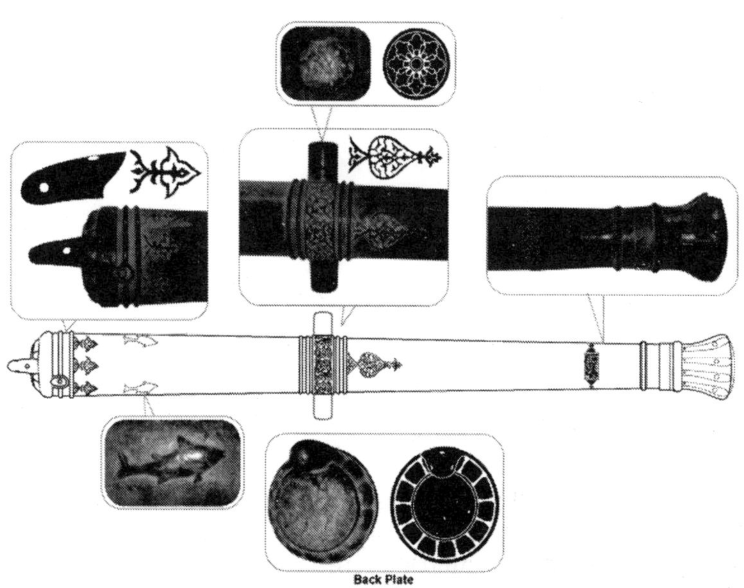

FIGURE 5.5A: ORNAMENTAL DESIGNS OF CANNON #2.

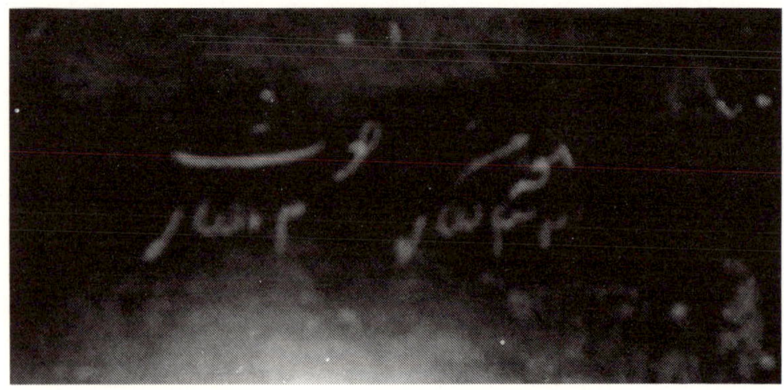

FIGURE 5.6: PERSIAN INSCRIPTION ON CANNON #2.

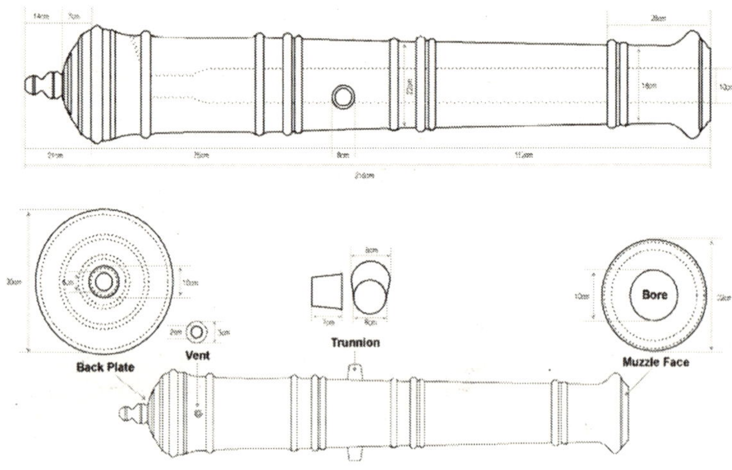

FIGURE 5.7: THE FORGE WELDED IRON CANNON #3.

CONCLUSION

In Bengal, a number of cannons which belonged to the reign of Shah Jahan still exist. One of the most important specimens is the *Jahankosha* (Conqueror of the World), now located at Murshidabad. This is a forged welded iron cannon, and was manufactured by a Bengali blacksmith, Janardan Karmakar, under the supervision of Daroga Sher Muhammad and Inspector

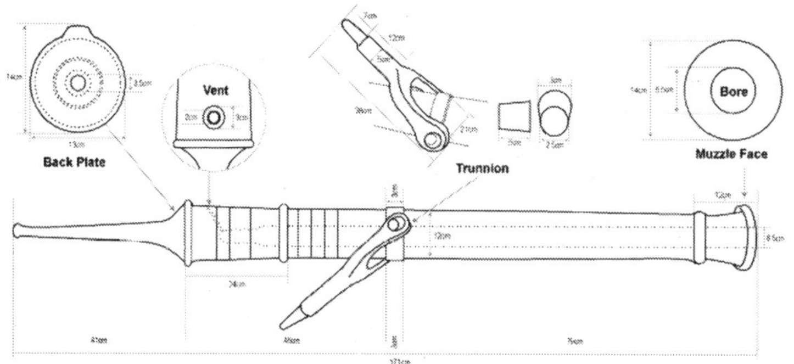

FIGURE 5.8: THE FORGE WELDED NAVAL IRON CANNON #4.

Haraballav Das. The year of manufacture was the month of *Jumadi al shani* in the eleventh regnal year of Shah Jahan. The weight of the cannon was 212 maunds. Another cannon was perhaps manufactured in Bengal as it included fish symbols and Bengali inscriptions. A number of small naval cannons are preserved now at the Bangladesh National Museum, Tripura State Museum, Agartala; and Tai Museum, Shibsagar in India. Most of those were developed or modified from *firingi* cannons. Our present knowledge is incomplete as regards the skills of the cannon makers of Bengal, like Janardan Karmakar, Brajakishore Das, Saiyad Ahmad Rumi or Muhammad Hussain Izzat. Detailed archaeometallurgical studies are needed to evaluate the manufacturing methods of those cannons.

ACKNOWLEDGEMENTS

We acknowledge the Chittagong University to permit us to study the cannons and to publish the same in this volume. Pratip Kumar Mitra and Abul Barkat Jilani have helped us in reading the Persian inscriptions.

NOTES

1. N.K. Bhattasali, *Nutan Badshahi Amaler Kaman, Prabashi*, Paus 1328 BS (1922), pp. 313-18; A.S. Beveridge, *Baburnama in English (Memoirs of Babur)* (London: Luzac & Co, 1922), pp. 671-2 (Downloaded from

the internet, digitized by Microsoft); For *bans*, see Kaushik Roy, 'Rockets Under Haidar Ali and Tipu Sultan', *Indian Journal of History of Science*, vol. 40, no. 4 (2005), pp. 635-55.
2. A.K. Biswas, 'Epic of Saltpetre to Gunpowder', *Indian Journal of History of Science*, vol. 40, no. 4 (2005), pp. 539-71.
3. Biswas, 'Epic of Saltpetre to Gunpowder', p. 558.
4. H.E. Stapleton, 'Notes on Seven Sixteenth century Cannons Recently Discovered in the Dacca District', *Journal and Proceedings of the Asiatic Society of Bengal*, New Series, vol. 5 (1909), p. 368.
5. P. Niyogi, *Iron in Ancient India* (1914, rpt., ed. R.S. Shukla, Delhi: Pratibha Prakashan, 2007), pp. 50-6.
6. N.B. Sanyal, 'A Note on the Additions to the Society's Museum During the Year 1927-28', *Annual Report Varendra Research Society*, pp. 1-8.
7. S. Sharf-ud-din, 'A Note on the Arabic & Persian Inscriptions in the Society's Museum', *Annual Report Varendra Research Society*, 1927-8, pp. 1-6.
8. S. Hossain, 'Cannon of Shah Jahan with the Title Buland Iqbal', *Journal of Asiatic Society of Bengal*, vol. 33, no. 2 (1985), pp. 9-11; *Art and the Vintage* (Chittagong: Chittagong University Museum, 1988).
9. *Baburnama* (*Memoirs of Babur*), tr. from the original Turki Text of Zahir-ud-din Muhammad Babur by A.S. Beveridge, vol. 2 (rpt., New Delhi: Saeed International, 1989), pp. 469, 536, 547.
10. R. Balasubramaniam, *The Saga of Indian Cannons* (New Delhi: Aryan Books International, 2008), p. 93.
11. Ibid., p. 84.

CHAPTER 6

The Idea of Navy and Naval Warfare Under Shivaji

AMARENDRA KUMAR

The sea and rivers have helped in fostering the development of the human civilization in many ways. They have been significant in facilitating the human requisites of transportation-communication since time immemorial. Depending on the natural conditions—viz., access to the sea, geographical position of a nation, and, more recently, the geo-political compulsions—the human interest in the sea, and his desire to use and possess it for his own benefit have been found to have gradually increased in the historical context. As a result, the concept of 'sea power' has evolved and gradually assumed larger significance, especially in the pre-modern (pre-industrial) era. The main aim of this paper is to examine the response of the Maratha Navy under Shivaji—the founder of the Maratha *Swarajya*—towards the superior challenges coming from the organized fleets of European nations (having rich maritime traditions and expertise) in the Arabian Sea. This paper will take into account geo-political surroundings of the Maratha *Swarajya* under Shivaji, the Konkan Coast and the Arabian Sea; beginning of the Maratha Navy; structure of the naval establishment under Shivaji; declared objectives of the Maratha Navy; threat perceptions and the naval wars of Shivaji—nature and intensity; and finally naval strategy and tactical appreciations.

I

The nineteenth-century military theorist A.T. Mahan identifies six fundamental factors which affect the development of a nation as a sea power. These are: geographical position, physical configuration, extent of territory, size and character of population, national character and the character of government/institutions.[1] The main mechanism through which this power is exercised is by the use of a naval fleet. It should be noted here that in the pre-modern era, there was a little or virtually no separation between the functions of a navy and a mercantile fleet. The real separation of functions in respect of the two could become more identifiable only after the introduction of cannon in the sea-worthy vessels. The use of cannons brought about a transformation in the manner in which the naval engagements were conceived. Heavy guns changed naval warfare from hand-to-hand combat to fighting from a distance. Subsequently, the difference in the function of the navy, which had hitherto been totally committed to the defence of the shore from an invasion or for the transportation of land troops (army) by sea for the purpose of invasion to other land, became more pronounced after the introduction of the cannon. The state agency now started fitting special types of vessels for different but specific purposes, and worked in the direction of raising an effective naval flotilla. It was, as if, to have a 'military command of the sea' in a manner, theoretically not different from the way the control over land was exercised. The policing of trade routes, exploration, charting, lighting, the establishment of bases of supply and their protection were all essentials, and were all made the functions of the navy. In this way, the 'use of sea' for a choice of purposes was explored by the state, and constituted the very basis of sea power.

In the absence of concrete evidences, it is difficult to postulate as to when Shivaji started ruling like an independent king.[2] He started his career of conquests from the surroundings of Pune as he had made himself familiar with the paths and defiles of the hilly country in and around the city while enjoying his father's *jagir* near the city.[3] He found the adjoining hill forts in a state of utter neglect as regards their maintenance and garrisoning. There, he realized his expansionist possibilities, and embarked upon a systematic acquisition of the most prominent of them. In 1656, the Mughal viceroy of Deccan, Prince Aurangzeb was on a military assignment to expand the Mughal imperial hold over the Muslim (Shia) principalities of the Deccan. Shivaji took advantage of the situation and tried to consolidate his hold over the recent conquests he had made. The departure of Aurangzeb to Agra to participate in the succession to the Mughal throne further provided Shivaji a free hand. He particularly selected the region of north Konkan[4] which, though was conquered by the Mughals (from Bijapur), still had an overlapping jurisdiction of the Mughals as well as Bijapur. The departure of Aurangzeb provided Shivaji (on the basis of his offer to serve the Mughals) the very pretext to his claim that he was seizing the Konkan territory for the Mughals.[5] Aurangzeb tried to ensure that Shivaji could not make a big gain out of the situation by signing peace with Bijapur in 1657, but it was too late. Shivaji was well placed by that time as he was firmly perched atop the Sahyadris overlooking the Konkan Coast and commanding strategic passes.[6] Hence, the next phase of his military action was destined to be in the Konkan plains.

The undulating lowland of the Konkan[7] roughly extends for about 330 miles (528 km) from north to south[8] and has an area approximately 12,500 sq miles.[9] Between the sea and the Sahyadri wall, which runs almost parallel and presents a steep face to the lowland, this littoral region has a variable width from about 45 to 75 km.[10] These are the widest near Bombay, in the amphitheatre-like basin of the river Ulhas (or river Kalyan). In this region, we indeed come across a flat, alluvial belt along the coast but this is only 4-8 miles wide; behind it lies a series of parallel ridges reaching 1,500-2,000 ft in which rivers like the Vaitarani, Ulhas and Amba have lower course, more or less parallel to the coast before reaching it transversely.[11] Further, south of Bombay, the pattern of landscape changes and the continuity is interrupted by deeply furrowed land and fast-flowing streams (which are shorter and directly transverse) from the Sahyadris, giving it a crisscrossed appearance with highly eroded remnant ranges and knolls.[12] Southwards again, in the south of Ratnagiri, the distinctive feature is the series of extensive laterite-capped residual plateau. And throughout this length, the Konkan is dominated by the tremendous scrap of Ghats rising to 3,000 ft in a mile or two, fretted into wild canyons at the valley heads.[13] In this way, the physiographic features of the Konkan suggest that this area is far from being purely a plain or even-land.

The river and drainage system of the area is also peculiar. None of the streams rising from the Sahyadri drain a large area to be called a river. They are short, varying in length from about 50 to 150 km, and they run mostly parallel to their respective small individual catchment areas. After receiving tidal water at a distance of 12 to 48 km from the sea, they wind along low, mangrove-covered salt marshes and undergo so much change in their character that many of them have two different names, one for the upper course as fresh water streams and the other for their salt-water creeks. Most of these rivers have a steep gradient in the sub-Sahyadrian tract, and are fast flowing and more intensely seasonal in their regime. These fast flowing and rain-fed rivers of the Sahyadri, towards the coastal side, therefore, could not develop estuaries at their mouths. They, however, formed creeks at their mouths—the ports developed along which obviously did not possess deep harbouring facilities. Moreover, the wide and shallow expanses of their mouths are often marked by sand-bars because of the action of the sea-waves. These rivers, though important at their mouths for the depth and tidal effects, are equally disadvantageous and useless further upstream for navigational purposes because of their variable nature of water and steep gradient. Such features, in particular, prevented the European ships, comparatively bigger in size, to make inroads from the western side into the interior. The Europeans, therefore, clung to some pockets along the coast. Even the small vessels anchoring near the river-ports up the creeks could not move upstream because of the steep upstream gradient. The great fluctuation in the water level of the rivers, because of their seasonal nature, was another restricting factor to reckon with if these were to be used for navigational purposes. In short, though the river-ports in the Konkan served large tributary areas, there was no free or easy access from the coast towards the interior, which, on the other hand, remained unaffected by minor developments along the coast. In this way, the rivers of Konkan did not permit any easy navigational use of them by man.[14]

The Konkan Coast is dotted with numerous creeks and inlets over which the influence of the sea water is felt in the form of tides. The movements of the ships and crafts were regulated by the tidal effects of the sea water. The smaller boats could utilize the incoming tides to enter inside and even move a few miles upstream. The tides that penetrate deep into the country (because of the inlets) have always favoured the growth of a number of littoral and estuary ports, like the ancient port of Suppara (Sopara), Kallinea (Kalyan), Semylla (Chaul), Byzantium (Vijayadurg), to name a few. In the medieval period, there were a few prominent ports which attracted the attention of the political powers. These ports fall under two categories: (i) coastal ports on sheltered bays and river mouths, for example, Bankot, Harnai, Devgarh, Dabhol, Anjanvel, etc.; and (ii) inland ports up tidal creeks, generally at the points where navigation ceases, for example, Chiplun, Sangameshwar, and Rajapur. The frequency of coming across a creek or a river bay along the coast is too frequent, particularly in north Konkan, where at every 6 or 8 km we can find safe harbours for indigenous crafts. The Maratha rulers, on the contrary, made a calculated and strategic use of them for political and military (naval) purpose. In addition, there were many fortified locations which had strategic value along the Konkan Coast. Most of them are found to be surrounding the important ports, or guarding the tidal mouths of the various Konkan rivers or, even sheltering dockyards.[15] Though, the date of construction of each of these forts cannot be determined with precision, it can be inferred nevertheless, taking into account their location, construction and distribution pattern, that fortification of the ports or commercial centres had been an integral part of policy matter of the coastal rulers.

The river valleys were the main arteries along which the trade moved from the creek-ports to the interior and vice versa. Protection to this trade and to the movement of the goods was provided by building sea-forts, preferably at the mouths of the creeks, near the ports, and hill forts at suitable points along the trade routes. M.S. Narwane, in a painstaking study, shows that a total of 88 forts, which constituted well over 19 per cent of the total number of forts located in Maharashtra, could be clubbed together and studied under the category of coastal forts.[16] The idea of providing protection to the coastal commercial settlements is not of recent origin or area specific. It has been a global phenomenon since ages, and has been proved through the findings of archaeological excavations. Excavations carried along the coast in the Indian Ocean region have brought to light the fact that many of the commercial settlements/ports coexisted with military/political/fortified settlements even during the ancient period.[17]

On the conditions prevailing in the Konkan Coast when Shivaji had started raiding it, S.R. Sharma writes,

> Despite its importance and value neither Bijapur nor the Mughals had bestowed on the Konkan the attention it deserved. They marked the earth with ruin, but their control stopped with the shore. As Muslim rulers they were anxious to protect the pilgrim traffic to Mecca; but, otherwise, their interest was confined to importing Arab horses and maintaining a few private ships for personal profit.[18]

It, therefore, seems that there was an absence of a powerful military presence or of a naval force of the Muslim rulers in the vicinity of the coast. Shivaji thus extended his military conquest to the plains of Konkan, which is indicated in the contemporary European documents too.[19] Once Shivaji got a foothold in the Konkan, he was quick to appreciate the military advantages of the coast which was characterized by numerous broken inlets and creeks. Moreover, the economic importance of the Konkan was also too obvious to have escaped his attention. Ports like Bassein, Chaul, Ratnagiri, Goa, and so on attracted foreign shipping and could be an excellent source of income. Shivaji's entry into the Konkan also brought him to the attention of the maritime powers on the west coast of India, notably the Portuguese, the British and the Dutch. The European companies in India, which till then were fiercely competing amongst themselves for the possession of the ports and other strategic locations on the Konkan Coast, found another contender—Shivaji. For Shivaji,,on the other hand, it was necessary to eliminate/subjugate all challenges if he had to realize his dream of Maratha *Swarajya.* He, therefore, had to think of a different strategy altogether if he was to survive in this struggle. In the coastal surroundings, it must have been the perceived threat from the potent quarters (particularly the Europeans) that impelled the great visionary to set the keel afloat immediately in the waters of Kalyan-Bhiwandi in 1657. This was indeed a practical and well-calculated decision that speaks volumes of a political genius who would not only like to scrupulously guard his political fortunes, but would also like to add to the economic fortunes of his nascent *Swarajya* by an organized use of the navy. The first ever definite information regarding Maratha naval activities comes from a Portuguese document of the year 1659 which, apart from other things, also mentions that the Maratha ship-building activities had commenced at Kalyan-Bhiwandi.[20]

At the same time, Shivaji must have come to know about the helplessness of the Indian coastal rulers (with the exception of the Mughals), who, in the matters of oceanic and coastal navigation, were constantly exposed to the merciless exploitations of the naval fleets belonging to some of these European nations. The Portuguese were claiming 'sovereignty over the sea' and enforcing their injunctions on the Indian Ocean as a whole since the early sixteenth century. The *Cartaz* system established by them had virtually brought the Indian shipping and navigation, particularly on the western coast, under their monopolistic control. This system served not only as a source of income to the Portuguese, but was also a well-conceived method adopted by them to project their uncontested naval supremacy over the Arabian Sea. It seems, therefore, that Shivaji's natural desire behind the foundation of the Maratha Navy was to resist such pressure-tactics of the Europeans and, at the same time, challenge it by the use of an organized fleet. The geo-political compulsions of the long stretch of the coast forming the natural frontier of the newly founded *Swarajya* in the west forced Shivaji to arrange for its proper defence. This could have been possible only with the help of a chain of coastal and island forts as well as a functional navy which could guard the coastal waters. Very soon a chain of coastal fortresses like Vijaydurg, Sindhudurg, Suvarnadurg, etc., were put into use; and the attempt to harness the sea followed thereafter.

II

That the idea of a state owned naval establishment was well conceived by Shivaji could be established from a perusal of the contemporary Marathi narrative titled *Rajniti* of Ramchandra Pant Amatya. The *Rajniti* clearly indicates that during the reign of Shivaji, the authority of the Maratha *Swarajya* was equally well exercised over the Arabian Sea abutting the Konkan Coast. This was done in a manner befitting to the status of a legitimate sovereign. The state's authority was exercised in matters of administration and safeguard of the rights over the usage of the sea, whereas, parity (with other powers—indigenous or foreign) was claimed in the matters of commerce and rights of navigation. This is amplified by Amatya's considerations: 'Navy is an independent limb of the state. Just as a King's fame for success on land is in proportion to the strength of his cavalry, so the mastery of the sea is in the hands of him who possesses the navy. Therefore a navy should necessarily be built.'[21] That, Shivaji had also organized his flotilla under the office of admiralty is further attested by *Sabhasad Bakhar*.[22] Hence, the objectives behind the raising of Maratha Navy by Shivaji become intelligible at the very outset. We have already noted that Shivaji made earnest efforts to raise a navy for the *Swarajya* immediately after getting a foothold in the Konkan. As a result, as soon as Kalyan-Bhiwandi was captured from the Mughals, it was transformed into a dockyard by the Maratha visionary. The brilliance with which Shivaji comprehended the geopolitics of the region is worth admiring. On the one hand, adequate good quality teak was available in the surroundings of Kalyan-Bhiwandi which could be utilized for the purpose of ship-building; on the other, the fortification of the creek of Durgadi would have afforded the much needed military and strategic protection to Kalyan which was destined to play the crucial role of a dockyard in the immediate future. The theory of warfare, it should be remembered, also states that secured bases are necessary prerequisites which facilitate the military leader the freedom of action against the enemy.

The Portuguese documents also confirm that by the year 1659, the construction of vessels had commenced at Kalyan-Bhiwandi by Shivaji.[23] Other Portuguese documents of similar nature further update us that the first 20 armed ships of Shivaji were constructed with the help of Portuguese artisans.[24] Specific reference on the ship-building activities of Shivaji in the Kalyan-Bhiwandi waters can also be found in the English Factory Records. A letter dated 22 October 1672 records, 'We have thought good to advise you that there is laid up in this harbour six small vessels, belonging to Shivagee with his consent, all new and lately built at *Culian Bhindi* (Kalyan Bhiwandi). . . .'[25] Such armed vessels, being the components of an organized/specialized navy, were surely planned to project the authority of the Maratha state on the coastal waters. Shivaji appeared well determined to secure the freedom of the seas along the coastal extent of his *Swarajya*. Hence, ship-building activities were undertaken, though initially in a cautious manner.

Shivaji was shrewd enough not to provoke any kind of suspicion in the minds of the Portuguese governor of the adjoining Thane/Bassein region when he began the boat-building enterprise at Kalyan-Bhiwandi. Not only did he declare that the main aim of the Maratha Navy was the destruction of the Siddi chiefs of Janjira, who, according to *Sabhasad Bakhar*, were like 'rats in the house',[26] but also convinced the Portuguese to supply ammunitions and equip his vessels. After all, the newly fitted Maratha vessels from Kalyan-Bhiwandi were to pass through the channel skirting Thane before they made into the open sea. Even at a later stage many craftsmen of Portuguese origin were found working for the Maratha Navy, particularly assisting in ship-building activities. It is also believed that through his minister Pitambar Senvi's influence (who was well posted in the affairs and rivalries of the European trading companies, and also well acquainted with the knowledge of their languages), Shivaji borrowed the services of a few naval and artillery experts from Goa and established his own ship-building yards and arsenals at Malwan.[27] Logically, therefore, it follows that the vessels which were being constructed by the Marathas could have adopted some of the European designs and compatibility. Moreover, since they were being purpose built for naval warfare (apparently to be fought against the Siddis of Janjira), they must have been designed to withstand damage, and to inflict the same on the enemy too.

Though it is difficult to tell precisely as regards the quantum and quality of vessels being constructed in the Maratha dockyards, particularly in the formative years of the Maratha Navy, the contemporary references do facilitate us to outline a rough estimate of the same. *Sabhasad Bakhar*, the most reliable of the various *Bakhars* on Shivaji, gives us a rough estimate of the numerical strength of the Maratha Navy along with details regarding the different types/categories of vessels of the Maratha fleet. This helps us to outline the spirit of the Maratha naval enterprise under Shivaji. Some other contemporary Marathi chronicles also list the different types of Maratha vessels along with their nomenclature. This is indicative of the fact that vessels for specialized purposes, keeping in mind the naval-strategy of the state and varied requirements thereof, were built by Shivaji. For example, flat-bottomed boats would have suited the specific purpose of being used in the shallow and rocky waters of the various creeks; small rowing boats would have been employed for towing the heavy vessel (particularly from the coastal forts and bases) which could attain speed once it used sail after reaching out in the open waters, etc., and deliver attacks on the enemy vessels. *Sabhasad* writes, 'Ships of various types, such as *Gurabs* and *Tarandes*, *Tarus*, *Galvats*, *Sibads* and *Pagar* were fitted and organized under two *Subedars*. . . . A *Suba* (naval district) comprising of two hundred ships.'[28] As per this estimate, the total number of vessels (of various types as enumerated above), at that particular point of time seems to have been somewhat around four hundred. Another piece of information of similar nature comes from a letter of the British East India Company, dated 21 November 1670. It reads, 'The General and Admiral of the fleet (which consists of 160 small vessels counted by my own servant, who I sent as a spy) is one Ventji Sarung with whom I have a correspondence these 7 or 8 years. . . .'[29] It is interesting to note that the number of vessels under the admiralty, as reported in the English correspondence, is not very different from the number specified by *Sabhasad* under one *Subedar*. The perusal of some Portuguese documents further confirms that different kinds of vessels were being constructed at a good pace for the Maratha fleet. A letter written by the Governor of Goa in April/May 1662 reads, '. . . Ragho Ballal, the (Maratha) Subedar of Dabhol, had requested him to allow five warships (*Sanguiceis*) and one *Pataxo* interned in the Karanja creek to go out to the sea. . . .'[30] Another Portuguese letter written by the captain of Chaul to the governor of Goa in August 1664 mentions that Shivaji was building fifty ships in upper Chaul and that seven of them were ready to set out to sea.[31]

Simultaneously, with the ship-building activities at Kalyan, Shivaji also worked to cater for a logistical system for his naval enterprise. His priorities were now to provide naval bases for safe anchorage of the various kinds of ships where their repair and maintenance works could also be undertaken. For an effective operation of a navy, naval bases are the key requirements. Such bases were, therefore, built at carefully selected places along the stretch of the Konkan Coast under his possession. In this way, Shivaji worked on the principle that secure bases constituted the necessary foundations for power upon the sea. Construction works had commenced at the most prestigious marine fort christened *Sindhudurg* (the Sea Fortress) under the personal supervision of this great military visionary as early as 1664. Ramchandra Pant Amatya highlights the importance of the naval forts in providing strength to the indigenous fleet and offering protection to the ships and sailors. He writes, 'Whatever may be our strength . . . our fleets should be brought under the protection of sea-forts. The safety of ships and sailors should never be risked.'[32] In consideration of the above, therefore, many naval forts were constructed along the coast of Konkan; and those which already existed were occupied or captured and then suitably strengthened.[33] The curiosity of the Portuguese, particularly at the Maratha vessels and marine forts, becomes quite manifest by the year 1667. In a letter to the king of Portugal, the Portuguese Vice-Rei Conde de San Vincente, expressed his apprehensions regarding the Maratha naval establishment in the following manner, 'I am afraid of Shivaji's naval ships. We did not take sufficient preventive steps and so he has built many forts on the Konkan Coast. Today he has several ships and they are large ones. . . .'[34] *Sabhasad*, on the other hand, also justifies Shivaji's decision of undertaking the construction of marine forts in the light of his constant preoccupation against the Siddis of Janjira. The latter could not be subdued despite strenuous military attempts of Shivaji, as he (the Siddi) held the island fort of Janjira under his control. Sabhasad writes:

> Rajapuri (or Danda-Rajapuri as the fort of Janjira was referred to in the contemporary literature) was left (unconquered by Shivaji) in the sea and on that account the name of the Nizamshahi still continued there. As that place had to be conquered, the *Raje* (Shivaji) built forts selecting (for their site) rocks at various places, as these commanded the sea and (these forts) would weaken the sea kings. Realizing this, he fortified some submarine rocks and constructed forts in the sea. Building such sea-forts or *Janjiras* and uniting ships with forts, the *Raje* saddled the sea.[35]

The most significant aspect of Sabhasad's narration is 'uniting ships with forts', which indeed highlights the importance of naval bases for an effective operation of naval fleet. The Europeans (the Portuguese, in particular), had been working on the same principle ever since their arrival in India. It should be remembered that the absolute control of the Portuguese over the Indian waters in the sixteenth and seventeenth centuries was very much facilitated by their naval bases at Gujarat, Chaul and Goa. The Portuguese records, in fact, further suggest that Albuquerque, the governor of Portuguese Goa, planned to have 'a garrison of 500 to 1,000 men in each fortress; 300 men for each factory and 1,600 for the fleet'.[36] This was deemed necessary to maintain total dominance over the coastal navigation on the stretch of the Arabian Sea mingling with the Konkan coast.

The observations of the contemporary French traveller Abbe Carre further highlight the strategic significance of the sites which were selected for his marine forts by Shivaji. Carre writes, 'He (Shivaji) selected such maritime places as were easier to defend and more difficult to attack.'[37] Amatya's narration further upholds the keen observation of Carre as he writes, 'On account of the fear of the fort, the enemy should not enter the creek of the sea.'[38] It must have been the prime endeavour of Shivaji, therefore, to ensure that suitable arrangements to strengthen the defenses of such marine establishments, which acted as naval bases also, were made on a priority basis. It is equally significant to note that in principle, there was no difference as regards the rules of construction, pattern and layout of the marine forts with respect to the hill or other forts of the *Swarajya*. Many of Shivaji's coastal forts commanded the mouth of the creeks, and were built on the banks of rivers (not in the open sea), or, on a cliff or a spot of land (headland) more than half surrounded by the sea, if possible.[39] Shivaji was aware of the fact that military strategy owes much more to the politico-military perceptions of operational effectiveness. Hence, at many of the strategic marine locations, Shivaji strengthened the defenses of the already existing structures as per the operational requirements. In addition, his initiative to fortify the off-shore islands, and bring them within the sphere of operational control of his flotilla was also strategically justified. When we analyse these structures carefully, we find that most of them had ample defensive features. We find gun holes and port holes on the ramparts of most of the marine forts. Guns and canons were fitted on most of these forts to keep at bay the enemy ships, approaching by sea/creeks. A large number of the coastal forts had sea facing walls from three sides, and from the land side the protection was provided by digging ditch or having a moat around the periphery. The gates opening on the sea sides also had protective features and were, at times, also regulated by tidal patterns. In addition, these forts had sufficient provisions to keep the garrison well fed and defiant to the enemy diktat for a longer time. This was in accordance with the policy directives of Amatya who suggested that apart from water and food, the naval forts must have sufficient ammunitions to keep the enemy at bay.[40] Last, but not the least, in the most important of these forts, facilities of a fully functional dockyard also existed. The most notable examples, to name a few, were those of Kulaba, Vijaydurg, Sindhudurg and Jaitapur. It should always be kept in mind that apart from lucrative economic prospects (which could be grasped by providing military protection to the ports and coastal mart), the freedom of escaping to the sea—particularly at a time when faced with a determined attack originating from the land side—was also one of the added and the most practical of the advantages of having such fortified bases. The East India Company in India resorted to this practice quite often, as that time they were devoid of any territorial base on the coast.

It ought to be kept in mind that the Portuguese armada operated from their bases and compelled the Indian ships to purchase Portuguese passes for navigation in the coastal waters. M.N. Pearson has, in fact, argued that 'there was no evidence of any use of force at all (by any Indian powers) in the Indian waters. No warships patrolled forcing ships (of any nationality) to call at any of the ports', whereas the Portuguese armed vessels could enforce their authority by compelling the Indian ships to touch Daman or Diu, pay taxes, and then proceed to their onward journey.[41] The Portuguese, according to Pearson, achieved their above mentioned objective by establishing a string of fifty forts and fortified areas (acquired through conquests/ treaty or even newly constructed by them), and a total fleet of up to 100 ships of various sizes in different areas.[42] In a similar manner, Shivaji also tried to impose his sovereign authority over the coastal waters of Konkan by asking even the European ships to buy the Maratha passes, and thus earned money. Here, it may be noted, that by doing so, he not only tried to disallow the loss of revenue to the *Swarajya,* but also augment his own collections at the same time as the 'native' merchants were until then, buying passes from the Europeans. Amatya's *Rajniti* echoes such a kind of policy by a sovereign, and notes, 'If the ships of foreign merchants, besides those of the enemy, not possessing permits, are coming and going, then they should not be allowed to move without inspection.'[43] He further writes, '. . . if any merchant ships coming from the hostile territory are found in the sea, they should be captured by making great efforts (and) brought to the port. . . .'[44] For the projection of sovereignty over the coastal waters, a strong and effective flotilla was a prime requisite. Moreover, the requirement of naval bases at appropriate locations would have further helped in doing so, as 'any hostile ship' had to be captured and taken to the nearby friendly port or base so that it could be dealt with suitably. It is not very clear whether Shivaji was in a position to ask European/foreign ships to compulsorily touch any particular Maratha port or base in the absence of an effective flotilla and strong naval bases, at least in the initial phase of his navy building enterprise. However, contemporary European documents give enough examples to show that the capturing of each others' ships for the purpose of bargain was not an uncommon practice.

Construction of warships in large numbers required, after all, time, financial resources and professional expertise. Hence, for projection of force and authority, it was imperative to maintain the required number of ships. It was rather a policy obligation of the Marathas to prevent any harm/loss to it because of enemy activities. Amatya writes that even when the naval ships were sheltering in a creek, well defended by a fort at its mouth, 'the fleet should not be kept in one place, it should be distributed in various places; and in the night—patrolling both by creek and by land—should be done round about the fleet'.[45] This tactic (of distributing the vessels in various places) of Amatya finds its proper usage during the time of Shivaji and has been confirmed by English sources. A letter from a factor in Bombay addressed to the governor of the English Factory at Surat says, '. . . we are certainly informed that Shivaji, with a considerable fleet, is at Nagaon (creek) about seven leagues off of us, where he has made provisions for a siege or storm. . . .'[46] The reference to the presence of the Maratha fleet at Nagaon could also be found in the contemporary Portuguese documents. A letter written by the Portuguese Captain of Chaul, dated 25 March 1677, states, '. . . Shivaji's fleet was stationed at Nagaon: there were ten *Galvats* and thirteen other ships.'[47] The creeks could have served the purpose well only if the land adjoining it was under the control of the friendly naval power. Though the undefended coast was always exposed to the depredations of a hostile navy, there was a bigger risk of the vessels placed in such creeks being damaged or destroyed by way of bombing or fire caused by the power which controlled the land adjoining the creeks. Nagaon or Nagothana, or for that matter any such creek was threatened from land; therefore, it was important to occupy them, as Shivaji did. It was the absolute control of the coast-line by Shivaji that discouraged the Siddi (who was in Mughal pay) to winter his fleet in the Konkan creeks. Siddi, under such a situation, was left with no option but to be always dependent on the British Bombay for the same.

III

Every state formulates a military doctrine of its own which is in harmony with its political doctrines and geographical realities. The geographical location of a state affects its accessibility, vulnerability, bargaining capacity, strategic worth, political sensibility and national policy. In terms of terrestrial location, the Maratha *Swarajya* under Shivaji could be categorized as 'maritime location', i.e. a territory having access to sea and also connected with land. A large stretch of the *Swarajya*, in fact, was coastal terrain, as it was the actual zone of contact between the land and sea. The coastal terrain, so far as strategic considerations go, offers at least two distinct military advantages to the state:

1. Offers opportunity for the deployment of naval forces for defensive purposes.
2. Offers opportunity of employing maritime commerce to sustain its defensive effort.

A cautious appreciation of the hydrographical character of the coastline (viz., the presence of gulfs, bays, river estuaries, large lagoons with sufficient entrance) and the character of the ocean/sea bed (viz., the nature of the favourable tides and streams, presence of sand-bars or submerged reefs stones) further helps in extracting the military advantages offered by the coastal terrain. Shivaji had been working to raise an effective flotilla to exercise his authority over the Konkan waters right from the beginning of his political career. The tasks for the Maratha Navy were as follows:

(a) *Military objectives:* To destroy enemy navies.
(b) *Political objectives:* To enforce political demands on the adversary.
(c) *Geographical/geo-strategic objectives:* To overthrow the enemy from strategically important areas.

A better understanding of the military strategy and tactics is possible only in the background of the task which is to be accomplished through military undertakings. Strategy may be defensive or offensive, or, at times a combination of both, depending on the circumstances under which the troops operate. Hence, to have a better comprehension of the Maratha naval strategy and tactics, let us start with the defined aim or objective of the Maratha naval establishment. Sabhasad puts forward the view that the Maratha Navy was created to fight and defeat the Siddis of Janjira. The Portuguese sources also support this hypothesis. Hence, particularly with regard to the Siddis of Janjira, the strategy of the Maratha Navy outrightly appears to be 'offensive'. This is also evident from the long-drawn conflict between Shivaji and the Siddi chiefs both on land and water. Here, the 'military objective'—of the destruction of the enemy and its war machine—was being served with the navy as one of the 'means'. In fact, the prolonged naval war between these two parties could also be perceived as a 'war of attrition'. The question now is, was the Maratha naval policy under Shivaji the same with respect to the Portuguese, the British or, for that matter, any other European power? A study of the naval encounters between Shivaji's fleet and the Europeans gives us the impression that Shivaji was not completely hostile to these nations and that his military actions were mixed with caution. Moreover, the European trading companies, too, were not keen in pursuing a policy of altercation with the Marathas or even the other 'native' powers so long as their commercial objectives were being satisfied by other means. Skirmishes between the European and the Maratha navies indeed took place, but, by nature, they were not very intensive. The usual practice was to capture each other's vessels for the purpose of bargain and then return to normalcy. Under such a situation, it may not be grossly inaccurate to infer that Shivaji, by means of his naval policy with respect to the Europeans, was trying to achieve 'political objectives', i.e. to enforce political or associated demands on the adversary in the interest of the *Swarajya*. As regards his endeavours to oust the Siddis from Janjira, or his occupation and fortification of the desolate off-shore rocks like Suvarnadurg, or even the forcible occupation and fortification of the uninhabited island of Khanderi, they could well be understood in the light of his desire to fulfil the geographical objectives, i.e. to overthrow the enemy from strategically important areas and to prevent them from occupying one.

The contemporary documents appear deficient as regards definite information on the details of strategy and tactical dispositions of the Maratha Navy during its engagements with other powers, especially the Europeans, on the Konkan waters. However, it could not be denied that with whatever meagre naval means Shivaji had under his disposal (viz., comparatively smaller vessels, poor quality of fire power and related technology), he would have tried to make the best possible use of them, particularly by resorting to superior strategy and tactics. It would be inappropriate to presuppose that smaller, 'native' war-vessels (variously called as *Galvat* or the rowing vessels in the contemporary Marathi chronicles) were of no value as compared to bigger, higher, and more manoeuvrable European sailing ships. The post-sixteenth-century European ships were indeed modified warships with multiple (and complex) sails which afforded them more manoeuvrability, particularly in battle formations. They were comparatively bigger and higher, hence more defensible and secure against enemy attack comprising boarding actions. The Portuguese, the Dutch, the British and the French had such ships. The Marathas under Shivaji, on the other hand, had smaller rowing vessels of small or medium size. Shivaji apparently made an efficient use of them to his advantage in his naval engagements with the Europeans. It had been his strategy to operate his war vessels from the vicinity of the shore, under the protection of the guns of his marine forts.[48] This was a good ploy as it deterred the enemy ships from engaging there without seriously undermining their own safety. Even Admiral Mahan recommends that in a situation where there is the presence of numerous lines of inland water communications and strong fortresses along the estuaries, a weaker navy should 'always fall back to elude pursuit or to receive protection' of the coastal guns and avoid a direct confrontation in the open seas.[49] The most perceptible advantage of the tactics of 'clinging on to the shore', put into repeated use by the Marathas in their naval engagements could be comprehended in terms of the appreciation of the geomorphologic peculiarity of the Konkan Coast by the Maratha vessels. As noted earlier, there was always a possibility of submerged rocks or reefs, sandbars, or even shallow water (depending on the nature and occurrence of tide) in the environs of the shore. The Marathas with their, smaller, lighter and flat-bottomed rowing boats could have exploited such a surrounding at ease when pitched against the bigger and heavier European ships. It should be noted that the Europeans were still oblivious of the geomorphology of the Konkan Coast. The contemporary European records contain a few references of their efforts to chart the Konkan Coast for a better (economic) use of the coastal surroundings. The English records make us believe that the French Company designed to take the island of Karanja, which was in the close proximity of Bombay, as early as 1671. The French fleet had also taken the soundings of the water near the Karanja island to study the possibility of establishing a base there. The 'efforts' seem to have continued for a couple of years after which the 'project' was abandoned by them.[50] The European ships, on the other hand, could not follow the 'native' crafts in their hot chase because they did not have the correct details of the submerged rocks or reefs in the coastal surroundings. They seemed to have understood this problem. Hence, we find the East India Company administrators directing the factory officials to map the routes of the coastal water and also to take 'soundings' of the different locations on way.[51] One of the prominent causes of the failure of the British to prevent Shivaji from occupying and fortifying the island of Khanderi appears to be their ignorance about the submerged coastal rocks as the 'soundings' of the Bay of Khanderi had not been done.

Another, and more decisive utilization of the coastal surroundings for the purpose of naval engagements during the period of Shivaji, has been visibly highlighted by Amatya. Amatya writes, 'Ships of medium size should necessarily be built which should not be dependent on wind (sail).'[52] Though Amatya does not elaborate his contention, but a lot could be inferred from his generalized statement. The European ships, which were capable of undertaking long distance and risky sea-voyages, were largely dependent on deep oceanic waters and strong winds to sail fast and manoeuvre properly. Once the wind dropped completely or if a wind shift occurred in the vicinity of the coast, it was liable to render a sailing ship (in this case a European vessel, principally dependent on sail) helpless as it would be unable to manoeuvre, and hence, unable to engage the enemy (the Marathas in this case with their smaller and more manoeuvrable rowing vessels) in a profitable manner. Even if things did not happen as anticipated, the Marathas, well accomplished in the guerrilla tactics of warfare, could always 'disengage' and make good their escape through the numerous creeks and shallow bays, where it was difficult for the European vessels to follow them.

The matter of fire-power commanded by the Marathas and the Europeans, and its impact on the outcome of naval engagements between them also needs to be addressed here. Scholarly studies have, time and again, pointed towards the fact that by the sixteenth century, the use of guns and cannons had become common practice in the Indian Ocean region. The Ottoman Turks, in particular, manufactured cannons of good quality on a reasonable scale and had been instrumental in making the use of guns and cannons in the ships a common feature in the Indian Ocean region. The English Factory Records mention an incident of the year 1625 in which a 250-ton Indian ship with 20 pieces of artillery and shots (apparently to deal with the high-sea pirates), failed to defeat a comparatively smaller and lightly armed British ship.[53] It is a different matter altogether whether the Indian mariners were competent to use the artillery effectively.[54] In fact, many of the Indian merchant ships carried a surprisingly large number of guns, which were perhaps a necessity because of the pirates. But, what is really surprising, however, is not the number of their guns, but that they were never in a position to use them effectively.

As far as the Marathas under Shivaji were concerned, they suffered a severe handicap in terms of artillery. There were no foundries in the Maratha *Swarajya*, and Shivaji had to rely upon the Europeans for the supply of ammunitions and guns for his navy. There are numerous instances where the French, the British, the Dutch and the Portuguese alike had obliged Shivaji by supplying him with ammunitions and guns from time to time.[55] In fact, it would not be erroneous to argue that one of the reasons why Shivaji refrained from entering into serious hostilities against the Europeans (on land as well as on the sea) was the reality that the Europeans alone were capable of contributing meaningfully towards the enhancement of the naval capabilities of his *Swarajya*. Though the supply he obtained from them was too less and not even up to the standards to have significantly altered the prospects of Maratha Navy as a big player in the Konkan waters, the genius of Shivaji converted even this limitation to his advantage through sheer imagination. As on land, in the naval warfare too, in accordance to the guerrilla traditions, the element of surprise as a 'Principle of War' constituted the basis of his naval operations. To inflict surprise on his opponents, Shivaji's vessels would invariably exploit the advantage offered by the numerous creeks and coves along the Konkan Coast. Such recesses were used not only as forays for quick offensives, but were also used as secret hide-outs for their vessels. Hence, by resorting to the principle of surprise, the Maratha Navy not only assimilated the geographical realities in their naval policy, but also made the best use of the available resources. We have already noted that it was a part of his naval strategy that Shivaji's naval vessels would offer war in the vicinity of the shore rather than in the open sea, mainly on account of manoeuvrability. The 'medium sized' or small Maratha vessels powered by oars, even if fitted with a single gun/cannon over their prows were bound to be very effective because of the rapidity in manoeuvring in shallow water. On the other hand, the bigger European vessels which had fixed port-holes cut for accommodating their cannons on their broadsides were less effective because it was difficult to get the ship to its broadside (in the calm winds of the coastal surroundings) every time the enemy (the smaller Maratha vessels) changed position. From the paintings and sketches of the Maratha vessels of the period of Kanhoji Angre and his successors (some three to four decades after the death of Shivaji), it could be ascertained that a large number of the Maratha fighting vessels had a few guns positioned near the bow or the head of the respective vessels so that they could be effectively used during rapid tactical manoeuvres.[56]

It is indeed true that the Marathas suffered from severe limitations in terms of resources, finances and even technology which made their programme of naval expansion difficult. It should also be remembered at the same time that the combat and control of the sea could not offer those 'visible advantages' as it offered as a consequence of the conquest over a land mass. Shivaji, thankfully, did not suffer from this typically 'continental outlook'. As such, he was hard pressed on land because of the military campaigns against him by the likes of Afzal Khan, Jai Singh and Shaista Khan, etc. Hence, the control over the Konkan water essentially offered him a safe refuge in case of exigency arising from the land. He had extended his control over the entire stretch of the Konkan coast[57] and to supervise and secure them, he had constructed a reasonably large flotilla as noted earlier in this paper. Superiority of numbers allowed him a few crucial advantages. In the first place, it enabled him to stretch and extend the deployment of his vessels in the numerous bays and creeks and guard them intensively. At the same time he had the option of simultaneously undertaking naval campaigns directed at different points.[58] During the time of naval engagements, he could have ensured the 'concentration of force' through swift manoeuvre and gained advantage against enemy vessels. In fact, the method of naval engagements of the Marathas (with respect to bigger and lesser number of European vessels) was simple. The Maratha vessels would surround the bigger vessels and would try to dismast them first so that it could not sail or manoeuvre. Once that was done, boarding action followed from different sides and finally culminated in a 'close quarter battle' with the enemy on board their ships. The Marathas thus engaged in daring action in naval engagements too. Since the entire act was risky as the Maratha vessels were exposed to the superior artillery fire of the European vessels, and there was a possibility of the loss of Maratha vessels too, for Shivaji it was the risk worth taking. He knew that losing a few vessels would not make a big difference to the prospects of his naval establishment. His vessels were small with hardly a gun or two fitted on them. As compared to that the loss the Europeans were expected to experience because of the destruction of well equipped and specialized ships[59] was going to be enormous. This must also have been a practical consideration why the Europeans refrained from engaging with the Marathas on a large scale.

Another tactic being put into practice by Shivaji in his naval wars was that of 'naval blockades'. The contemporary Marathi and English sources frequently refer to the military contest between Shivaji and the Siddi of Janjira. Shivaji was aware of the fact that the control over the island fortress of Janjira was the mainstay of the Siddi strength. After having successfully reduced the authority of the Siddi on the coast, Shivaji, therefore, set out to capture Janjira. He knew that his vessels and artillery were incapable of doing any harm to the sea fortress of Janjira, hence he decided to carry out a naval blockade of the island fort. His strategy was somehow to push the Siddi garrison of Janjira on to the brink of starvation, thus leaving them with no other substitute but to surrender Janjira (to the Marathas). To achieve this objective, Shivaji deployed the Maratha naval ships around Janjira in the best possible way. The island fort of Janjira was thus virtually surrounded and blockaded every year. Shivaji could anticipate some success in breaking the resistance of the Siddis, but the military assistance of the Portuguese at the eleventh hour prevented the Siddi from succumbing to Maratha pressure. It seems that the Siddi was so hard-pressed (as no military assistance was coming from the Portuguese) that he opened channels of negotiations with the British as well as the Mughals. The British administrators of the nearby territory curiously kept a track of the Maratha–Siddi conflict as the outcome of this conflict was surely going to affect their trading fortunes on the Konkan coast. The British factors at Bombay acknowledge the receipt of a message from Fateh Khan, the Siddi of Danda-Rajapuri, 'whose island fortress of Janjira was being closely besieged by Shivaji'.[60] The same letter contains an explicit hint of the Maratha naval strategy against Janjira. The letter reads, '. . . Though he (Shivaji) cannot storm the place (Janjira), only thinks to starve him (Siddi) out. . . .' It also appears, on the basis of the above-mentioned letter, that Aurangzeb (on receipt of a request for help from the Siddi) had also asked Shivaji to withdraw his force from the surroundings of Janjira.[61] This letter further reveals the very mindset of the besieged Siddi admiral of Janjira. The letter records, '. . . so the Siddi resolves to hold it (Janjira fort) to the last, and then has thoughts of delivering it up to the Mughals, who have an army about Kalyan, commanded by Lodhi Khan.'[62] It seems that the Mughal Emperor Aurangzeb was also prepared to pay a sum of Rs. 3 lakh to the British to bring Danda-Rajapuri (Janjira) under the Mughal control.[63]

The Bombay authorities, on the other hand, tried to create an opportunity for themselves out of this situation. One of their letter reads,

> . . . the Portuguese Captain of Chaul has dealt falsely with him, in seizing one of his vessels after he has assured him contrary; so he (Siddi) will not trust them, and rather desires our friendship. . . . This place is doubtless of great concern, almost invincible if not blocked up by sea. . . . The Company formerly have had an eye towards it, and now may be the very nick of time to compass it. . . . If two or three ships were sent down from Surat and we empowered to appear with them (Siddi fleet) at Danda Rajapuri, and advantageous treaty might be made with the Siddi. . . .[64]

The policy of naval blockade seems to have worked fine for the moment for Shivaji, as the Siddi was apparently desperate to enlist himself in the services of anyone who would have relieved him from the ceaseless Maratha pressure. This also brings into light the crucial fact that mere possession of an isolated island base was not enough to sustain the naval operations and related expenditures. A base could serve its purpose only if its requirements and supplies can be continually guaranteed. The Siddi was deprived of any landed possession on the coast which could have supplemented the food requirements of his garrison at least. The island of Janjira, at the same time, was too small to have produced any food grains or met any other requirement (i.e. wood) related to the military/naval apparatus. It was, thus, necessary to have a territorial base for the Siddis to operate efficiently in the coastal waters. At the same time, so long as Janjira was with the Siddi, he was difficult to get to, and could not be brought to task. Under such circumstances, to cut-off any kind of contact or supply to the Island of Janjira with the Konkan coast was the only way out to coerce the Siddi into submission. For that, it required a strong naval fleet which could have defied the Siddi fleet. Shivaji, thus, continued with his policy of naval blockade against the Siddi. The next year (i.e. 1670) the Siddi resistance seems to have finally shown the sign of cracking.

Being exhausted by the constant struggle (now seeming hopeless too), the Siddi Admiral Fateh Khan allegedly had made up his mind to come to terms with Shivaji. This decision came in the wake of acute scarcity of food and other provisions in Janjira. Moreover, the Portuguese, their new allies, failed to come to their timely rescue this time because of bigger compulsions. The contemporary Mughal chronicler Khafi Khan throws more light on this event. As per Khafi Khan's narrative, Fateh Khan, the admiral (chief) of Janjira, was almost going to hand over Janjira to Shivaji[65] and settle down with a rich *jagir* as compensation, but somehow, three of his lieutenants got an inkling about this plan. They immediately swung into action in order to prevent Janjira from going into the hands of an 'infidel'. In a bloodless coup, the Siddi admiral was put behind the bars. Siddi Khairiyat, one of the three subordinates as well as the second in command of the Siddi fleet, was proclaimed the chief. The Mughal help and intervention was immediately invoked. The Mughals also proved to eager to step in and oblige—after all, it would have enabled them to put a close watch on the 'anti-imperial activities' of Shivaji.[66] The control over this unsubdued island fortress helped the Siddi to successfully contest the sovereignty of the Marathas on the Konkan waters for years to come.

It was because of his repeated failure to grab Janjira that Shivaji decided to occupy and fortify the desolate island of Khanderi having an area not more than even 2 sq miles. The location of Khanderi was indeed strategic from the point of view of navy. In fact, the Portuguese, in the past, had also worked on the idea of using Khanderi for their own purpose but that idea was shelved by them as they could find no source of fresh water on that island.[67] Moreover, this island was uninhabited and was being used only as a source of fuel (wood). What made Shivaji eager for its possession and subsequent fortification was his sheer strategic vision. He knew that by possessing it, he could keep an eye over the movement of ships (including those of the Siddis as they wintered at Mazagaon every monsoon) to and from the British port of Bombay. The British could also visualize the danger posed to the navigation of Bombay because of an enemy post at Khanderi, hence they immediately protested against the action of Shivaji. The Siddis were equally alarmed and even tried to resist Shivaji in fortifying Khanderi by joining hands with the British.[68] But the alliance failed to achieve any success against the firm determination of Shivaji. The fright of Siddis could be presumed from the fact that they captured another small island (that too inhabited only by rats!) in the vicinity called Underi.[69] It was the turn of Shivaji now to voice protest. He asked the British to direct (their ally) the Siddis to vacate Underi,[70] but nothing came out of this effort and the *status quo* was maintained. Shivaji, in any case, did not live long after the Khanderi-Underi episode, which had altered the situation significantly in favour of the Marathas. But, so long as this great visionary lived, he tried to exercise his full authority over the Konkan waters with the help of his fleet.

CONCLUSION

What becomes evident from the foregoing discussion is the fact that the Maratha naval establishment under Shivaji had assimilated nearly all the features which we find as the basic elements of the naval establishment of the modern era. Shivaji had a keen understanding of the very concept and foundations of sea/naval power, and it had been his cherished desire to gain equality in the matters of navigation with respect to the other powers present on the western coast of India. To achieve this objective, he devised his naval strategy in such a manner that the reliance on foreign marine technology or his deficient fire power was minimized to a great extent in the naval expansion he followed slowly but steadily.

NOTES

1. For a general discussion on the foundations of sea power, see, A.T. Mahan, *The Influence of Sea Power Upon History (1660-1783)* (1890, rpt., London: Methuen, 1965), pp. 25-89.
2. D.V. Apte, 'When Did Shivaji Start His Career of Independence', in *The Indian Historical Records Commission* (Proceedings of Meeting), XVII Session, Baroda, 1940, pp. 44-6. Apte has tried to establish that the year 1656 marks the beginning of the independent career of Shivaji. His conclusion is based on the assumption that scores of letter written by

Shahji to the officers of the *jagir* of Pune could be found pertaining to the period 1636 to 1655 but not a single letter written after 6 December 1655 could be found.

3. His father Shahji Bhonsle was a *jagirdar* of good repute in the court of the Sultan of Bijapur. For his services he was granted a *jagir* in the region near the modern city of Pune in the Maharashtra state. See J. Sarkar, *House of Shivaji* (3rd edn., Calcutta: M.C. Sarkar & Sons, 1955) pp. 33-4. (Chapter III of this book gives a good detail of Shahji Bhonsle's early career.)
4. The Konkan refers to a coastal tract which is one of the major geographical divisions of western India. In simpler terms, Konkan can be described as a narrow strip of land lying between the base of the Sahyadri and the Arabian Sea.
5. It is interesting to note that Aurangzeb 'really listened to the overtures of Shivaji, assented to his keeping what he had wrested from Bijapur, and, with the alleged right of the emperor to dispose of that kingdom, consented to a proposal from Shivaji of taking possession of Dabhol and its dependencies of the sea coast'. See Grant Duff, *History of the Mahrattas*, 3 vols. (n.d., rpt., Jaipur: Aavishkar Publishers, 1986), vol. 1, p. 118 (Duff has based his statement on the basis of the original letter from Aurangzeb).
6. For a general discussion on Shivaji's military action in the Konkan during 1656-7, refer, Jadunath Sarkar, *Shivaji and His Times* (New Delhi: Orient Longman, 1973), pp. 53-8.
7. At present, there are four districts, in addition to that of Mumbai or Bombay, namely Thane, Raigarh, Ratnagiri and Sindhudurg which cover the Konkan Coast from north to south in the state of Maharashtra.
8. R.L. Singh, *India: A Regional Geography* (Varanasi: The National Geography Society of India, 1971), p. 911.
9. W.W. Hunter, *The Imperial Gazetteer of India*, 13 vols. (London: Trubner & Co., 1885-7), vol. 8, p. 289.
10. C.D. Deshpande, *Geography of Maharashtra* (New Delhi: National Book Trust, 1971), p. 18.
11. O.H.K. Spate and A.T.A. Learmonth, *India and Pakistan: A General and Regional Geography*, 3rd rev. edn. (London: Methuen & Co. Ltd., 1967), p. 654.
12. Deshpande, *Geography of Maharashtra*, p. 18.
13. Spate and Learmonth, *India and Pakistan: A General and Regional Geography*, p. 654.
14. Deshpande, *Geography of Maharashtra*, p. 24.
15. Recently, Wing Commander (Retd.), Dr. M.S. Narwane has come up with a valuable publication on the coastal forts, titled *The Heritage Sites of Maritime Maharashtra* (Mumbai: Maritime History Society, 2001).

16. M.S. Narwane, *Forts of Maharashtra* (New Delhi: APH Publishing Corporation, 1995), p. 270.
17. H.P. Ray, 'Shipping in the Indian Ocean', in David Parkin and Ruth Barnes (eds.), *Ships and Development of Maritime Technology on the Indian Ocean* (London: Routledge, 2002), p. 17.
18. S.R. Sharma, *Foundations of Maratha Freedom* (Bombay: Orient Longman, 1964), p. 222.
19. A Portuguese letter dated 28 November 1657, quoted by P. Pissurlencar, *Portuguese-Marathe Sambandh (arthat Portugezanchya Daftaratil Marathayancha Itihash*, in Marathi language) (Pune: Pune University Press, 1967), p. 41, mentions that, 'Shahji's son has taken possession of Upper Chaul'. Another document from the Factory Records of the English suggests that Kalyan-Bhiwandi were under the control of Shivaji—see B.G. Paranjape (ed.), *English Records on Shivaji* (1659-82), Shivaji Tricentenary Memorial Series, 6 vols. (Pune: Shiva Charitra Karyalaya, 1931), vol. 1, no. 160.
20. Quoted by P. Pissurlencar, *Portuguese-Marathe Sambandh*, p. 41.
21. *Rajniti* of Ramchandra Pant Amatya, tr. S.V. Puntambekar (Madras: Diocesan Press, 1928), p. 48.
22. Kirshnaji Anant, *Sabhasad Virachit Chhatrapati Sri Shivaji Raje Yanchi Bakhar*, ed. U.M. Pathan (Pune: Snehwardhan Publishing House, 2000), p. 87.
23. Pissurlencar, *Portuguese Marathe Sambandha*, p. 41 (fn. 9).
24. Ibid., p. 42 (fn. 11). The Portuguese records refer to such Maratha boats as *Sanguiceis.* It is stated that two of Portuguese artisans namely Roe Leitao Viegas and his brother Fernao Leitao Viegas supervised some 340 workmen, Portuguese and others, and helped Shivaji to construct the fighting vessels.
25. B.G. Paranjape (ed.), *English Records on Shivaji (1659-1682)* (Shivaji Tercentenary Memorial Series) (Pune: Shiva Charitra Karyalaya, 1931), vol. 1, no. 333.
26. Pathan (ed.), *Sabhasad Bakhar*, pp. 84, 86.
27. G.S. Sardesai, *New History of the Marathas*, 3 vols. (Bombay: Phoenix Publications, 1946-8), vol. 1, p. 219.
28. Pathan (ed.), *Sabhasad Bakhar*, pp. 93-4. It is interesting to note that during the time Kanhoji Angre as many as 51 different types of Maratha boats were being used. A detailed list is provided by D.G. Dhabu, *Kulabkar Angre Sarkhel* (Bombay: Alibag, 1939), pp. 378-9.
29. Paranjape (ed.), *English Records on Shivaji (1659-1682)*, vol. 1, no. 238. The cause of this letter was Shivaji's apparent show of strength as a large number of Maratha vessels were noticed hovering around the mouth of Bombay harbour. This was enough to alarm the British settlement of Bombay, though no engagement seems to have taken place with the Maratha vessels.

30. Quoted by Pissurlencar, *Portuguese Marathe Sambandha*, p. 43.
31. Ibid.
32. *Rajniti* of Ramchandra Pant Amatya, tr. Puntambekar, p. 50.
33. Duff writes, 'He (Shivaji) built Sindeedroog or Malwan. . . . He rebuilt and strengthened Kolabah; repaired Severndroog and Viziadroog and prepared vessel at all these places. . . .' See, Duff, *History of the Marathas*, vol. 1, p. 137.
34. Letter to the King of Portugal from Vice-Rei Conde de San Vincente, dated 20 September 1667, quoted by Pissurlencar, *Portuguese Marathe Sambandha*, p. 43.
35. Pathan (ed.), *Sabhasad Bakhar*, pp. 89-90.
36. R.S. Whiteway, *The Rise of Portuguese Power in India (1497-1550)* (2nd edn., London: Suhil Gupta, 1967), p. 172.
37. Abbe Carre, in S.N. Sen, *Foreign Biographies of Shivaji* (Calcutta: K.P. Bagchi, 1977), pp. 154-5.
38. *Rajniti* of Ramchandra Pant Amatya, tr. S.V. Puntambekar, p. 51.
39. A.K. Nairne, *History of the Konkan* (1896, rpt., New Delhi: Asian Educational Services, 1988), pp. 73-4.
40. *Rajniti* of Ramchandra Pant Amatya, tr. Puntambekar, pp. 46-8.
41. M.N. Pearson, *The New Cambridge History of India*, vol. I.I: *The Portuguese in India* (Cambridge: Cambridge University Press, 1987), p. 29.
42. Ibid., p. 31. See also M.N. Pearson, 'The Portuguese in India and the Indian Ocean: An Overview of 16th Century', in Pius Malekandathil and Mohammad Jamal (eds.), *The Portuguese, Indian Ocean and European Bridgeheads 1500-1800* (Tellichery: Institute for Research in Social Sciences and Humanities of MESHAR, 2001), chapter 4.
43. *Rajniti* of Ramchandra Pant Amatya, tr. Puntambekar (Madras: Diocesan Press, 1929), p. 49.
44. Ibid., p. 50.
45. Ibid., p. 51.
46. *English Records on Shivaji (1659-1682)*, ed. Paranjape, vol. 1, nos. 235, 242.
47. Quoted by Pissurlencar, *Portuguese Marathe Sambandha*, p. 48.
48. Ramchandra Pant Amatya also suggested that all the creeks and coast should be fortified and guns be placed there so that the enemy ships could not dare come closer to those areas. See, *Rajniti* of Ramchandra Pant Amatya, tr. Puntambekar, pp. 49, 51.
49. A.T. Mahan, *The Influence of Sea Power upon History (1660-1783)*, p. 43.
50. Charles Fawcett, *The English Factories in India*, New Series, 3 vols. (London: Oxford University Press, 1936), vol. 1, pp. 36, 62, 215.
51. See, for example, *Shiva Kalin Patra-Saar Samgraha*, 3 vols. (1930-7) (Pune: Shri Shiv Charitra Karyalaya, 1930), Letter no. 1229. This

document suggests that the British wanted to take an estimate of the depth of the coastal waters in between Pen and Bombay, for navigational purposes. Another document (Letter no. 1528) gives us the idea that for trading purpose, the region surrounding Nagothana was to be charted and mapped by the British. This included ascertaining the depth of the river, its width, area of the small boats which could easily enter the creeks, levels of tide, etc. Similar information was to be obtained regarding Pen and Batty.

52. *Rajniti* of Ramchandra Pant Amatya, tr. Puntambekar, p. 48.
53. William Foster (ed.), *The English Factories in India* (London: Oxford University Press, 1921) (1621-3), p. 72. Even Fryer, a contemporary official (an English surgeon) observed that ships carrying 30 or 40 cannons on board were being built at Surat. (See, William Cooke (ed.), *A New Account of East India and Persia, being Nine Years' Travels, 1672-81*, 3 vols. (London: The Hakluyt Society, 1909), vol. 1, p. 267.
54. It seems that many of the vessels so constructed in western India (under the orders of the local rulers) catered for some sort of defense mechanism in view of rampant piratical activities. One such example is of the Mughal ship *Ganj-i-Sawai* which was mercilessly plundered by (British) pirates in the year 1694. It is a fact that the Mughals were never interested in escorting their mercantile fleet. It seems they were confident of the guns/canons, which were placed on board of the cargo ships. According to one version, the Mughal ship had 80 guns and 110 muskets on board (in Elliot and Dowson, *History of India as Told by its Own Historians*, rpt., Delhi: D.K. Publishers, 1996, vol. 7, pp. 350-5. But, what appears to be more shocking in this account is the absolutely callous attitude of the Mughal authorities towards the defense of the ship. Jan Jilliszon, a Dutch navigator who was lent to the Mughal imperial service and was on duty on board the *Ganj-i-Sawai*, in 1694, wrote to his superiors in wonder that the ship was so overcrowded with men and goods that scarcely a gun could be loaded or used (so that it could be successfully operated. See, Ashin Das Gupta, 'The Early 17th Century Crisis in the Western Indian Ocean and the Rise of Gujarati Shipping in early 18th Century', in B. Runachalam (ed.), *Essays in Maritime Studies* (Mumbai: Maritime History Society, 1998), p. 63.
55. The contemporary English letters contain references of Shivaji being supplied, though not on a regular basis, with cannons, powder and shots by them (see, for example, *Shiva Kalin Patra-Saar Samgraha*, 3 vols., Pune: Shri Shiv Charitra Karyalaya, 1930-6, Letter nos. 1584, 1592, 1603, etc.). At times, in order to escape the attention of the Mughals who were at war with Shivaji, the British would devise other ways to ensure the supply of ammunition to Shivaji. One such reference suggests that the planned sale of ammunitions was to take place with the help of the Portuguese who would act as intermediaries [see,

Paranjape (ed.), *English Records on Shivaji (1659-1682)*, vol. 1, no. 282]. The reference of the Dutch supplying ammunition could be found in *Shiva Kalin Patra-Saar Samgraha*, vol. 3, no. 2805, etc.

56. Some of the paintings/sketches have been reproduced in B.K. Apte's *A History of the Maratha Navy and Merchant Ships* (Bombay: State Board for Literature and Culture, Govt. Central Press, 1973), pp. 119-40.
57. The contemporary traveller Abbe Carre notes, 'In fact along the stretch of the sea where he (Shivaji) was the master, there never passed a ship of Europe to which the governors (of those coastal areas under Shivaji) did not send refreshments with all the good offices that could hardly be expected by an allied prince. I passed that way in 1668 with two ships of the company and we were treated in a manner which was beyond our expectation. It was an act of his policy, but it was also due to the preference he felt for the people of Europe. . . .' See, S.N. Sen, *Foreign Biographies of Shivaji*, p. 155.
58. Late in 1664, when Shivaji undertook the mission to Kanara, he apparently planned to march his army down the west coast, get on board his fleet waiting at Bhatkal (the southern point of North Kanara district), and raid the coastal towns. But Khawas Khan, a Bijapuri general, who was sent by the Adil Shah to prevent the Marathas from doing so, barred his path (see, Sarkar, *Shivaji and His Times*, p. 226; Sarkar has based his statement on the Surat Factory Records). The Basrur naval expedition of 1665 is another example of coordination between the navy and army to achieve a single objective. The Basrur expedition commenced with a contingent of about 85 frigates and 3 large ships from Malvan in the South Konkan with Shivaji aboard. By that time, the Portuguese had reconciled to the enlarged status of Shivaji in the affairs of Deccan. Hence the Portuguese, purposefully, did not interfere in the movement of the Maratha naval fleet so long as their own interests were not threatened. As a result, the Maratha convoy sailed past the Portuguese headquarters of Goa and proceeded further south to Basrur, an important port of Kanara under the small kingdom of Bednur. The Maratha fleet took the king of Bednur by surprise and the port of Basrur was at the mercy of the Maratha marine troops. The Marathas could lay hands on a considerably rich booty which was further utilized in augmenting their naval establishment. See, Paranjape (ed.), *English Records on Shivaji (1659-1682)*, vol. 1, no. 107; cf. Sarkar, *Shivaji and His Times*, pp. 231-3.
59. It has been pointed out that a number of large Dutch ships out from Europe were kept for long periods in Asia and they were supplemented by smaller 'yachts' permanently stationed in the east and often used as specialized warships. By 1626 there were 29 Dutch ships in Asia. By 1626, there were 26 specialized ships in Asia; 54 in 1632 and 76 in 1635

(quoted in P.J. Marshall, 'Western Arms in Maritime Asia in the Early Phase of Expansion', *Modern Asian Studies*, vol. 14, no. 1 (1980), p. 20 (fn. 23).

60. William Foster (ed.), *The English Factories in India (1668-1669)*, p. 242. The date mentioned in this letter is 16 October 1669.
61. Ibid.; see also Paranjape (ed.), *English Records on Shivaji (1659-1682)*, vol. 1, no. 171.
62. William Foster (ed.), *The English Factories in India (1668-1669)*, p. 243.
63. *Shiva Kalin Patra-Saar Samgraha*, 3 vols. (Pune: Shri Shiv Charitra Karyalaya, 1930-6), Letter no. 848.
64. Ibid.; see also *Shiva Kalin Patra-Saar Samgraha*, 883. It seems that the East India Company administrators were interested in having Danda-Rajapuri as an additional base because of its stratgeic location.
65. See also, Paranjape (ed.), *English Records on Shivaji (1659-1682)*, vol. 1, no. 479.
66. Sarkar, *Shivaji and His Times*, pp. 170-4, 236, 262, etc. The Siddi secured royal admiralty and a *mansab* of 900, in addition to a *jagir*, the yearly income form which was worth Rs. 3 lakh. The Siddi was also obliged to protect trade and *Haj* pilgrims frequenting from the Surat port. See also Robert Orme, *Historical Fragments of the Mogul Empire, of the Morratoes, and of the English Concerns at Indostan; From The Year MDCLIX* (London: F. Wingrave, 1805), p. 57.
67. B.G. Paranjape (ed.), *English Records on Shivaji (1659-1682)*, Shivaji Tercentenary Memorial Service, 6 vols. (Pune: Shiva Charitra Karyalaya, 1931), vol. 2, no. 418.
68. Ibid., vol. 2, nos. 364, 365, 367, 369, 373, 374, 377, 378, 379, etc.
69. Ibid. vol. 2, no. 472.
70. Ibid.

CHAPTER 7

British-India *versus* Portuguese Goa: Politics, Warfare and Diplomacy, 1799-1815

SHANTHA HARIHARAN

INTRODUCTION

The Franco-British rivalry in European politics in the eighteenth century is well known and both powers attempted to construct spheres of influence with the 'native' states in India. But Portugal was an enigma. She was a neutral nation, but there was ancient and centuries old alliance and friendship between England and Portugal, counter-balanced by a weak Portugal's need to keep peace with France to survive against Spain.[1] In this atmosphere of latent fear and mistrust clothed in patent professions of friendships and diplomatic adjustments, the population of Portugal were divided in their attachment to France and England. This lent credence to rumours of French pressures on Portugal to part with its possessions in the East. These attitudes were reflected in their Indian settlements also and in the conduct of the viceroy and his advisors, and the Portuguese Army in India.

An English residency was established in Goa in 1784 after the end of the Second Anglo-Mysore War with Tipu Sultan, because of its importance as a centre for gathering information from Pune, Seringapatnam, and the French Islands. It was identified as the part of India where agents from France to the 'country states' in

*The basic work for this essay was undertaken by the author as a post-doctoral research scholar at the Fundacao para Ciencia e a Tecnologia, Lisbon, Portugal. This essay is a revised version of a paper presented at the 20th Conference of the International Association of Historians held at New Delhi in November 2008.

western India will probably land.[2] The post of Resident was first held by Charles Crommelin, an ex-governor of Bombay and later by General John Carnac. A long-time Portuguese resident of Calcutta was admitted as agent of the Portuguese nation at Calcutta in 1784.[3] France and Portugal were at peace in Europe, but the French agents were very active in Tipu Sultan's Mysore.[4] The British, therefore, considered Goa as a potential threat point for their Indian possessions.

The Anglo-Portuguese alliance ensured cordial and friendly relations and mutual cooperation between the British and the Portuguese in India in matters of protection of their coastal trade in the Western Coast from attacks by the pirates. One of the most important events of this period was, however, the 'stationing' of British troops in Goa and later in Daman and Diu from 1799 to 1813, as 'auxiliaries' or additional help for the safety and protection of the Portuguese possessions in India from being taken over by France, either by forced consent of Portugal or by intrigues. Joshua Uhthoff was deputed in July 1799[5] as an envoy by the British government in India to Goa, for the purpose of accomplishing this arrangement and he succeeded in his mission. He was followed by Colonel Sir William Clarke, Colonel Colman, and Captain Courtland Schuyler as an envoys during the period from 1801 to 1813 when British troops remained in Goa, Daman and Diu as an auxiliary force. This paper examines the role and contribution of these envoys in Goa during this critical and delicate period of Anglo-Portuguese relations, with reference to the original unpublished documents of the English East India Company in the National Archives of India.

BRITISH TROOPS IN GOA

The period of the presence of British troops in Goa falls into two phases. The first phase lasted from 1799 to 1802, when the forces were assiduously built-up in a period of uncertainties, with the consent of the viceroy, for the defence of Goa, Daman and Diu. The troops were suddenly withdrawn when the Peace of Amiens was announced. The forces were needed also in Gujarat by the Bombay government in 1802, just before the beginning of the Second Anglo-Maratha War (1803-5), when the British supported the Peshwa (hereditary prime minister of the Maratha Confederacy) against the Scindias, Bhonsles and Holkars. The second phase started from 1803 after the European truce collapsed and war was resumed between France and England and lasted until 1813, when finally the British troops were all withdrawn from Goa. The long-standing alliance between the Portugal and England certainly acted as a check on either side, preventing them from adopting any rigid stand or coercive methods.

The proposition of stationing a British force in Goa originated from the British side, with the discovery of papers in the palace of Tipu Sultan, after his death. The papers revealed a project had been germinating between the French and Tipu for taking possession of Goa and Bombay, as a measure connected with the general plan of cooperation against British possessions in India.[6] As it was deemed probable that in pursuance of this project the French might dispatch an expeditionary force to India, directed in the first instance against Goa, the British government considered it of great importance to secure that post by inducing the viceroy of Goa to receive into it a British garrison as an auxiliary force, for the security of the Portuguese possessions against the threat of French invasion. Joshua Uhthoff was deputed to Goa in July 1799 by the governor-general of British-India Lord Mornington, as his envoy, to negotiate with the viceroy of Goa, Sr. Francisco Antonio da Veiga Cabral,[7] for admission of British troops into Goa for the security of the settlement.

The Viceroy at first showed some hesitation in acceding to this proposition, although he professed himself gratified and obliged by the attention that the British government showed to the interest of the Portuguese nation in India.[8] The treacherous but unsuccessful attempt made earlier in 1796 by the French to seize the fortress of Diu and the information about the deployment of the Brest Fleet of the French convinced the viceroy of the probability of a French invasion of Goa, and he accepted the succour of a British garrison to be stationed in Goa. In September 1799, British troops numbering some 1,300 men under the command of Colonel Sir William Clarke arrived from Bombay and were admitted into Goa by the viceroy.[9] The British troops were stationed at Goa as auxiliaries in the service of the Portuguese government for the special and exclusive purpose of protecting that settlement against an eventual attack on the part of the French and 'not generally for the purpose of being employed on any species of duty which might be requested by the Government of Goa'. It was also clarified by the governor-general at the time of introduction of the British troops that the commanding officer will be subject to the order of the viceroy of Goa in everything relating to the security of the place against the enemy or in the maintenance of its internal peace and tranquillity.[10]

Lord Mornington had succeeded Sir John Shore as governor-general in 1798. Taking into account all the factors, he had assessed that about 3,000 auxiliary British troops will be necessary for the defence of Goa, acting in concert with the Portuguese forces. But, because of other commitments of the three presidencies, such a force for Goa had to be assembled by detachments from Bombay, Fort St. George and Bengal as they became available in a piecemeal manner. Therefore, after the initial admission of some British troops, the viceroy of Goa had to be approached for his consent every time to augment this auxiliary British force to the projected level of defence strength. While the initial hesitation of the viceroy to accept a British garrison in Goa was overcome when he was satisfied with the threat of French invasion, difficulty and uncertainty was experienced by the envoy at Goa, at every subsequent stage to secure the Portuguese viceroy's consent for augmentations of the strength of the British troops. In November 1799, after persistent efforts the envoy was able to secure the viceroy's consent to receive another battalion of 'native' infantry in place of a detachment from the original force, which was sent to Mangalore.[11]

The viceroy of Goa reviewed the Portuguese troops numbering 2,117 persons (about 800 Europeans) at the end of December 1799. The total Portuguese Army numbered at 1,160 Europeans and 4,000 'native' troops in different parts of the Portuguese territories. Besides, there were about 2,000 indisciplined, badly armed sepoys who lacked uniform and received Rs.3.50 each per month. These sepoys were dispersed throughout the Portuguese territory. The British auxiliary force at the same time consisted of 134 European commissioned officers, 1,022 European rank and file, 1,004 'native' troops in addition to 182 men on the sick list.[12] While the Portuguese troops were dispersed all over their possessions, the British troops were all garrisoned at the capital Goa and its military implications were obvious to the viceroy of Goa. He was therefore careful to avoid any military confrontation with the British and instead adopted a policy of unpredictability, objections and delays. The envoy Colonel Sir William Clarke wrote to Admiral Pellew in December 1805 that the Portuguese regular troops numbered about 1,300 Europeans and 3,500 natives, of which about 800 of the Europeans including the Portuguese Grenadiers and artillery regiment were in good discipline. The 'Native' Regiment was a fine body of men, respectably clothed and armed and officered by Europeans, but in the envoy's words badly disicplined.[13]

On occasions, like the viceroy's annual review or special incidents, when all the Portuguese forces were usually brought to Goa, the envoy was very careful to avoid any dispute or showdown and was alert, apprehensive and defensive. Clarke writing to Governor Duncan in early December 1801 told him that the customary annual review of Portuguese troops will take place on 17 December 1801 and expressed his fear that the viceroy may even retain them at Goa afterwards, and surmised that the purpose may be 'to overawe us and endeavouring to stress his future pleasure to force us out of our present situation in Goa'. The viceroy had already asked for the withdrawal of the British troops on peace being concluded between France and Portugal, but bringing the Portuguese forces to Goa would also weaken the frontiers. Clarke requested for directions about the stand he should take in such an eventuality and wanted Bombay to return the three *pattamar*[14] boats to him urgently.[15] Again in 1806, when news reached Goa that a new viceroy had left Lisbon for Goa, Clarke informed Duncan that the acting viceroy was making preparations for the reception of the new viceroy and had directed the Portuguese troops at the provinces including the Grenadiers under Clarke's charge to be deployed at Goa for the reception. This created a dilemma for the envoy. Such a concentration of Portuguese troops would weaken the position of the British garrison and it could prove disastrous if the new viceroy turned out to be anti-British. At the same time a refusal would be construed as a hostile act by the viceroy.[16]

THE BRITISH ENVOY IN GOA

Joshua Uhthoff was appointed in 1799 as the Envoy of the Supreme Government in India, placed at Goa for discharging political and diplomatic duties with the Government of Goa and he directly reported to the governor-general. He was shortly joined by Colonel Sir William Clarke who was the commanding officer of the British auxiliary force in Goa, looking after the military aspects of the British presence. They formed a synchronous team, patient, persuasive, diplomatic and anticipatory; very well informed and communicative. They won over the confidence and cooperation of the viceroy of Goa, the Portuguese officers and the public in general. Uhthoff in his farewell letter to Richard Marquis of Wellesley, governor-general, attributed his success in the post to the

> liberal attachment to the British government of His Excellency the Governor and Captain-General of Goa and his conciliating conduct towards British subjects on all occasions, to the zealous and able cooperation and support received from Colonel Sir William Clarke and the exemplary good conduct of the British troops within the Portuguese territories in this quarter and the most kind and conciliating manner in which the Portuguese society in general conducted themselves to the British party.

He also acknowledged his pleasure at the conduct of the ADCs of His Excellency, who were his principal channel of intercourse with the viceroy and praised the services rendered to the British Army and naval vessels by one Antonio Pereira, a 'native' of Bombay settled as a merchant in Goa for some years past and participated in the campaigns against Tipu Sultan.[17] His Excellency the Viceroy also expressed his regret over Uhthoff's departure and made a spontaneous handsome offer to accommodate him with a passage to Bombay in one of the Portuguese frigates.[18]

Initially, the office of the envoy looking after the political and diplomatic duties and that of the commander of the auxiliary force looking after the military duties were separate under two different officials who cooperated with each other. When Uhthoff handed over charge on 1 December 1800, Colonel Clarke who had so far looked after the military duties was appointed as the British envoy also, thus combining the political, diplomatic and military duties as commanding officer of the British forces in Goa in a single office. The viceroy also welcomed the appointment of Colonel Clarke, 'whose character and conduct had rendered him so entirely satisfied'.[19] The governor-general considered that a good envoy should be an officer who by his address, character and qualifications shall have gained the goodwill of the viceroy and of the principal authorities at Goa and shall have establisned that general influence, which was necessary for preservation of the British interests in that quarter of the peninsula.[20] The offices of the political duties as envoy and military duties as commanding officer of the British forces in Goa were united in Colonel Clarke and proved to be an efficient arrangement for the next six years.

Colonel Sir William Clarke understood the moods of the viceroy of Goa, his compulsions, limitations and predicaments through reliable local sources, and always showed respect and offered felicitations to the viceroy, which were his due. When the viceroy recovered from a serious illness, Clarke organized a military parade to celebrate his recovery[21] and the viceroy rendered every mark of respect to the memory of Lord Cornwallis on his death. Without standing on prestige or protocol or assuming superior airs, Clarke discharged the combined military and political duties with great advantage and was known to possess a great degree of influence over the viceroy of Goa.[22] Even when it appeared that the viceroy might refuse a proposal, he made his moves patiently in such a way that finally he got a positive response, without straining relations. The taking over of the Portuguese Grenadiers battalion into British pay[23] and the viceroy entrusting the joint command of the Portuguese and British Forces to Clarke, with a special rank of Major-General of the Portuguese Army, showed the extent of the viceroy's confidence and trust in this British officer.[24] Very often he would confide even his administrative problems with the envoy as a friend. For example, when Clarke proposed relocating the Portuguese troops from their present quarters and encamping them near the beach for better defence, the viceroy refused it on account of his financial problems and explained to Envoy Uhthoff how the insanity of the Queen, the inexperience and limited powers of the Prince Regent and the weakness of the 16 member council had involved the Portuguese government in Europe in considerable disorder and debility, resulting in their neglecting to send cash and stores, etc., to their colonies, which were formerly usual and by which the prosperity of those colonies were maintained. The viceroy suggested an alternative that some irregular native soldiers should be attached to the artillery as gun lascars and a party of British troops under

Lieutenant-Colonel Grimstone could be stationed near Shapra River to cover the northern beach. It was decided that some good field artillery pieces with adequate ammunition should be attached to the European regiment stationed at Agoada, Margong, Bardes and Calwal.[25]

On promotion as major-general in the British Army, Clarke had to go back to England according to the demand of the regulations and his place was taken over in February 1806 by Colonel Colman, who was commanding the British forces at Goa at that time.[26] Colman also combined in himself both the diplomatic and military duties.[27] Although the viceroy welcomed the appointment and the arrangements, the command of the Portuguese forces was taken back and restored to the Portuguese military officers. His services were also acknowledged by the viceroy, who preferred to deal with British officers whom he had come in contact with and not new ones. Thus, when the question of replacing Colman by his senior officer, Colonel Dickenson in the regiment was mooted, the viceroy objected and the governor-general of British-India accommodated the wishes of the viceroy by replacing the regiment itself where the questions of seniority did not arise. The governor-general appreciated that the viceroy would not readily transfer his confidence and cordiality to any successor, while a compliance with his declared wishes in the selection of an officer as envoy could not fail to afford him gratification and consequently confirm and augment the disposition which he had manifested to consider the interests of his own nation and those of the British government as inseparably combined.[28]

Colman's tenure at Goa was comparatively short and he had to leave due to health reasons.[29] Captain Courtland Schuyler who had served under the previous envoys in the diplomatic side was appointed as envoy in his place and remained in that post until the withdrawal of British troops in 1813. Lieutenant-Colonel Adams was to command the British forces in Goa. There was thus a reversion to the original position of different officers looking after the diplomatic and military duties in Goa. But they worked as a good team and when Adams proposed to accompany his corps in 1811, the envoy praised his knowledge of the local country and cordiality between them on both public and private concerns and how his conciliatory manners endeared him to all the Portuguese. Schuyler proved to be quite effective and negotiated many important conventions with the viceroy in the wake of the Prince Regent initially aligning with Napoleonic France against Great Britain and within a few days fleeing from Lisbon to Brazil under British escort. Unlike Clarke, Schuyler's relations with the new viceroy were not very close or cordial as he suspected the new viceroy as being biased towards France and expressed this suspicion quite frequently to the governor-general and others. His communications generally described the viceroy as acting in a dubious manner and not in the interest of friendship with Britain. Even at the time of withdrawal of British troops from Goa in 1813, this prejudice of Schuyler was noticed by the Court of Directors who regretted allusions made by Major Schuyler to the viceroy's supposed predilection for the Government of France and held that in the altered state of affairs, no material inconvenience could arise to British interests in Goa from the viceroy's political bias. They directed that the future intercourse between the viceroy and the envoy will conform to the spirit of the respective courts.[30] Clarke was a prodigious writer judging from the frequency and variety of his correspondence. Colman comparatively was quite restricted in this respect. Schuyler was again quite prolific in the information he fed to higher authorities.

A lot of tact and discretion was involved in discharging the political and diplomatic duties in a foreign possession. Sensing the local sentiments and situations, based both on facts and circulating rumours or private information and the ground realities, the envoy kept the governor-general and the governors of Bombay and Fort St. George informed of the developments. Because of the time required for exchange of letters with Bombay or Calcutta, he could not get required instructions urgently and had to act in his discretion in anticipation of approval.[31] The Secretary of Fort William wrote to Clarke that 'Governor-General is confident that your action will have anticipated the guidelines and under any circumstances you will have adopted measures conducive to the honour and interest of the British Government'. This shows the trust and relationship between the governor-general and the envoy.[32] He had to keep good relations with the Portuguese Army officers, the viceroy and his ADCs and collect information discreetly without being obtrusive. He had to temper instructions from higher authorities which might create problems locally if implemented blindly. In other words his judgements, actions and moves had to be more careful, sensitive and responsible than to blindly implement the instructions of the higher authorities. He had to cultivate and maintain relationships, win the confidence of contacts, avoid confrontations and at the same time achieve the results demanded by the higher authorities. His was a balancing act simulating that of the governor-general. The governor-general also had to carry out the orders of the directors of the East India Company as well as that of the British government with due consideration for the actualities of the Indian situation. The time factor for correspondence between Britain and India affected him equally, making him depend more on discretion than actual orders.

THE BRITISH POSSESSIONS IN INDIA AND THE FRENCH THREAT

The secret committee of the Court of Directors in their letter of July 1801 warned the governor-general about the probability of an attack by French ships displaying Portuguese colours and a secret convention between Portugal and France, facilitating execution of French designs of attacking British possessions in India. The letter justly demanded extra vigilance to repel any such attempts.[33] The governor of Bombay in a letter dated 20 October 1801, consequently requested the viceroy of Goa through Clarke to issue orders to the governors of Daman and Diu to receive British garrisons 'the more effectually to enable these fortresses to withstand any attack or attempt, whether open or insidious from the French'.[34] Clarke with great difficulty prevailed upon the viceroy to accede to this order, reversing his earlier orders of refusal and transmitted the viceroy's new orders dated 26 October 1801 to the governor of Bombay, in which the governors of Daman and Diu were ordered to receive into their garrisons a body of 200 British European auxiliaries and artillerymen. Forwarding the viceroy's orders to the governors of Daman and Diu to admit a British garrison each of 200 British auxiliaries, Clarke explained to Governor Duncan of Bombay that he thought it better not to press for induction of more troops because he found it difficult to obtain even so much. Adding that he learnt from the viceroy that orders had earlier been issued against admission of any British troops into Daman and Diu and even retaining the force in Goa was contrary to his instructions from court.[35] The viceroy in his reply to the governor-general stated that he had sent the orders requested by him in respect of Daman and Diu 'in spite of great objections' and that he was 'stimulated by the intelligence and energy of the above mentioned Colonel (Clarke) and the support of the well disciplined troops under his command' and assured that they will both discharge their duty to their August Sovereign.[36] The British troops were admitted into Daman on 9 November 1801 and into Diu on 14 November 1801. On reports from the senior commanders that these forces[37] were not adequate to offer effective resistance to any French attack, the viceroy was requested to allow an addition of 700 'native' troops at these two posts, which, however, was refused by him.[38]

With news coming in from Europe of a convention concluded by France and Spain with Portugal and the hazard of the Portuguese settlements being seized by the French under the terms of that convention, the governor-general decided not to depend on the doubtful disposition of the viceroy of Goa or the governors of Daman and Diu and sent a letter on 20 November 1801, seeking the cooperation of the viceroy for admitting the reinforcement of British troops to all the Portuguese possessions in India and Macau, for the protection of those possessions. In the governor-general's instructions to Clarke dated 20 November 1801, he was informed of the details of deployment of British forces for protection of the possessions, so that he can act in concert with the viceroy to obtain his consent for augmenting the British garrisons, use discretion, share what was absolutely the minimum, and assure the viceroy of all exertions on the part of the British. He was, however, to avoid giving details of the force deployed to the general points of danger. To preclude the execution of any agreement concluded between Portugal and France relating to India and China, Clarke was asked to convince and carry the viceroy with him; but in case ultimately the viceroy should raise any objections or obstacles to reinforce the garrison of Goa, Clarke was authorized to surmount those difficulties and at all events to establish the British contingents in possession of that place.[39] A battalion of 'native' troops was also sent to Goa. Clarke adopted a strategy of fast action and delayed communication with the viceroy and quickly disembarked and encamped the detachments from Bombay in a commanding position on the island of Goa and postponed replies to the viceroy's objections.[40] Between 30 December 1801 and 4 January 1802, the position changed dramatically and Clarke reported to the governor-general on 4 January 1802 that a proposal laid by him before the viceroy on 1 January 1802 was accepted the next day and 'the whole of the Portuguese settlements in the East are placed without reserve under the protection of Your Lordship'. By 3 January 1802, the various forts and posts like Agoada, Cabo, El Reis, and so on, which were strategically situated were occupied by additional British troops with the approval of the viceroy himself and British troops were augmented in Daman and Diu. A senior Portuguese officer left to confer with the governor of Bombay and settle matters regarding Daman and Diu amicably. The command of all the allied troops in the settlement and their distribution was also vested in Clarke. The arrangement thus amicably settled, obviated the use of any open force authorized by the governor-general to achieve the same objective.[41] The viceroy in his reply to the governor-general dated 2 January 1802 explained that his earlier refusal was based on orders of his court, brought by the annual *Carreira* Registership. Since the governor-general decided to send the auxiliary force in anticipation of any emergency, the viceroy was convinced that the governor-general's action was based on information of later developments which he was not aware of and, therefore, consented to receive the British troops and cooperate with them as desired by the governor-general.[42]

When the Peace of Amiens was announced, the viceroy wanted the British troops to be withdrawn from Goa. The bulk of the troops were withdrawn. In July 1803 after the resumption of hostilities in Europe following the Peace of Amiens, Clarke explained to the commander-in-chief his conviction that judging from the late conduct of the viceroy, he felt that the latter will not give offence to France, will not voluntarily admit a British squadron into Goa and will yield to intimidation if a French force is sent to occupy this settlement. The Portuguese people having influence with the viceroy were un-inclined to the British and would certainly urge him to submit to French measures and the Portuguese soldiery would easily duped by French intrigue.[43]

However, the viceroy wrote to the governor-general on

1-2 September 1803 that he had requested Clarke to ask for the auxiliary troops to Goa due to the appearance of a large French fleet before Pondicherry. On 3 September 1803, the British troops marched into Goa. The viceroy admitted that 'surrounded with the valorous troops of the illustrious British nation, I feel myself invincible . . .',[44] showing a vast change in attitude. Clarke described the viceroy as follows: 'with an unaccountable inconsistency which marks his character',[45] 'his aptitude on slight grounds to change opinions',[46] 'attachment to the British nation and hatred of France'.[47]

The governor-general was against 'the hazard of the precarious dependence upon the doubtful disposition of the Viceroy'. The other occasion when the envoy was authorized discretionary powers for forcible occupation of Goa was in 1808 (Schuyler), when the secret committee of the Court of Directors ordered such a course in their despatch of 11 December 1807, on receipt of information that the Prince Regent had joined France against England and had closed all the Portuguese ports for British vessels. The governor-general was ordered to occupy Goa if necessary by force.[48] Though the latter information of the removal of the royal family to Brazil, repressed the alarm caused in Goa by the previous information of rupture between Great Britain and Portugal,[49] doubts were expressed by the envoy about the new viceroy's amicable disposition towards the British. There were reports of the viceroy giving asylum to some Frenchmen secretly and there were some signs that the former was suspecting that the British would occupy the settlement. In the situation of diminished confidence and indications of hostile position of the new Viceroy Dom Bernardo J.M. Silveira Lorena, Conde de Sarzedas (1807-16),[50] Schuyler was vested with discretionary authority to proceed with the occupation of Goa if open and decided hostility was met with. This authority like the earlier one was also not invoked due to subsequent developments.[51]

THE ENVOY AND THE GOVERNORS OF THE PRESIDENCIES

Although the envoy reported directly to the governor-general, the governors of Bombay and Fort St. George had also a stake in Goa. The British garrison in Goa was made up of detachments of European as well as 'native' infantry, collected according to availability from the three presidencies. Each presidency had its own commitments for troops and was reluctant to spare the full complement of troops for garrisoning Goa. The troops deployed in 1799 and later additions in Goa were withdrawn in October-November 1802 almost completely when the Peace of Amiens was announced. It is also true that the troops were required by the Bombay government to reinforce the detachments guarding the Gujarat frontier.[52] With the British forces spread over different parts of India according to local needs and confrontations on the borders of the British possessions in India, everyone of the commanders felt that his forces were inadequate and opposed any detachment moving to Goa. Bombay, for example, repeatedly expressed fears for the safety of Bombay itself, as the forces available were much less than what had been laid down as the necessary minimum in the Report of Sir Archibald Campbell—quoted often as the *Bible* of the armed forces. Requirements of Colonel Wellesley in Mysore, and Canara[53] or of the Government of Bombay for safeguarding the Gujarat frontier or of Fort St. George to meet the needs of Travancore/Malabar were often sought to be met by diversions from Goa, where the forces were not actually fighting, but were deployed strategically. The temptation was too difficult to resist. It was, however, the experience of the envoy at Goa that while initial deployments were difficult in itself, there were refusals and resistance from the viceroy for every stage of addition to the garrison. Even the return of detachments temporarily sent out became matters of uncertainty and controversy and the reactions of the viceroy to such proposals were unpredictable and wavering. The envoy therefore tried to retain the garrison without reductions, for fear that they may not be readmitted by the viceroy at a later date. At critical junctures when threat perceptions of the British and Portuguese coincided, there was cooperation but resistance and reluctance was apparent at every subsequent attempt to reinforce the garrison to a strength considered adequate for repulsing an enemy attack.

The envoy kept the governors of the presidencies of Bombay and Madras informed of his activities and took advantage of their support in getting his proposals approved by the governor-general. He also sought their instructions in anticipated difficulties or emergencies. In a letter to Governor Duncan in October 1801, Clarke wanted four *pattamar* boats to be sent to Goa, ostensibly to cruise with the cruiser *Princess Augusta* for protection of the coastal commerce, but actually as a logistic support for the easier movements of British troops both for offensive and defensive action, in case of necessity. The British troops were concentrated in the island of Goa and to meet any emergency due to Portuguese adversity, Clarke also stocked food and military stores for six weeks in Goa.[54] Clarke also got a guardship stationed in Goa 'to prevent intrusion of Jacobins whether French or Portuguese into Goa'.[55]

A crisis developed between the British envoy at Goa and the governor of Bombay in October 1805. There was a rumour that a French fleet was heading towards India and the governor of Bombay wanted to strengthen the city by withdrawing from Goa the detachment sent earlier by the Bombay Presidency. The British Envoy at Goa, William Clarke appealed to the governor-general. Clarke was supported by the governor of Madras. The governor-general informed the governor of Bombay that the French fleet was heading towards West Indies and refused removal of the Bombay detachment from Goa.[56]

The envoy also assisted the British resident at Pune in his negotiations with the Peshwa regarding piratical activities in the ports of the Raja of Kolhapur and Sawant Waree and the Malabar Coast[57] and for delivering up some forts on the coast to the East India Company for this purpose. The negotiations were conducted outside Portuguese territory for diplomatic reasons.[58]

THE EYES AND EARS OF THE SUPREME GOVERNMENT

The envoys functioned as the eyes and ears of the supreme government and reported every piece of information gathered discreetly during private discussions. The envoys also collected data on the administrative, judicial and ecclesiastical set up as well as the revenue sources of Goa. They also informed the governor-generals about their personal observations and inferences regarding certain scenarios.[59] Clarke also mentioned in a communication that in general the Portuguese at Goa were extremely jealous, if not disaffected to the English, that only by making it more the interest of the Portuguese troops than they appear at present, could the British expect their help to resist any French moves. He apprehended their intrigues and treachery and proposed to take these Portuguese soldiers under the more attractive British pay, and thereafter organize them to acquire their fidelity.[60]

The Portuguese Grenadier Battalion which had been carved out of four companies of the Portuguese Army was disbanded twice by the viceroy to severe the British hold on them.[61] Just as the Portuguese mistrusted the intentions of the British, the British also mistrusted the loyalty and dependability of the Portuguese troops even under British pay, to cooperate and support them in case of an actual French invasion.[62] In fact, it was even argued at times that improving the fighting discipline of the Portuguese troops may become counter-productive and detrimental to British interests in case of any conflict and therefore, some contradictory policies were advocated.

It is on record that Clarke had friends among the Portuguese officers whom he utilized to collect information or intelligence on the happenings in the Portuguese Court, *carreira* movements, and information brought by Portuguese ships and the consultations in the viceroy's palace.[63] It is quite probable that Clarke might discreetly have also let the viceroy know that the result of refusal may entail the use of force for which he had authority. Particularly with the additional forces already landed in Goa, Clarke was negotiating from a position of strength. It is otherwise difficult to explain the total change in the viceroy's disposition to various issues, from time to time. Clarke even helped to recruit and send from Goa 250 strong, well-built kafris as soldier recruits to the governor of Ceylon.[64] The instructions to the envoy from the governor-general while forwarding letters to be delivered to the viceroy often included directions to convey additional oral information not contained in the letters but calculated to influence his decision. In one such instance the envoy was asked to intimate the viceroy that the governor-general was requesting HM's ministers (at London) to convey to the Portuguese ambassador in England the undesirable conduct of the public functionaries at Macau against the British.[65]

Reporting about the information gathered in private conversation with the Portuguese adjutant-general, Envoy Schuyler informed the governor-general that the reason given by the

Portuguese officer for the honours heaped on the *Disembargador* (the chief judge) of Macau (who thwarted two British attempts to garrison Macau in 1802 and 1808) was that he helped the Chinese to fight pirates, although the British always suspected that it was for his role in preventing the garrisoning of Macau with British troops in the 1808 expedition during the Napoleonic wars. He stated that both the former viceroy and the present viceroy had confided in him that they considered the chief judge of Macau as an 'improper' person with an unfriendly disposition to the English. Schuyler also sent a paper purporting to be a representation made to the Prince Regent by the citizens of Goa against Viceroy Count Sarzeda for his involvement in anti-people activities.[66] In his letter dated 1 November 1811, he informed the vice-president about the receipt of the treaty between the British government and the Prince Regent at Brazil in 1810 and the objection supposed to have been raised by the minister of colonies at Brazil and about dissatisfaction with the treaty in Goa itself. He sought instructions whether he should request the viceroy to give the treaty terms publicity, if he withholds its promulgation.[67]

Even in matters like the withdrawal of troops from Portuguese territory in 1813, the governor-general took the envoy into confidence and sought his assessment of the ground reactions.[68] The envoy replied that the viceroy would not throw any material obstacle in the way, but would rather be pleased with the absence of British troops from their territories. He again repeated his view that in spite of the outward profession of the viceroy and his advisers of strongest attachment to the British nation and abhorrence of the rulers of France, he distrusted them and considered them as having a bias towards the French nation. He gave this opinion to enable the governor-general to judge, how far political it would be to leave the settlement in an unprotected state under the direction of these potentially disloyal persons, 'in the event of any reverse of fortune befalling the allied nations'.[69]

CONCLUSION

In fact, during the last decade of the eighteenth century, the British troops were in Gujarat, Bombay, Goa, Canara and Mysore and there were forces under subsidiary alliances with 'native' states in Hyderabad, Pune, Travancore and also in Ceylon. Thus, British detachments practically covered the entire west coast of India. In that region only Goa was in the possession of another European nation in alliance with Britain. In the assessment of the East India Company and the British government, Goa was the point most vulnerable to a French attack. Portugal as a neutral nation was friendly with France in order to check Spain and therefore, susceptible to French influence in the Court of Lisbon. With the claim made to a part of the Indian revenues by the British Parliament, the Indian possessions were indeed very valuable for the English East India Company and the British Crown. They stood to lose most by a French invasion of India.[70] No wonder the English East India Company and the Government of Great Britain took the French threat to its Indian possessions very seriously and took all possible preventive measures.

When news of a compulsory convention between France and Portugal was received about cession of Portuguese possessions in the East to France in 1801, tension developed in Anglo-Portuguese relations over the issue of garrisoning the forts of Diu and Daman with British troops and again in 1807, when the Prince Regent of Portugal joined Napoleon's continental blockade against Britain and issued a proclamation barring British ships from Portuguese ports. On these two occasions the envoys were authorized to use force to secure possession of the Portuguese settlements in India under certain circumstances without prior approval of the governor-general in council. The diplomatic manoeuvrings of the envoy in the first case and the flight of the Portuguese royal family to Brazil under escort of a British squadron in the second case saved the situation and avoided any such contingent action. But certainly the hold of Britain in Goa was strengthened during each of these crises.

Sr. Francisco Antonio da Veiga Cabral assumed charge of the post of governor and captain-general of Goa in 1794 and had earlier been the commander-in-chief of the Portuguese armed forces for almost a decade. The Portuguese viceroy had very limited access to information from Europe usually through the annual registership from Lisbon[71] and from private Portuguese ships calling at Goa, or Bombay and from private correspondence of Portuguese merchants with their friends and partners in Lisbon and Rio de Janeiro.[72] He was painfully conscious of his weak military strength and strained financial resources and had difficulty in maintaining the national pride and prestige against British and French pressures and the internal problems and intrigues of his advisors.[73] In the absence of clear indications of court policy in Lisbon (due to the madness of the queen and the functioning of the regency), he was unsure of the reception to his actions in India in the Court of Lisbon. Additionally, Viceroy Cabral who assumed the post of governor of Goa in 1794 had long past completed his tenure of three years even before the British troops were introduced into Goa (1799). And, with rumours floating in Goa, he was expecting a replacement in the registerships which were coming from Lisbon. By his own admission, he admired the British Army and the nation and was very chary about French designs on his country, but was so loyal to his Crown that he would not have disobeyed orders from Lisbon to surrender Goa to the French, even if he knew they were issued under French pressure and the order was patently disadvantageous to Portugal. His viceroyalty coincided with the French Republic and the rise of Napoleon Bonaparte, a very difficult period when no one could easily predict the outcome of those wars and its effect on Portugal.

The rising power of the British in India was apparent to the viceroy who was aware that adequate reinforcement of Portuguese troops from Lisbon for the defence of the Portuguese possessions in India would not come. He must therefore have felt that a temporary superimposition of British troops as an auxiliary force to check the French designs was a far better alternative to supercession or annexation by the British.

In contrast, his successor the Conde de Sarzedas, who joined in 1807 held charge at a time when France had invaded Portugal, the court and royal family of Portugal had moved to Brazil and Portugal received military support from Great Britain. The war was also slowly turning in favour of the allies. But, he also had the same problems about communication, court approvals to his decisions, internal intrigues and in addition was looked upon with suspicion by the British envoy as a person biased in favour of the French.

The British professed that their intention in placing an auxiliary force in Goa was only to protect the Portuguese possessions from

French invasion and many of their actions tended to support this assertion. It was, however, natural for the Portuguese viceroy and his officers to suspect British designs, particularly during a period when the British were annexing territories in India at every opportunity and the Portuguese forces were no match for the well-disciplined British troops. This apprehension was reinforced by a section of the military and the general public who were pro-French and anti-British. The reactions of the Court of Lisbon under the weak regency were also unpredictable due to pressures from the French government. In defence of the viceroy's wavering policy towards the British troops in Goa, all these aspects would be relevant. In spite of his admiration for the British Army, his loyalty to his sovereign was such that he would obey the orders even against his best convictions. The threat of a French attack on Goa was more imaginary than real, with the British prepared to take the brunt of it, if it happened. An envoy sitting in Goa with a disciplined armed force at his command and adequate authority from the governor-general himself had to be handled very sensitively and diplomatically, lest any move should result in rupture of relations, exposing the Portuguese possessions to a British attack.

The conduct of the envoys required great tact, negotiating skills, resilience, patience and diplomacy in executing the orders of their superiors without offending the national pride and sentiments and wishes of Portuguese viceroys and officials. The envoys had to mesh fully their own orders, the local ground realities, and limitations and restrictions placed on the viceroys. It was not an ordinary or easy matter to entrust the command of the Portuguese defence forces in Goa to a foreign military officer, but even when this happened, the Court of Lisbon advised the viceroy against its repetition. Understanding the moods and compulsions of the viceroys, the envoys achieved by persuasion and compromise, what the supreme government thought might need use of force. The fact that during the period 1799 to 1813, when British troops were present in the Portuguese possessions in India, there were no unpleasant incidents, embarrassments or sour relations, speaks volumes of the practical approach and spirit of cooperation of both nations and appreciation of their mutual needs and limitations. In contrast every attempt for similar arrangements in Macau, handled by the Select Committee of Supra-cargoes[74] in Canton, China, ended in failure and unpleasantness.

NOTES

1. Christopher D. Hall, *British Strategy in the Napoleonic War: 1803-15* (Manchester: Manchester University Press, 1999), p. 167.
2. *Fort William India House Correspondence* (*FWIHC*), vol. 15, *Foreign & Secret, 1782-86*, ed. C.H. Philips and B.B. Misra (New Delhi: National Archives of India [NAI], 1963), Introduction, pp. xxiv, 545, 539.
3. *FWIHC*, Letter to Court, 8 December 1784, pp. 305-6.
4. Sydney J. Owen, *A Selection from Despatches, Treaties and other Papers of the Marquess Wellesley KG During his Government of India* (Oxford, 1877) (hereafter *Owen, Wellesley Papers*) selection no. 81, pp. 567-77; Report of Lieutenant-General James Stuart, Commander-in-Chief, of the Bombay government dated January 1800, submitted to Henry Dundas on his arrival in England in July 1800, dealing with the danger of foreign invasion by land and sea, importance of securing the whole seaboard and commercial policy (hereafter *Stuart's Report*) stated that 'the French character is more popular in India than ours. This is however true and individuals of that nation are always sure of a more favourable reception from the native princes', p. 570; about their power at the Nizam's court, he added that 'it was by the instrument of force and agency of fear that we succeeded in expelling them from Hyderabad', p. 570; again 'before the reduction of Seringapatam, had the French reached this country, they would have landed in Tipoo's states or at Goa, which is in their neighbourhood and considered an easy conquest', p. 574.
5. Joshua Uhthoff was Assistant to the British Resident at Pune and was appointed as the Civil Administrator of Kanara in 1799 and was then sent to Goa to negotiate with the viceroy. Secret and Political Diary, no. 56, pp. 669-70, Diary no. 81, pp. 5013-14, Records of the Bombay Government, Maharashtra Archives, Mumbai.
6. *Owen, Wellesley Papers*, selection no. 24, pp. 130-1, despatch dated 19 May 1799 to the Secret Committee of the Court of Directors of the East India Company, London; Robert Montgomery Martin, *Despatches of the Marquess Wellesley*, vol. 1 (London: n.d.), no. 7, pp. 591-2; Celsa Pinto, *Goa Images and Perceptions* (Goa, 1996), p. 26; and P.P. Shirodkar, *Researches in Indo-Portuguese History* (Jaipur, 1998), vol. 1, p. 233, quoting Lieutenant-Colonel Mark Wilks, *Historical Sketches of South India in an Attempt to Trace the History of Mysore*, ed. Sir Murray Hammick (Mysore: Mysore Govt. Branch Press, 1932), vol. 2, p. 640.
7. Francisco Antonio da Veiga Cabral da Camara Pimentel was the 80th Governador da India, 1794-1807, the longest term for any holder of the post. The term viceroy has been used in this article to denote the governor and captain general of Goa for easier differentiation from the English governors of provinces and the governor-general of Bengal, who was the head of the British Supreme Government in India. The title of Vice-Rei was given to '*os grandes fidalgos, com Dom e titulo*' (the great

noblemen with Dom and titles). Of the 113 *governadores da India*, up to the proclamation of the Republic, there were only 52 who had the title of Vice-Rei. See *Tratado de Todos os Vice-Reis e Governadores da India* (Lisboa, 1962), p. 62, 194.

8. *FWIHC*, vol. 17, Foreign, Secret & Political, 1792-95, ed. Y.J. Taraporewala (Delhi: NAI, 1955), p. 502.
9. *Foreign Department, Secret Branch, Proceedings and Consultations of the Governor General in Council, Fort William, Calcutta* (hereafter *FSC*), New Delhi: National Archives of India, FSC no. 52 dated 27 March 1806 regarding the system followed by the viceroy, Goa in his relations with the British and no. 3 dated 28 December 1807, memorandum dated 23 December 1807 recorded by the Secretary Fort William, about the circumstances in which troops were stationed in Goa.
10. FSC no. 3 dated 8 December 1807 from Governor-General (GG) to Viceroy Goa, copy to envoy Clarke. But, when treachery or imbecility or ignorance of the viceroy was sufficiently proved, the commander could waive obedience to the authority of the viceroy.
11. FSC no. 7 dated 18 December 1800. Envoy Uhthoff's letter dated 30 November 1799 to Commander-in-Chief, Bombay.
12. FSC no. 106 dated 18 December 1799, letter from the British commander at Goa, Colonel William Clarke to GG, enclosing three lists FSC nos. 107 to 109 giving this information with details.
13. FSC no. 78 dated 9 January 1806, Envoy Clarke to Admiral Pellew dated 9 December 1805.
14. A *pattamar* boat was a swift sailing vessel used to carry despatches, generally a runner or courier. See Henry Yule and A.C. Burnell, *Hobson-Jobson, Being a Glossary of Anglo-Indian Colloquial Words and Phrases and of Kindred Terms* (1886, rpt., Delhi, 1968), p. 687.
15. FSC no. 72 dated 4 May 1803 letter dated 4 December 1801, from Envoy Clarke to Governor Duncan, Bombay.
16. FSC no. 57 dated 10 April 1806, Envoy Clarke's letter to Duncan dated 11 September 1805.
17. FSC no. 42 dated 14 March 1800, Envoy Uhthoff to GG dated 18 February 1800.
18. FSC no. 105 dated 18 December 1800, letter from Viceroy to GG.
19. Ibid.
20. FSC no. 44 dated 19 June 1806, letter dated 19 June 1806 from Secretary, Fort William to Secretary to the Government of Bombay.
21. FSC no. 8 dated 26 December 1805, Envoy Clarke to GG dated 28 November 1805.
22. Para 37 of letter to the Hon'ble Secret Committee of the Hon'ble Court of Directors, dated 20 September 1808 from GG. Col. Clarke's distinguished military talents, general abilities and long local experience

were such that the GG wanted to appoint him to the post for a second time on his return from Europe.

23. FSC no. 345 dated 29 November 1804, letter from Envoy Clarke to Governor Fort St. George, dated 9 December 1802 and FSC no. 57 dated 14 March 1805 Governor's reply dated 9 November 1801.
24. FSC no. 94 dated 14 March 1805 Viceroy's Commission dated 2 January 1602.
25. FSC no. 42 dated 14 March 1800, Envoy Uhthoff to GG dated 18 February 1800.
26. FSC no. 51 dated 27 March 1806, Envoy Clarke to GG dated 24 February 1806
27. FSC no. 55 dated 23 January 1806, letter dated 17 January 1806 from Secretary, Fort William to Envoy Colonel Colman.
28. FSC no. 44 dated 19 June 1806 letter from Secretary, Fort William to Secretary Government of Bombay.
29. FSC no. 94 dated 4 December 1806 letter dated 5 November 1806 from Viceroy to GG.
30. *Secret Letter from Court,* 1811-16, vol. 6, letter from Court dated 18 November 1814 to GG.
31. FSC no. 65 dated 4 March 1802, Envoy Clarke to Governor Duncan asking for the stand that he should take if a new governor were to land in Goa, as rumoured.
32. FSC no. 9 dated 1 November 1805, Letter dated 24 October 1805 from Secretary, Fort William to Envoy Clarke.
33. Holden Furber, *John Company at Work: A Study of European Expansion in India in the Late Eighteenth Century* (Cambridge: Harvard University Press, 1951), fn. 8, p. 12; FSC no. 4 dated 4 March 1802, circular letter dated 17 November 1801 from GG to Governor Duncan, Bombay, Envoy Clarke and others.
34. FSC no. 3 dated 4 March 1802, letter dated 20 October 1801 from Governor Duncan to Envoy Clarke.
35. FSC no. 13 dated 4 March 1802, Proceedings of Jonathan Duncan, Governor of Bombay, dated 1 November 1801.
36. FSC no. 101 dated 4 March 1802, letter dated 29 October 1801, Viceroy to GG.
37. FSC no. 13 dated 4 March 1802 from Senior Commander of British troops at Daman to the Governor Duncan.
38. FSC no. 64 dated 14 March 1805, Envoy Clarke to Governor Duncan dated 26 November 1801.
39. FSC no. 23 dated 4 March 1802, letter to Envoy Clarke from GG dated 20 November 1801.
40. FSC no. 88 dated 14 March 1805, letter from Envoy Clarke to GG dated 30 December 1801.

41. FSC no. 95 dated 14 March 1805, letter dated 4 January 1802 from Envoy Clarke to GG, and FSC no. 2 dated 25 March 1802 from envoy Clarke to GG dated 15 February 1802 seeking GG's approval to his withholding the letter to the Viceroy and not acting on those instructions.
42. FSC no. 99 dated 14 March 1805 letter dated 2 January 1802 from Viceroy to GG.
43. FSC no. 344 dated 29 November 1804, letter of Envoy to Commander-in-Chief, dated 4 July 1803.
44. FSC no. 200 dated 2 November 1803, letter dated 12 September 1803 from Viceroy to GG.
45. FSC no. 59 dated 14 March 1805, Envoy Clarke to Lord Clive, Governor, Ft. St. George dated 22 November 1801.
46. FSC no. 352-354, dated 29 November 1804, Envoy Clarke to Col. Wellesley, Chief Commander, Canara , copy to Governor Duncan dated 22 September 1803.
47. FSC no. 15 dated 4 March 1802, Proceedings of President Duncan dated 1 November 1801 referring to Envoy Clarke's letter dated 21 October 1801.
48. Secret Letters from the Court of Directors dated 11 December 1807 to the GG.
49. Letter to Court from GG, 13 May 1808, para 3.
50. The Conde de Sarzedas, was the 81st Governador and 48th Vice-Rei da India, from 30 May 1807. See *Tratado de Todos os Vice-Reis e Governadores da India* (Lisboa, 1962), pp. 195-6, which credits him with the expulsion of the English from national territory (*Alem da expulsão dos ingleses da território nacional. . .*).
51. Letter to Court dated 20 September 1808, para 25.
52. FSC no. 70 dated 21 February 1803, letter dated 22 October 1802 from Dhillon, temporary Envoy at Goa to Governor Bombay. While Bombay government wanted the troops to be embarked from Goa, Colonel Wellesley and the Governor Fort St. George stopped it for reconsideration of the GG.
53. In 1800 a detachment was required by Col. Wellesley for his expedition against Dundia Waugh and it was temporarily spared by Col. Clarke from Goa.
54. FSC no. 15 dated 4 March 1802, Proceedings of President Duncan dated 1 November 1801.
55. FSC no. 72 dated 4 March 1802, Governor Duncan to GG dated 27 November 1801.
56. FSC, dated 9 January 1806.
57. FSC no. 29 dated 19 April 1812 and no. 3 dated 10 April 1812, Envoy to Vice-President.
58. FSC nos. 22-27 dated 5 November 1812. Envoy to GG.

59. FSC no. 3 dated 25 March 1602, enclosed to Envoy Clarke's letter dated 15 February 1802 to GG.
60. FSC no. 13 dated 4 March 1802 Envoy Clarke to GG.
61. FSC no. 363 dated 29 November 1804, letter dated 6 September 1804 from Envoy Clarke to GG.
62. FSC no. 352-354 dated 29 November 1804, Envoy Clarke requested Governor Duncan to defer sending the 68 pieces of ordnance, gun-powder, etc., to Goa, fearing that to place the greater part in the hands of men whose fidelity is by no means to be relied upon. FSC no. 355 of same date to Duncan and GG from Clarke.
63. FSC no. 346 dated 29 November 1804 letter of Clarke dated 9 December 1801 to governor Ft. St. George, about Major Brieston of the Portuguese Engineers and recommending a moderate stipend to him.
64. FSC no. 349 dated 29 November 1804, letter from Governor Ceylon to Envoy Clarke dated 23 July 1803.
65. FSC no. 13 dated 8 July 1809, letter dated 8 July 1809 to Viceroy from GG and FSC no. 14 dated 8 July 1809, covering letter to Envoy from Secretary, Fort William.
66. FSC nos. 17-18, dated 27 September 1811, Envoy Schuyler to Vice-President Fort William.
67. FSC no. 1-2 dated 1 November 1811, letter dated 25 September 1811 from Envoy to Vice-President Fort William.
68. FSC no. 2 dated 4 December 1812, letter of Chief Secretary to Envoy dated 25 September 1812.
69. FSC no. 3 dated 4 December 1812, letter dated 27 October 1812 from Envoy to Chief Secretary.
70. Furber, *John Company*, p. 9.
71. FSC no. 53 dated 27 March 1806, Viceroy to GG dated 26 February 1806.
72. FSC no. 21 dated 21 August 1806 and no. 93 dated 4 December 1806.
73. Furber, John Company, p. 7; M.N. Pearson, *Coastal Western India: Studies from the Portuguese Records* (New Delhi: Concept Publishing Company, 1981), p. 21.
74. The select committee of supracargoes was the equivalent of the governor-in-council of the British provinces in India, and looked after the China trade of Britain. Shantha Hariharan, 'Macau and the English East India Company in the Early Nineteenth Century: Resistance and Confrontation', *Portuguese Studies*, vol. 23, no. 2 (2007), p. 139.

CHAPTER 8

The Anglo-Gorkha War and the Territorial Structure of the Anglo-Gorkha Frontier, 1750-1814

BERNARDO A. MICHAEL

One of the persistent problems encountered by the English East India Company (EIC) since its formal acquisition of territory in South Asia in the eighteenth century was that of territorial illegibility. Company administrators inherited territories whose actual geographical location, layout, and internal divisions were vague and ill-defined, possessing blurred edges and interlocked bodies. Without knowledge of the architecture of the territories it had acquired, and lacking the means to employ cartographic surveys, the early colonial state's capacity to tax its subjects, impose law, and establish order was severely blunted. This situation assumed grave proportions all along the Company's forested frontier with the Himalayan state of Gorkha (present-day Nepal), especially prior to the Anglo-Gorkha War of 1814-16.

This paper examines the variables that informed the structure of territory along the Gorakhpur-Butwal section of the frontier—entitlements such as service (*birt*) and rent-free tenures (*maafi*), the territorial histories of relevant *parganas* and *tappas*, and the shifting relationships between petty kingdoms (such as Butwal, Gulmi, and Bansi), and their regional overlords (like Awadh, Gorkha, and the EIC). The focus will be on the spatial or territorial implications of these entitlements and relationships; for example, how rent-free landholdings tended to create islands of semi-autonomous authority within the territorial framework of pre-colonial polities. Consequently, the territories of these principalities were often made up of fluctuating, ill-defined, and even overlapping jurisdictions.

This paper is part of a larger project on the Anglo-Gorkha War of 1814-16. This project rejects received notions about the causes of the Anglo-Gorkha War to argue that questions pertaining to the geography of the two states informed these disputes over territory and revenue administration. Using an interdisciplinary approach, and tapping into the holdings of archives spread over India, Nepal and the United Kingdom, this study aims to trace the connections between state formation and spatiality in early colonial South Asia. Eventually, in the nineteenth and twentieth centuries, the colonial state resolved the question of its illegible territories through projects of surveying and territorial reordering. New administrative divisions were constructed over the debris of the older Mughal divisions to create the geographical template of the modern state.

The Anglo-Gorkha War (1814-16) between the EIC and the Himalayan kingdom of Gorkha has been studied from divergent perspectives—diplomatic, nationalist, and military.[1] Military accounts of the war have usually portrayed the conflict in terms of the key engagements and tactics, along with personalized accounts of fighting highlighting bravery, courage and honour displayed by combatants.[2] Understanding the Anglo-Gorkha War in terms of diplomatic manoeuvrings, nationalist sentiment and military action certainly has its place in historiography, but there is also much to be gained by remembering that armies, military policy and operations are invariably embedded within a wider world shaped by the forces of culture, power, history, society, and space. Such an insight has already been proposed and developed by what has come to be called the 'New Military History'.[3] In this connection, this chapter will pay particular attention to questions of spatiality, something that has been often ignored in writings on war.[4] Space then is not to be viewed as a container or stage within which social life unfolds. Rather, space refers to the layout and organization of territory produced by the ebb and flow of dynamic social forces.[5] Such forces were at work along the Gorakhpur-Butwal frontier leaving many territorial divisions fluid in their organization and layout. The argument of this paper then is that the Anglo-Gorkha War actually encoded struggles over the geographical construction of the state. A careful examination of these disputes for their spatial implications yields valuable insights into the organization and knowledge of territory along this frontier which contributed in no small measure to the outbreak of the Anglo-Gorkha War in 1814.

The Gorakhpur-Butwal frontier was formed by a stretch of territory that straddled the northern reaches of *sarkar* Gorakhpur and the plains (terai) of Butwal (see Map 8.1). In November 1801, the EIC acquired territories along this frontier when the Nawab of Awadh ceded the *sarkar* of Gorakhpur. Gorakhpur encompassed an enormous and fluctuating tract of land comprising 48 administrative divisions known as *parganas*. Each *pargana* was typically further sub-divided into smaller divisions called *tappas*. A few years later, in 1806, the Gorkhalis arrived on the frontier when they acquired territories in the area after they incorporated the little kingdom of Palpa, setting the stage for future conflict.

From the very beginning of their administration, questions of spatiality crept into matters of everyday governance on this frontier. The boundaries of the northern reaches of Gorakhpur were undefined, and some of its northern *parganas*—like Tilpur and Binayakpur—were territorially entangled thanks to the complicated histories of agrarian entitlements involving rights to taxable and tax-free lands and the levying of tribute. This made it difficult for Company officials to discern the actual extent and boundaries of the *parganas* under their control. After 1804, the ebb and flow of such territorial dynamics increasingly drew the two states into a series of protracted disputes that culminated in the outbreak of war in November 1814. These territorial disputes are diagnostic of deeper questions of spatiality or territorial organization that can only be understood in terms of existing agrarian entitlements pertaining to land tenure, taxation, and tribute. These entitlements produced the mobile and shifting administrative spaces that made up the Gorakhpur-Butwal frontier.

THE SOURCES OF SOCIAL POWER ON THE GORAKHPUR-BUTWAL FRONTIER: RAJAS, *TALUKDARS*, *BIRTIAS* AND *MAAFIDARS*

On the Gorakhpur-Butwal frontier, centuries of state making had resulted in the strengthening of highly localized systems of social power. The dominant landholders in Awadh belonged to the Rajputs (Bhumihar) and Brahmin castes, as well as some Muslim

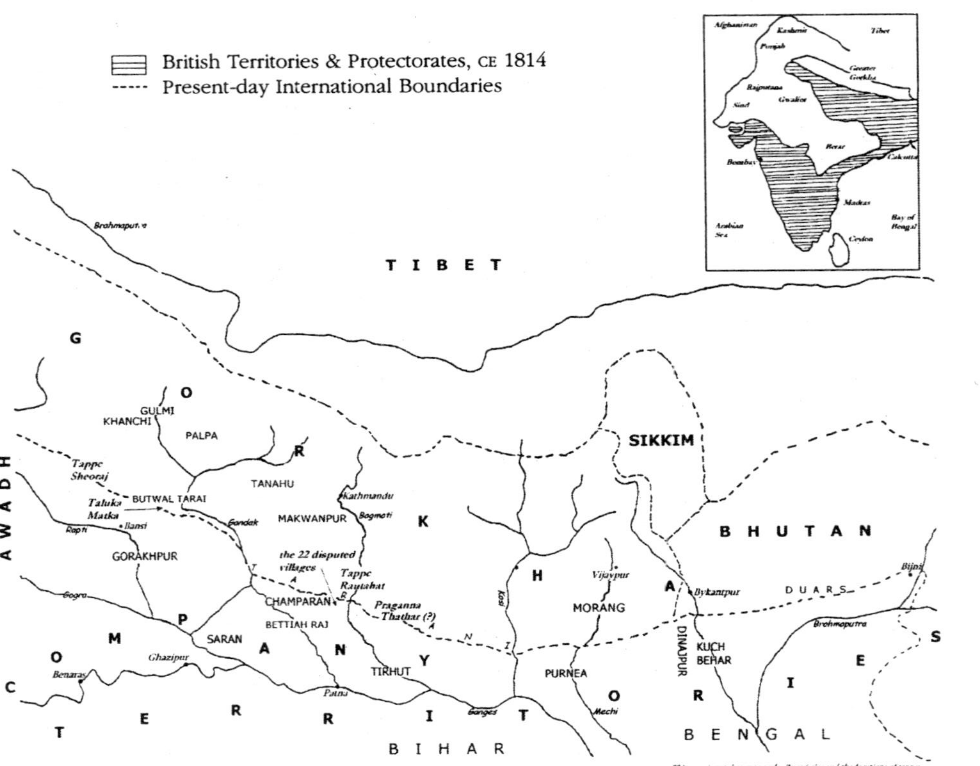

Map by Saramma & Sharon Michael

MAP 8.1: THE ANGLO-GORKHA FRONTIER, CE 1814

families.[6] Variously termed as rajas, zamindars, and *taluqdars*, they exercised considerable local power and influence. Such landholding groups became key local players and enjoyed a range of entitlements concerning political authority, taxation, and land tenure. Regional and trans-regional rulers had to take into consideration the authority and influence wielded by such local elites in the course of making arrangements for governance. Alongside these dominant groups were a number of lower, and in some cases dominant, service castes such as Ahirs, Kurmis, Koeris, Banias, and Chamars. Taken together, these local and often opposed elements constituted a diffused terrain of power in the post-Mughal world. When the EIC took over the district of Gorakhpur in 1801, it had to confront these dispersed configurations of local power.[7]

Another source of local power in eighteenth-century Gorakhpur was the holders of *birt* grants (or *Birtia*). A *birt* signified a tenure originally granted by a chieftain to a dependent on account of kinship or services rendered. It could apply both to assignments of money or land, and be either absolute or conditional. A varied class of landholders, the *birtias* enjoyed certain rights and privileges. In the eighteenth century, such grants were widespread throughout the district of Gorakhpur.[8] Large areas of Butwal were held in such grants, often conferred by the Palpa raja or high-ranking members of his immediate family. *Birt* grants complicated the tenurial structure of the region by creating fresh layers of tenurial rights and new loci of power.

Maafi or rent-free grants constituted another tenure which was usually conferred on a subject by a superior landholder or sovereign in lieu of loyalty displayed or meritorious service performed. Over the course of the eighteenth century these rent-free grantees became a numerous class of landholders and a powerful social prop for the ruling regime in Awadh.[9] Determined to cling to their entitlements they always constituted a substantial class of landholders who would persistently resist attempts by central authorities to resume their lands or convert on a taxable basis. The nawabs had made numerous rent-free grants in the district of Gorakhpur, particularly within its northern *parganas* of Bansi and Rasulpur Ghaus. For instance, in *pargana* Bansi over 50 per cent of the villages were held on a *maafi* basis. By 1811, nearly 5½ million *bighas* of land were *maafi* in Gorakhpur, many being granted not just by the nawabs but by local elites (landholders and officials).[10] The presence of substantial rent-free lands along the Anglo-Gorkha frontier (especially in the Butwal terai made up largely of the *parganas* of Tilpur and Binayakpur) is confirmed from EIC's records. In 1804, approximately 41-6 per cent of the Butwal terai's land value was subsumed under rent-free tenures.[11] Within the *pargana* of Binayakpur an estimated total of 36 per cent of income-yielding lands were given out on rent-free basis.[12] In those portions of the *pargana* of Binayakpur held by the Gorkhalis (some time after CE 1800), as much as 70 per cent of the land was alienated under rent-free tenures.[13] About 33 per cent of the nawab's rent-free grants in the Butwal terai were made to the rajas of Palpa and their dependents who in turn issued further rent-free grants of their own. In fact, 86 per cent of the rent-free grants made in the Butwal terai were made by the kings of Palpa. Most of these were conferred in the years between 1789 and 1801.[14] Of this 48 per cent of Palpali *maafi* lands were held by the raja's family and immediate relations.[15] In this sense, the granting of rent-free lands in the Butwal terai reflected processes of Palpali state making whereby land was redistributed among Palpa's clients to create a body of loyal supporters, promote agrarian colonization, and reconstitute the authority of its kings as well.[16] It is highly likely that the Palpali practice of issuing rent-free land grants was further replicated by their beneficiaries as they too sought to extend their influence. After 1801, the threat posed by the expanding kingdom of Gorkha to Palpa's own survival might have accentuated this process. Consequently, after the Gorkhalis occupied parts of *pargana* Binayakpur in the first decade of the nineteenth century, they too issued numerous rent-free grants in the area which may have amounted to as high as 70 per cent of the total value of lands under their control.[17] This they may have done in order to establish their authority in the area, especially after the incorporation of the Palpali kingdom in 1806. Such practices created hierarchies of authority and dependency while enhancing the power and initiative of local elements. Territorially this meant that lands were held under multiple regimes whose political authority and power overlapped and shifted. The territorial space of the frontier presented a patchy mosaic of tax-yielding and rent-free lands with its holders owing political loyalty to various overlords.

For the EIC, the large body of rent-free tenures meant a loss of revenue and low returns. While nearly 41 per cent of the value of the Butwal zamindari was alienated in the form of rent-free tenures, the EIC was only able to collect 61 per cent of revenues due from the remainder in 1805.[18] The situation was further exacerbated by the fact that many *maafi* holders (including the Palpa raja) did not possess any documents to support their claims. To cite just one example, in 1817 the former Palpa Raja Rattan Sen claimed 103 rent-free villages in the *parganas* of Tilpur and Binayakpur. These were originally granted to his grandfather Mahadat Sen by Nawab Asafuddaulah (1775-97) of Awadh. However, there was little documentary evidence to be found to support his claim. The only *sanad* (royal decree) from the nawab to Prithvipal Sen (Rattan Sen's father) and his wife (dated 1200 *hijri*), however, made no mention of the concerned villages. Repeated investigations did not yield any further information in the form of documents. When the British concluded settlements for the northern *parganas* of Tilpur and Binayakpur in 1802, no mention was made of the considerable rent-free or assessed estates which they were reluctant to recognize anyway. Ultimately, the EIC resumed these lands in 1817 on the grounds that there was no proof of their ownership.[19]

The EIC's officials also tried to sift through the plethora of claims to rent-free lands and discern if they were hereditary or non-hereditary. As far as the frontier district of Gorakhpur went, in 1813, Company officials recognized the right of the nawab of Awadh to issue rent-free grants but were reluctant to recognize similar rights for the lesser rajas and zamindars. In their investigations in the lands of Butwal, they found no evidence of the nawabs of Awadh having made rent-free grants to the Palpali kings. The British were unwilling to consider any claims by the Gorkhalis to the lands of Butwal and the raja of Nepal was considered only a 'pretended sovereign'. In making such assertions, the Company state was only simplifying a complex and active picture of landed entitlements that involved numerous parties, sometimes without any documentary evidence in sight. Company officials such as Major Paris Bradshaw who was stationed along the frontier to inquire into the Anglo-Gorkha territorial disputes often resorted to such simplifications which had the net effect of easily justifying the Company's claims to these lands.[20]

Rent-free land grants have implications for questions of state-making and spatiality. Traditionally, these grants provided patrons a means to create and reward loyal clients. Such grants, especially along the Gorakhpur-Butwal frontier were not always documented. The verbal commitment of a patron was the only source of the grant's legality. Rent-free landholders themselves were never passive recipients of royal largesse. Their rent-free entitlements tended to diminish the landed resources of the patron in addition to presenting an alternative source of social power that could potentially rise up in rebellion. Many *maafidars* would try to ensure that their rights were secured in perpetuity. Their loyalties could fluctuate, especially during moments of political turmoil, regime change, and rebellion. Thus, the presence, absence, increase or decline of rent-free grants over time is usually indicative of variations in patterns of state making. There were spatial consequences for all this as well. Agrarian entitlements such as rent-free lands were like little islands of perceived autonomy within the larger tax-yielding territories of a kingdom. The picture was further complicated given the shifting political relationships that landholders invariably found themselves enmeshed in. Where overlords and patrons exercised firm political control, rent-free landholders usually fell in line as loyal subjects. But during times of political uncertainty and regime change, there emerged a possibility for renegotiating their status. For instance, the distinction between tax paying (*malguzari*) and tax-free lands (*maafi*) could be used as an excuse by a rent-free landholder to declare autonomy from one overlord and seek the protection of another.[21] Alternatively, unique historical circumstance could emerge where a conquering kingdom would claim the rent-free lands belonging to another on the grounds that they did not belong to the taxable resources of a kingdom, and therefore could be alienated (see below). There is evidence to suggest that tax-free lands were often viewed as a separate and distinct category of lands that were taken off the tax paying register. They constituted islands of autonomy within the revenue paying lands of a kingdom. Such lands could, under certain historical conditions open up new kinds of claims for its holders and emerging overlords, and these set-off political struggles that characterized processes of state making.

This is what seems to have happened along the Gorakhpur-

Butwal frontier, especially in the case of the Kingdom of Palpa. In 1806, it was unclear whether the raja of Palpa was a subject of the nawab of Awadh (by virtue of the latter's overlordship of the Butwal zamindari) or of the raja of Gorkha (who claimed the same zamindari on the grounds of conquest). When the EIC claimed the territories of Palpa by virtue of its treaty with Awadh in 1801, Gorkhali officials would make distinctions between Palpa's *malguzari* and *maafi* lands. The Gorkhali found it expedient to concede that the EIC had a right to the taxable land of Palpa, but would present reservation about Palpa's rent-free lands. They argued that the EIC had no rights to the *maafi* grants conferred by the nawabs of Awadh and the rajas of Palpa because they were for *all practical purposes separated from the fiscal accounts of the pargana to which they were attached*[22] (italics added). The Gorkhali authorities seem to have reconfirmed Palpa's rent-free lands under their authority. And to prove this claim, the Gorkhalis presented an *ikrarnamah*, dated 1798, from the Palpali raja to the Gorkhali raja stating that no revenue for the Butwal zamindari had been paid to the nawab of Awadh, rather those lands were held in *maafi* (rent-free) *from the raja of Gorkha* (emphasis mine).[23] At least one former *qanungo* of the *parganas* of Tilpur and Binayakpur, one Ram Nawaz, was recruited to support the Gorkhali case in the Anglo-Gorkha boundary investigations of 1813. He would state that the lands in the Butwal terai were all *maafi* grants made to the Butwal raja by the nawabs of Awadh, thereby confirming the Gorkhali claim that these lands were never attached to the fiscal resources of Gorakhpur (which they conceded had been granted to the EIC by Awadh in 1801). This being so, these rent-free lands never formed part of the district of Gorakhpur, their new rightful possessor being the raja of Gorkha.[24]

In sheer spatial terms political, taxation, and tenurial relationships, and localized cultures of governance conspired to allow conflicting parties to make all kinds of claims on lands, tenures, and agrarian entitlements. Such claims were often articulated using all kinds of arguments leaving pre-colonial territories interlocked and entangled in all sorts of complicated ways. In 1801, along the Gorakhpur-Butwal frontier, the lands of the Butwal terai were drawn into a dynamic of territorial representation and conflict. One of the implications of this is that *maafi* lands were distinct from tax-paying lands and amenable to transfer from one kingdom to another. The Gorkhalis laid claim to the *maafi* lands granted to the Palpali kingdom by the nawabs of Awadh on the grounds that they did not form part of the tax-yielding lands of Awadh. Such rent-free tenures were perceived as autonomous of the authority of the nawabi of Awadh and now formed an integral part of Gorkha kingdom by virtue of their conquest of Palpa. Such *maafi* lands could be discontinuous patches of land scattered all over the body of a *pargana*. This did not matter to the Gorkhalis who were more concerned about their inheritance of these landed entitlements without reference to territorial contiguity or linear boundaries—variables the British were trying to uphold. On the ground what this entailed was that the space of Gorakhpur could have patches of land belonging to two distinct states that were highly intermixed.

In 1814, the EIC became unwilling to compromise its claims to the lowlands of Butwal. Faced with the loss of revenue arising from this dispute and the perception that Gorkha's claims violated its territorial sovereignty and honour, the EIC treated Gorkhali actions along the frontier as 'encroachments' that needed to be resisted with military force, if needed.

CULTURES OF GOVERNANCE AND ILLEGIBLE LANDSCAPES ON THE GORAKHPUR-BUTWAL FRONTIER

When the British acquired Gorakhpur in 1801, they possessed little knowledge of its internal resources and administrative divisions. The administrative units organized by the EIC were erected on the remains of the older Mughal divisions, which were themselves constructed out of a shifting medley of political, taxation, and tenurial rights. Revenue administration under the nawabi government in Awadh, prior to colonial rule, was made up of loose and flexible arrangements and this only compounded the situation of territorial illegibility.[25] This is clearly reflected in a letter from the Acting Collector of Etawa to the Board of Commissioners of the North-West Provinces in October 1807, when he observed:

> The objections alleged by several of the collectors, as well as by the late Commissioners, against the immediate conclusion of a permanent

settlement, are principally the imperfect knowledge yet acquired of the resources of the country, the inequality of the present assessment, the great proportion of the uncultivated lands, estimated at general at a fourth of the arable lands, the deficiency of population, and want of capital to extend the cultivation, the existing restrictions on commerce, land yet unascertained [...] the general uncertainty, [...] the doubtful value of the standard coin [...].[26]

Furthermore, there remained the question of determining the exact limits of hill kingdoms like Palpa which were along the frontier and whose lands had now fallen under the control of the EIC. These kingdoms had entered into political relations with regional kingdoms like Awadh in the plains and Gorkha in the hills. This left unclear their status *vis-à-vis* the EIC. While Palpa owned lands in the plains of Butwal for which it was willing to pay taxes to the EIC, in the early nineteenth century it also became increasingly subject to Gorkha's authority.[27] This meant that Gorkhali authorities began claiming greater influence in Palpali entitlements concerning authority, land, and taxes, something which the EIC was loath to surrender given the fact that these lands had been granted to it by Awadh under the terms of the Treaty of Cession of 1801.[28]

The only way the EIC state and the Gorkhalis could have accessed the territorial realities of their lands on the Gorakhpur-Butwal frontier was through a detailed examination of revenue and related records on agrarian entitlements. However, such a body of reliable records did not exist. This situation itself was a product of processes of state making whereby local power holders and officials had for long combined to evade the taxation demands of central authorities. Everywhere practices of forced exaction, revenue evasion, and misinformation seem to have characterized the arrangements taking place between revenue collectors and those engaging to pay revenue. The same could be said for the account books of *patwaris* (village accountants).[29] In a similar vein, Buchanan-Hamilton noted officials at Lucknow were very good at engineering all kinds of usurpations, defalcations, and forgery in the records at their disposal.[30] Forced to depend on these records, neither Gorkha nor the EIC was able to gain any accurate insight into the actual physical extent, boundaries, and internal resources of the districts that lay on the Gorakhpur-Butwal frontier. Numerous villages and wastelands never found mention in the records.[31] There was a frequent tendency for landholders in connivance with local officials to represent rich agricultural lands as waste or rent-free in order to secure tax concessions from the state, or even usurp the rights of another landholder.[32] P. Warden Grant, the EIC's revenue surveyor in Gorakhpur would note in 1821 that more than half the villages noted as *maafi* in Gorakhpur were actually in the hands of *qanungos* and their relations, suggesting the involvement of these officials in such activities. To add to this, cultivators were often enticed from revenue paying lands to work in such waste and rent-free lands, resulting in a considerable loss of government revenue.[33] Furthermore, the EIC was unable to discern the structure of tenurial rights in the region. For instance, it refused to recognize land grants that were made by authorities other than the nawab of Awadh, whom it considered to be the real sovereign of Gorakhpur district.[34] Again, in 1813, the EIC's commissioner on the Nepal frontier, Major Paris Bradshaw when investigating the status of *maafi* lands in the disputed Butwal terai made the following observations. He considered *maafi* grants to be of two important types. First were those made by a raja or zamindar to some of their relatives in lieu of cash payments for services rendered. The second class of *maafi* tenures were those made for the advancement of agriculture and bringing waste and barren lands under cultivation. Such grants were made to compensate the holders for the labour and expense incurred. Bradshaw considered both grants as non-hereditary, the former reverting back to the superior lord on the death of the zamindar and the latter on the expiration of the zamindar's *kabuliyat* (deed consenting to the terms for the engagement of revenue). Bradshaw further felt that a hereditary gift of land could only be made by the sovereign (in this case the nawab of Awadh). Thus, Bradshaw concluded that since there was no evidence of the nawab of Awadh having made any *maafi* grants to the Butwal raja, the claims of the Gorkhali commissioners were to *maafi* tenures of the second type. The attempts by colonial officials to discern the complicated structure of agrarian entitlements often ended up as gross simplifications which did not take into consideration the complicated histories by which land rights were constituted and reconstituted.

Adding to this situation was the presence of a large number

of land disputes. Such disputes took place between landholders of varying sizes. Many *birtia* tenures, which in 1801 numbered about 10,000, were steeped in such long-standing disputes.[35] These disputes were not easy to resolve.[36] A number of them revolved around water sharing rights.[37] There was a high incidence of boundary disputes as well which left ill-defined the boundaries of many *tappas* and *parganas* on the Gorakhpur-Butwal frontier.[38] In the 1800s, a few years prior to the Anglo-Gorkha War, the raja of Palpa was engaged in multi-cornered boundary disputes with the rajas of Satasi and Bettiah.[39] Ultimately, such disputes also spilled over into the territories of Awadh, the EIC, and Gorkha. The Gorkhalis were involved in a number of disputes with the British. In addition to the one over the Butwal terai, they also laid claim to other lands. In 1812, the Gorkhalis 'encroached' in two areas—*taluqa* Sheokhor and *tappa* Dhebrua which were jointly attached to *pargana* Ratanpur Bansi. *Taluqa* Sheokhor while belonging to *pargana* Bansi had for many years been in the possession of the raja of Butwal. This led Gorkhali authorities to lay claim to it after Butwal was incorporated into the Gorkhali empire. Such disputes, given their old, intricate histories, left most administrative divisions on this frontier illegible—deeply entangled in all kinds of territorial disputes. In 1814, when war broke out between the EIC and the Gorkha kingdom, the Gorakhpur-Butwal frontier had from the viewpoint of effective governance become an epistemological puzzle for both states. Both tried to render the territories legible, rein in local forces, and collect revenues, albeit in remarkably dissimilar ways. Such territorial illegibility and the conflicted constitution of land rights that characterized the cultures of governance of the Gorakhpur-Butwal frontier most certainly created conditions necessary for the outbreak of the Anglo-Gorkha War in 1814.

THE PALPA *RAJ*: A YAM BETWEEN THREE BOULDERS?

The Palpa *Raj* whose terai land in the vicinity of Butwal became of the focus of Anglo-Gorkha disputes in 1814, merits further scrutiny.[40] Like other hill kingdoms on the Anglo-Gorkha frontier, the Palpali rajas too sought to control the forest and agrarian resources of the terai. But for this purpose they had to fend competition from other little kingdoms (such as Bansi, Gulmi,

Argha-Khanchi, and Bettiah) while forging relationships with powerful states based in the north Indian plains, such as Awadh. In the eighteenth century Palpa submitted to the overlordship of Awadh, paying tribute both in cash and in kind.[41] This relationship allowed the Palpali kings to consolidate their territorial interests in the terai—getting their lands reconfirmed on favourable terms by the nawabs, building forts, and collecting taxes.[42] Palpa's political status became uncertain by the end of the eighteenth century with the decline in Awadh's influence and the emergence of Gorkha and the EIC in the region.

In 1784, Palpa strengthened its ties with an expanding Gorkha kingdom. In 1784, the Palpali ruler Mahadat Sen gave his daughter in marriage to the Gorkhali regent, Bahadur Shah and some time in 1786(?) a treaty was signed between the two kingdoms.[43] Palpa supported the Gorkhali military campaigns against the hill kingdoms of Gulmi, Argha, Khanchi, Dhurkot, Isma, Parbat, Pyuthana, Dang, and Rolpa. In return for the assistance provided, the Gorkha kingdom conferred the overlordship of Gulmi, Argha and Khanchi on Palpa.[44] Thus, by 1780, Palpa became a dependent of *both* the nawabs of Awadh and the Gorkha rajas. This fact about Palpa's double dependency is confirmed nearly two decades later by W. Scott, the Company's Resident at Lucknow.[45] Between 1790 and 1806, the growth of Gorkha's influence resulted in the imprisonment of the Palpali raja in 1804 and his execution in 1806.

With this, the Gorkhalis began to lay claim to Palpali lands lying in the Butwal tarai. In 1804, they sent letters to the governor-general seeking the right to farm the Butwal zamindari on the same terms that the former rulers of Palpa had done.[46] The Gorkhalis claimed that their conquest of the Palpa *raj* had made them heirs to all its entitlements to land and tax collection.[47] Such Gorkhali requests outraged Company officials. They treated it as an affront to their sovereignty, and affirmed the legitimacy of the Company's claims to these territories. The Company state was unwilling to share sovereignty with the Gorkhalis. Since the Gorkhali claims were based on conquest, the EIC viewed them as inferior to its own territorial claims that had emerged from legal contract (the Treaty of Cession concluded with the nawab of Awadh in 1801). Such pronouncements were indicative of the Company's vision of the juridical basis and clear-cut boundaries of its territorial possessions.

Thus, till 1794, Palpa's status was like that of a yam (or *tarul*, a Gorkhali metaphor for a weak state) placed between two boulders (*dhungo* or powerful states)—Awadh and Gorkha.[48] By 1801, the arrival of a third 'boulder'—namely, the British—on its southern flanks, and realignments within the Gorkhali court created a new political equation, which gave little room for Palpa's rulers to manoeuvre. The Gorkhalis, while willing to concede the Company's claims to these territories were not willing to relinquish their own territorial claims arising out of conquest. Squeezed between these two powerful states like a yam pressed between two boulders, Palpa's days as an independent and powerful kingdom were numbered. By 1806, it had succumbed to more powerful forces emerging along the Anglo-Gorkha frontier.

THE STRUCTURE OF TERRITORY ON THE GORAKHPUR-BUTWAL FRONTIER: *TALUQAS*, *PARGANAS*, AND *TAPPAS*

The lands straddling the Gorakhpur-Butwal frontier were carved into a number of *parganas*, *taluqas*, and *tappas* that had traditionally been subject to the overlordship of the nawab of Awadh (see Maps 8.1 and 8.2).[49] The organization, layout, and boundaries of these divisions were formed out of a shifting ensemble of multi-layered relationships that revolved around entitlements or rights to overlordship, taxation, and land-based resources. Since these rights were subject to frequent changes, prone to disputes, and shared between multiple overlords, they left their impress on the layout and organization of territorial divisions, especially along the Gorakhpur-Butwal frontier. This spatial illegibility would set the stage for the Anglo-Gorkha territorial disputes of the early nineteenth century that would culminate in the outbreak of war in 1814. It is for this reason that a brief excursion into the histories of these territorial divisions needs to be conducted.

The Butwal terai possessions of the Palpa *raj* were largely constituted by the two *parganas* of Tilpur and Binayakpur. The *pargana* of Binayakpur comprised three divisions—the *pargana* of Binayakpur itself, and the *taluqas* of Matka and Khajahani (Bhandar).[50] There is no unanimity in the written record concerning the organization, layout, and boundaries of these

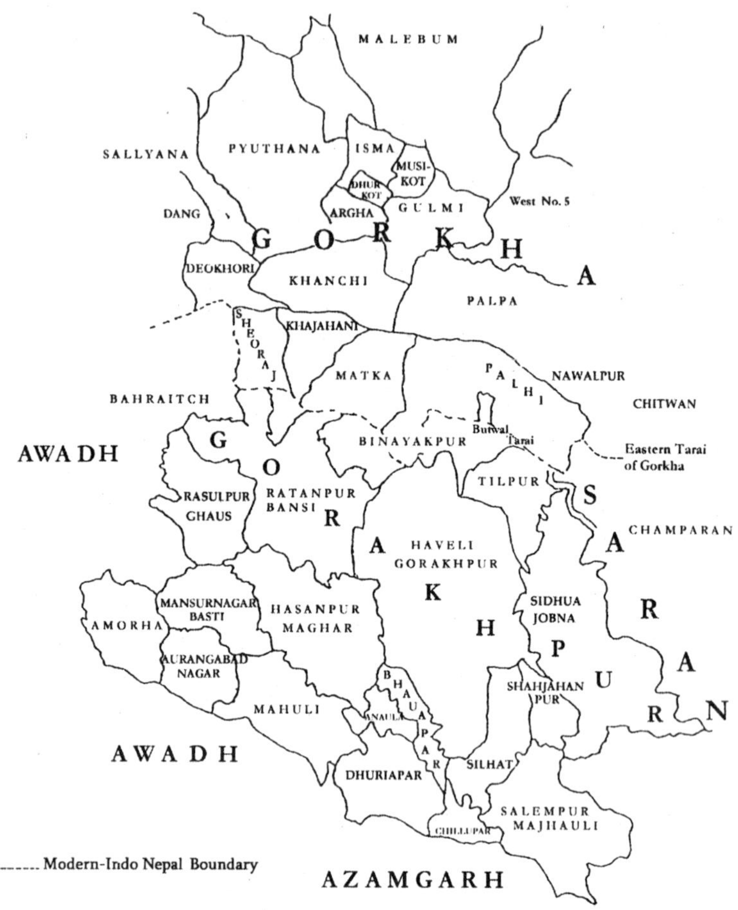

Map by Bernardo A. Michael and Sharon Michael

MAP 8.2: LITTLE KINGDOMS AND PARGANAS ON THE GORAKHPUR-BUTWAL FRONTIER, 1814.

divisions. For instance, in 1802, Lal Ran Bahadur Sen (the *diwan* of the Butwal raja) represented Matka as being a *tappa*.[51] On the other hand, in 1804, Alexander Ross, the Acting Collector of Gorakhpur noted that Matka comprised 18 *tappas*, indicating that it was a large division, possibly a *taluqa*.[52] British officials themselves were confused about the exact location, layout and structure of most of the fiscal divisions straddling the Gorakhpur-Butwal frontier. This was especially true for large *taluqa* divisions that were usually made up of loose collections of villages spread over a large area in a non-contiguous fashion.

The case of *taluqa* Matka merits further scrutiny. This *taluqa* was attached to the *pargana* of Ratanpur Bansi. It had been made up of 18/19 *tappas*[53] and had belonged at various times to little kingdoms of Gulmi, Palpa, and Bansi.[54] In 1786, Matka was transferred to the Palpali kingdom along with the hill kingdoms of Gulmi, Khanchi, and Argha. This was as a reward for the Palpali assistance provided to Gorkha's western campaigns.[55] This arrangement would have included the terai possessions of these kingdoms as well, including Matka. The 1786 grant to the Palpalis severed the fiscal connections that till then had existed between the *taluqa* and the *pargana* of Ratanpur Bansi. From 1786 onwards, the *taluqa* of Matka remained attached to the resources of the Butwal terai, more specifically the *pargana* of Binayakpur. Some time between 1786 and 1793, the Palpa Raja Mahadat Sen got his rights to the *taluqa* of Matka confirmed by Nawab Asafuddaulah (1775-97) of Awadh. In 1799, the Palpali raja's possessions were confirmed by the Gorkha raja.[56] At the time of the cession of Gorakhpur to the EIC (in 1801), the *taluqa* was included in the Butwal raja's *kabuliyat* (terms of agreement) for the *parganas* of Tilpur and Binayakpur.[57] In 1803, the Gulmi raja, seeking restoration of Matka, petitioned his sister who was in charge of the Gorkhali government at Kathmandu.[58] Kathmandu honoured the Gulmi raja's request.[59] The Palpali raja, unable to resist this move, sullenly consented by giving a *razeenamah* (certificate of surrender) to that effect. He even sent word to Company officials that from henceforth the revenues for Matka would be the responsibility of the Gulmi raja.[60] But it would seem that he never gave up his claim to this *taluqa* and actively resisted the efforts of the Gulmi raja's officials to take control of it. Later, when the ex-raja of Gorkha (Ran Bahadur Shah) returned to Nepal in 1804, he reinstated Prithvipal Sen as possessor of this *taluqa*.[61] No formal grant was issued, and it appears the order given was a verbal one.[62] By 1804, in a reversal of policy, the Gorkhalis detained the rajas of Gulmi (at this time a minor) and Palpa in Kathmandu, and incorporated their kingdoms into the Gorkhali kingdom.[63] In the meantime, even though the Gorkhali raja had reinstated the Palpali raja's right, the *taluqa* remained in the hands of the EIC under *kham* (direct) management setting the stage for a fresh round of disputes—this time between the EIC and the Gorkha kingdom.

Tappa Sheoraj, had for long been attached to the hill principality of Pyuthana, and formed a part of its possessions in the plains.[64] As in the case of most administrative divisions along the Anglo-Gorkha frontier, there is confusion as to whether Sheoraj was a *tappa*, *pargana* or *taluqa*. In November 1786, Pyuthana was forcibly incorporated into the territorial possessions of Gorkha.[65] From then on, the Gorkhalis began to take greater interest in managing the territorial possessions of the Pyuthana raja, including *tappa* Sheoraj. In 1795, Gorkhali officials were deputed from Kathmandu to survey lands mortgaged by the Pyuthana raja.[66]

At the time of the takeover of Sheoraj by Gorkha in 1786, the *tappa* was for purposes of revenue collection attached to the *pargana* of Ratanpur Bansi, lying in Gorakhpur district. In addition to this, the *tappa* itself lay mired in longstanding disputes. The seventeenth-century French traveller Tavernier observed an intense struggle over this *tappa* between the rajas of Pyuthana and Butwal.[67] It is also possible that the raja of Palpa had laid some claim to this *tappa*.[68] In 1790, when the Gorkhalis took possession of the *tappa*, a dispute arose between Gorkhali officials stationed there and the *amils* (revenue collectors) of *pargana* Ratanpur Bansi.[69] In 1791, information about these disputes over the *tappa* of Sheoraj was sent to the Court of Directors by Company authorities at Calcutta. But, the Governor-General's Council at Calcutta observed that the dispute was a trifling one, 'not likely to lead to any serious consequences or embarrassing consequences [since these lands were not under the authority of the Company].'[70] However, once the EIC acquired territories in this region after the treaty of 1801, they got drawn into the logic of these disputes. When this happened, as it did ten years later, the tone of the EIC's officials would turn into righteous indignation.

In August 1804, when Sheoraj's taxation arrears (payable at Bansi) began to mount, Bansi's revenue collectors demanded payment of these outstanding dues from the Gorkha raja. The nawab of Awadh too had sent a complaint to Company authorities that outstanding arrears were due from the authorities of Sheoraj for the years prior to its cession.[71] The Gorkha raja rejected this demand arguing that since 1765 (1786?), Sheoraj had been incorporated into Gorkhali territory, following the conquest of the hill kingdom of Pyuthana, and being a dependent of that kingdom. There seems to be some truth in this, as Gorkhali records after 1797 refer to Sheoraj as belonging to *zilla* Pyuthana.[72] In these Gorkhali documents, Sheoraj finds mention as a *pargana* and not a *tappa* as traditionally held. In yet another Gorkhali document, Sheoraj is combined with the neighbouring district of Deokhori to form the *pargana* of Deokhori-Sheoraj.[73] Company officials also conducted an investigation into the matter and concluded that *tappa* Sheoraj by virtue of being attached to *pargana* Ratanpur Bansi, actually formed a part of the EIC's possessions. However, Company officials also admitted that the *tappa* of Sheoraj had been in the possession of the Gorkhalis, prior to the cession of Gorakhpur to the EIC in 1801.[74]

Company records reveal that fiscal ties existed between the *tappa* of Sheoraj and its 'parent' *pargana* of Ratanpur Bansi between 1786 (when the Gorkhalis first obtained control of it) and 1801 (when the EIC first began to claim it).[75] In addition to the Anglo-Gorkha disputes over Sheoraj, the *tappa* was also wracked by a number of local disputes along its common border with the neighbouring *tappa* of Dhebarua lying in *pargana* Bansi. In the first decade of the nineteenth century, the *malguzar* of *tappa* Dhebarua, a Tharu magnate named Jurawan Chaudhury was involved in a protracted multi-cornered dispute with Gorkhali officials, Rajputs and other Tharu magnates.[76]

The recurrence of such disputes along the Gorakhpur-Butwal frontier worried Company officials. In late 1811, one Captain Williamson headed a military force to locate the boundary between the *tappas* of Sheoraj and Dhebarua (and by extension the *pargana* of Bansi from the territories of Gorkha). During the course of the boundary demarcation Williamson received a 'strange' request from the local Gorkhali *faujdar*, one Maniraj Bhaju. The *faujdar* had noted with some concern Williamson's tendency to draw the boundary in a straight line because this line cut through some lands belonging to the *faujdar*. The Gorkhali *faujdar* had evidently planted rice on a few *bighas* of land and wanted the boundary line to circumvent it.[77] Williamson found the request 'irregular' and denied it![78] Williamson, like many of his colleagues was unwilling to recognize the multiplicity of similar claims that peppered the Anglo-Gorkha frontier leaving discontinuous the boundary between the two states. Maniraj was not alone in this. A number of senior Gorkhali officials were granted lands in this *tappa*.[79] High-placed Gorkhali officials like Ranganath Pandit and Dalabhanjan Pande also held lands on or near the Gorakhpur-Butwal frontier, as well as among the twenty-two disputed villages on the Champaran-Tarriani frontier.[80] These Gorkhali officials too had a vested interest in preserving their claims and opposing the EIC's attempts to resume these lands. Local agency was critical in the historical development and sustenance of such disputes. Thus, such multi-cornered disputes often drew their social energies from localized systems of power and influence that were enmeshed with bundles of entitlements that were connected to questions of land control, property rights, political authority, and taxation claims. The local intricacies that fed into the Anglo-Gorkha disputes have been largely ignored by historians of the Anglo-Gorkha War.

In fact, local initiative was so unchecked that even Kathmandu was uncertain about what its officials were doing on the frontier. The Gorkhali commander on the Gorakhpur-Butwal frontier, Kaji Amar Singh Thapa discovered this when investigating the question of Sheoraj in 1813.[81] The Gorkhali *chaudharis* of Sheoraj claimed that they had never made any official remittances to authorities at Bansi.[82] Instead, they claimed that they had only offered gifts (*mamilat hoina bhetghat hunda sar saugad diyako ho*). The Gorkhali commander was unsure about the exact context for such 'gift' exchanges and whether it meant that the *chaudharis* were subordinated to officials at Bansi.[83] He then asserted that these 'gifts' were never sanctioned by Kathmandu and made without the knowledge of the Gorkha raja (*khamid*). In the final analysis, by 1814 it was clear that senior Gorkhali officials were unwilling to consider Sheoraj nothing less than a part of Gorkha's territorial possessions (*ajasamma aafnu amval bhayako jaga*).[84]

Thus, in 1813, *tappa* Sheoraj had become an illegible landscape—wracked by powerful local forces that eluded the centralizing thrust of both the Gorkha and the EIC. This left the *tappa* of Sheoraj with an unstable political, fiscal, and tenurial history. Between 1780 and 1814, it lay entangled with a number of *parganas* (Bansi, Deokhori), little kingdoms (Pyuthana, Butwal, Bansi), and more powerful state formations (Awadh, Gorkha, and the EIC). In 1814, both the Gorkhalis and the Company were unwilling to relinquish their claims to Sheoraj and other lands in the Butwal terai. In 1814, the raja of Gorkha sent two letters to firmly reassert Gorkha's claims to this *tappa*.[85] It was only after the Anglo-Gorkha War and the formal demarcation of the Anglo-Gorkha boundary that *tappa* Sheoraj was handed over to the Gorkhalis. The claims of the rulers of Pyuthana, to whom this *tappa* originally belonged, were forever silenced.[86]

CONCLUSION

The developments outlined in this chapter form the ingredients of a spatial history of a colonial frontier. On the Gorakhpur-Butwal frontier, the terai lands of Butwal stood at the centre of the Anglo-Gorkha territorial disputes that preceded the War of 1814. Entangled within a network of institutional arrangements for governance, and shifting inter-state relationships, these territories became the products of distinctive processes of state making. The open-ended jostling and situational adjustments made by local and supra-local agents (as they struggled to create, confirm, and protect their entitlements concerning landed properties, taxes, tributes, and political authority) constituted the cultures that marked governance along the frontier. This produced a multi-levelled layer of interlocking entitlements that when battened onto a single patch of territory resulted in the creation of multiple loci of authority and jurisdictions. By 1814, officials (local and supra-local), *pahikasht* cultivators, Tharus, rajas, traders, Company officials, Gorkhali commanders and others came together to produce a fluid world, argue about it, and ultimately to fight violently over it.

In many crucial respects, these arguments were about the produced space or territories of the state. The territorial coherence of administrative divisions, such as the *parganas* and *tappas* was rendered uncertain by the presence of discontinuous and overlapping jurisdictions, cross-cutting flows of political authority and dependence, multi-directional movements of taxes, and varied shifting tenurial arrangements. Cultures of governance on the Anglo-Gorkha frontier were shaped by the variables of local ecology, agrarian enterprise, patterns of land-use and the high incidence of *birt* and *maafi* landholders. These variables informed each other to produce the shifting spaces of many *pargana* divisions along the frontier. Consequently, the territories of the little kingdoms and states that straddled the frontier, in their layout and organization, lay mutually entangled and interlocked with discontinuous and fuzzy boundaries. Caught in the middle of this unfolding picture, British officials were never fully able to grasp the shifting geographies of these administrative divisions as they emerged out of multiple narratives of power and local agency. In trying to resolve the geographical puzzles of pre-colonial polities and render them legible, the EIC sought to impose its own vision of how a state should look.

By 1814, too much was at stake for both for the Gorkhalis and the British, for different reasons. With no resolution in sight, a violent conflict seemed imminent. On 29 May 1814, Gorkhali troops in a surprise dawn attack struck at the *thanas* of Biswarea, Chitwa and Sourah in the Butwal terai, killing a number of Company policemen stationed there. The *thanadar* (police inspector) of Chitwa, one Kashinath was captured, tied to a tree and shot to death with arrows. This became the immediate cause of the war and very soon the questions of spatiality that had plagued the relations between the two states receded into the background as EIC officials responded to this attack by declaring war against the Gorkhas on 1 November 1814.

ACKNOWLEDGEMENTS

Earlier versions of this paper were presented at the Annual Meeting of the Society for Military History, Center for the Study of War and Society, University of Tennessee at Knoxville, 1-4 May 2003 and the 20th International Association of the History of Asia

Conference in Delhi in November 2008. I am grateful to the participants in the panels and audience for their inputs and especially to Kaushik Roy for his comments and suggestions. The usual caveats apply.

ABBREVIATIONS

APAC	Asia Pacific and Africa Collections
BL	British Library
BOR	Board of Revenue
BS	Bikram Sambat, the official Nepali calendar usually arrived at by adding 57 years to the Gregorian
Coll.	Collector
Consl.	Consultation
CPC	*Calendar of Persian Correspondence*
FP	Foreign Political
FS	Foreign Secret
GCR	Gorakhpur Collectorate Records
IOR	India Office Records
NAI	National Archives of India
NAN	National Archives of Nepal
Procs.	Proceedings
RRC	Regmi Research Collection
RRS	Regmi Research Series
RSA	Regional State Archives, Allahabad
UPSA	Uttar Pradesh State Archives

NOTES

1. Nationalist accounts can be found in the works of Netra Rajya Laxmi Rana, *The Anglo-Gorkha War 1814-1816* (Kathmandu: Jore Ganesh Press, 1970), pp. 26-40; D.R. Regmi, *Modern Nepal*, vol. 2, *Expansion: Climax and All* (Calcutta: Firma K.L. Mukhopadhyaya, 1975), p. 291; Rishikesh Shaha, *Modern Nepal: A Political History, 1769-1955*, vol. 1: *1769-1885* (New Delhi: Manohar, 1990), p. 108. Among Nepali works see Maheshraj Pant, '1871-72 ko Nepal Angrez Yuddhama Nepal le Harnema Euta Thulo Karan', *Purnima*, vol. 1 (1964 [2021 BS]), pp. 47-58. See also Dineshraj Pant, 'Lahorema Nepali Viraharu', *Purnima*, vol. 5 (1965 [2022 BS]), pp. 63-70. Diplomatic treatments of the war can be

found in Asad Husain, *British India's Relations with the Kindgom of Nepal, 1857-1947* (Oxford: Clarendon Press, 1970), p. 36; Kanchanmoy Mojumdar, *Anglo-Nepalese Relations in the Nineteenth Century* (Calcutta: Firma K.L. Mukhopadhyaya, 1973); K.C. Chaudhuri, *Anglo-Nepalese Relations* (Calcutta: Firma K.L. Mukhopadhyaya, 1960); and B.D. Sanwal, *Nepal and the East India Company* (Bombay: Asia Publishing House, 1965).

2. See Chandra Khanduri, *A Rediscovered History of the Gorkhas* (Delhi: Gyan Sagar Publications, 1997); John Pemble, *The Invasion of Nepal: John Company at War* (Oxford: Clarendon Press, 1971); Tony Gould, *Imperial Warriors: Britain and the Gurkhas* (London: Granta Books, 1999), pp. 31-68.
3. John W. Chambers, 'Conference Review Essay: The New Military History: Myth and Reality', *Journal of Military History*, vol. 55, no. 3 (1991), pp. 395-406. For a fuller historiographic treatment of the subject of war see Anna Simons, 'War: Back to the Future', *Annual Review of Anthropology*, vol. 28 (1999), pp. 73-108.
4. But see Ludwig Stiller, 'The Principle of Limitation', *The Nepalese Perspective*, vol. 9 (1974), pp. 4-21; *The Silent Cry: The People of Nepal, 1816-1839* (Kathmandu: Sahayogi Prakashan, 1976), pp. 216-27. See also Mary Des Chene, 'Relics of Empire: A Cultural History of the Gurkhas, 1815-1987', unpublished Ph.D. dissertation (Stanford University, 1991), pp. 25-33. See especially p. 30. For a similar exploration of spatiality as in the case of the kingdom of Siam (present-day Thailand) see Thongchai Winichakul, *Siam Mapped: The History of the Geo-Body of a Nation* (Honolulu: University of Hawaii Press, 1994).
5. The literature on spatiality is too numerous to cite here. A sampling would include Derek Gregory and John Urry (eds.), *Social Relations and Spatial Structures* (New York: Palgrave Macmillan, 1985); Phil Hubbard, Rob Kitchin, Brendan Bartley, Duncan Fuller (eds.), *Thinking Geographically: Space, Theory and Contemporary Human Geography* (London: Continuum, 2004); Henri Lefebvre, *The Production of Space*, English translation by Richard Nice (London: Blackwell, 1991); Setha Low and Denise Lawrence-Zuniga (eds.), *The Anthropology of Space and Place: Locating Culture* (London: Blackwell, 2003); Doreen Massey, *For Space* (Thousand Oaks, CA: Sage Publications, 2005); John Pickles, *A History of Spaces: Cartographic Reason, Mapping and the Geo-Coded World* (New York: Routledge, 2004); Edward W. Soja, *Postmodern Geographies: The Reassertion of Space in Contemporary Social Thought* (London: Verso, 1989); Nigel Thrift, 'Space: The Fundamental Stuff of Human Geography', in Sarah L. Holloway, Stephen P. Rice and Gill Valentine (eds.), *Key Concepts in Geography* (London: Sage Publications, 2003), pp. 95-107. For feminist understandings of space see Doreen Massey, *Feminism and Geography: The Limits of Geographical Knowledge* (Minneapolis: University of Minnesota Press, 1993).

6. See, Francis Buchanan-Hamilton, 'An Account of the Northern Part of the District of Gorakhpur', Eur Mss D 91-93 and G 22-23, European Manuscripts, IOR, APAC, BL (hereafter *The Gorakhpur Report*), vol. 1, ch. 4, pp. 98-243; Meena Bhargava, *State, Society and Ecology: Gorakhpur in Transition, 1750-1830* (New Delhi: Manohar, 1999), p. 125.
7. The Company would over a period of time try to define and regulate the complex character of political, taxation and tenurial rights. Thus, Company officials took recourse to scrutinizing the *sanads* or deeds by which *taluqdars* had acquired control over the rights of the zamindars. In the event of such deeds being conferred by local revenue collectors (*amils*), and not by the nawab himself, they were deemed void. See, Bhargava, *State, Society*, p. 140.
8. For example, in 1810, Raja Sarabjit Singh of Bansi made a number of *jiwan birt* grants to his illegitimate sons in the northern *tappas* of Khajahani and Dhebarua (*pargana* Ratanpur Bansi) along the Gorakhpur-Butwal frontier. Cited in Bhargava, *State, Society*, p. 143. A *jiwan birt* was given as an appanage or an allowance usually granted to younger members of the grantor's family. The Satasi *raj* too was well known for the large number of *birt* grants its rajas had conferred—many of their *birtias* being located in the northern *tappa* of Lehera (*pargana* Haveli). During Nawab Asafuddaulah's rule, many of these *birtias* and zamindars placed themselves under the protection of the Satasi raja for fear of harassment from revenue collectors (*amils*).
9. The nawabs of Awadh had come to the early realization that they could not afford to alienate this powerful class of landed interests, and therefore made every attempt to co-opt them into the framework of the kingdom as allies. For rich details about the role of these *madad-i-mash* or rent-free landholders in Awadhi statemaking, see Muzaffar Alam, *The Crisis of Empire in Mughal North India: Awadh and Punjab, 1707-1748* (New Delhi: Oxford University Press, 1986); Bhargava, *State, Society*, pp. 177-83; and Meena Bhargava, '*Madad-i-mash* and *Maafi* Grants in Gorakhpur During the late Eighteenth and Early Nineteenth Century', *Proceedings of the Indian History Congress*, 52nd Session (1991-2), pp. 338-44.
10. Bhargava, *State and Society*, p. 182.
11. Statement of the Assets of Bootwul in year 1212 *fasli* (CE 1804), Letters Issued Register, GCR, *basta* 16, vol. 4, 17 July 1805, pp. 158-62, RSA.
12. These figures include lands held both by the Company state as well as Gorkha. See 'Statement exhibiting the resources in the year 1212 *fasli* (CE 1804) of each of the three divisions comprising the *pargana* of Binayakpur', Letters Issued Register, *GCR*, *basta* 16, vol. 4, October 1805, pp. 294-6, RSA.
13. See 'Statement exhibiting the resources in the year 1212 *fasli* (CE 1804) of each of the three divisions comprising the *pargana* of Binayakpur', pp. 294-6.

14. Ibid. The Palpali rulers during this period were Mahadat Sen (1782-93) and his son Prithvipal Sen (CE 1793-1806).
15. Statement of the Assets of Bootwul in year 1212 *fasli* (CE 1804), Letters Issued Register, GCR, *basta* 16, vol. 4, 17 July 1805, pp. 158-62, RSA.
16. The idea that the kingly authority of pre-colonial 'little kingdoms' was constituted and reconstituted through gifts of land has already been adequately emphasized in the work of Nicholas B. Dirks. See Nicholas B. Dirks, *The Hollow Crown: The Ethnohistory of a Little Kingdom* (Cambridge: Cambridge University Press, 1987).
17. The information has been gleaned from the following sources: Statement of the assets of Bootwul in year 1212 *fasli* (CE 1804), Letters Issued Register, GCR, *basta* 16, vol. 4, 17 July 1805, pp. 158-62, RSA. Reports and Observations of the Company's Political Agent on the Gorkha frontier, Lt. Col. Paris Bradshaw following his meetings with Gorkhali Commissioners, April-May 1813, FP Procs. 18 June 1813, nos. 18-24, NAI.
18. See A. Ross, Acting Coll. of Gorakhpur to BOR, 16 September 1805, Letters Issued Register, GCR, *basta* 16, vol. 4, p. 245, para 6, RSA.
19. See Translation of Proceedings of April 1817, Gorakhpur Collectorate, Letters Issued Register, GCR, *basta* 17, vol. 122, RSA.
20. See Encroachments of the Nepaulese, Boards Collections, F/4/422, no. 10381, pp. 293-382, See especially p. 378, IOR, APAC, BL.
21. In the 1780s, such a stratagem was attempted by one Mirza Abdulla Beg along the Anglo-Gorkha frontier in the Champaran region (present-day Bihar). Details about this can be found in Bernardo Michael, 'Statemaking and Space on the Margins of Empire: Rethinking the Anglo-Gorkha War of 1814-1816', *Studies in Nepali History and Society*, vol. 4, no. 2 (1999), pp. 247-94.
22. This is clear from the arguments of Gorkhali Commissioners like Gaj Singh Khatri and the depositions of Gorkhali witnesses in the 1813 Anglo-Gorkha border investigations. See Reports and Observations made by Paris Bradshaw, the Company's Agent to the Anglo-Gorkha frontier, April-May 1813, FP Procs. 18 June 1813, no. 18-24, NAI. Gaj Singh Khatri would clarify that the Gorkhali raja was claiming the *maafi* lands in 'right of his feudal superiority'. See, Encroachments of the Nepaulese, Boards Collections, F/4/422, no. 10381, pp. 293-382, IOR, APAC, BL. The quotation is from p. 298. It is unclear as to when and by what act such a 'feudal' authority was constituted as it would have enabled us to better understand the *locus standi* of the Gorkhali commissioner's claim.
23. Ibid. It is interesting to note that while the Gorkhalis admitted that the Palpali raja had received his lands from the nawab of Awadh, these very lands were now reconfirmed by Gorkha raja after Palpa (at least from a Gorkhali standpoint) became its tributary, probably some time

in 1787. In this manner, it can be suggested that the Gorkhalis re-constituted the older tenurial rights of the Palpali kingdom following its political subordination to Gorkha. See Encroachments of the Nepaulese, Boards Collections, F/4/422, no. 10381, pp. 293-382, IOR, APAC, BL.

24. See T. Balfour, Coll. of Gorakhpur to BOR, 20 April 1807, Letters Issued Register, GCR, *basta* 16, vol. 116, pp. 522-3, RSA. It appears that the Gorkhalis rewarded Ram Nawaz's services by giving him some land under *jagir* (resumable service) tenure. See Shankarman Rajbamshi (ed.), *Shahkalin Chittipatra Samgraha* (Kathmandu: Bir Pustakalaya, 1966 [2023 BS]), vol. 2, p. 6. See Bernardo A. Michael, 'Separating the Yam from the Boulder: Statemaking, Space, and the Causes of the Anglo-Gorkha War of 1814-1816', unpublished Ph.D. dissertation, University of Hawaii at Manoa, 2001, Table 29, Appendix 4, nos. 6 and 7, pp. 569-70.
25. For information on these and other matters pertaining to the Ceded and Conquered Provinces see Michael Mann, *British Rule on Indian Soil: North India in the First Half of the Nineteenth Century* (New Delhi: Manohar, 1999), pp. 22-43; B.B. Misra, *The Central Administration of the East India Company* (Manchester: Manchester University Press, 1959), pp. 200-12; and Sulekh Chandra Gupta, *Agrarian Relations and Early British Rule in India: A Case Study of Ceded and Conquered Provinces (Uttar Pradesh) 1801-1833* (Bombay: Asia Publishing House, 1963).
26. Cited in Michael Mann, 'A Permanent Settlement for the Ceded and Conquered Provinces: Revenue Administration in North India, 1801-33', *Indian Economic and Social History Review*, vol. 32 (1995), pp. 245-69. See especially p. 260. Colebrooke's reply provides a neat summary of the host of obstacles the Company would face during the early years of its rule in the Ceded and Conquered Provinces of north India.
27. The Company had entered into a revenue settlement for the Butwal zamindari in 1802. See letter from John Routledge, Coll. of Gorakhpur to Charles Boddam, Coll. of Saran, 5 March 1803, Letters Issued Register, GCR, *basta* 16, vol. 2, RSA. For further details see also Letter from A. Ross, Acting Coll. to Graeme Mercer, Secy. to Affairs of Ceded Provinces, 30 April 1803, in Letters Issued Register, GCR, *basta* 16, vol. 2, RSA; Translation of Proceedings of April 1817, Gorakhpur Collectorate, Letters Issued Register, GCR, *basta* 17, vol. 122, RSA; Thomas Balfour, Coll. of Gorakhpur to BOR, 20 April 1807, Letters Issued Register, GCR, *basta* 16, vol. 116, pp. 522-3, RSA.
28. For details see Bishnu Prasad Ghimire, *Palpa Rajyako Itihas*, part 1 (Varanasi: Aryabhushan Press, 1988 [2055 BS].), pp. 59, 60, 235, 249; Order authorizing Ram Chander Tiwari to reclaim wastelands in Palhi, AD 1811, RRC, vol. 41, p. 36. At the same time, one Mahadev Upadhyaya

was appointed as *subba* in the Butwal terai. See Boards Collections, F/4/185, vol. 2, p. 326, IOR, APAC, BL.

29. Bhargava, *State, Society*, p. 174. The Company over time after 1815, attempted to remedy the situation by converting local functionaries such as the *patwari* and *qanungo* into state employees. However, this measure failed to eliminate the problem of reliable records. Ibid., pp. 174-7.
30. Hamilton, *The Gorakhpur Report*, vol. 2, pp. 79-81.
31. The question of villages not finding mention in *qanungo* records was a serious one. Company officials in the first two decades of the nineteenth century just never knew the exact number of villages within their respective jurisdictions—waste or otherwise. Such kinds of information would emerge only when the first revenue surveys were conducted in the 1830s. These surveys would reveal the extent to which local cultures of governance had succeeded in evading the taxation net of the state. For instance, the surveys revealed as many as 520 villages in the *pargana* of Hasanpur Maghar had been concealed from the revenue demands of the Company. See J. Hooper, *Final Report of the Settlement of the Basti District* (Allahabad: Government Press, 1891), p. 46.
32. For instance, in one case, the *maafi* lands of one party were included as the *nankar* lands of another. See Petitions in Letters Received Register 1810, GCR, *basta* 2, vol. 12, RSA.
33. See Board of Commissioners to Vice-President-in-Council, Fort William 11 September 1809, GCR, *basta* 2, vol. 11, Letters Received Register, RSA. See also, R.M. Bird to J. Armstrong, Coll. of Gorakhpur, 18 June 1829, GCR, *basta* 6, vol. 43, pp. 21-2, RSA.
34. See Letters Received by Coll. (12 May 1802-25 June 1803), GCR, *basta* 1, vol. 1, Letter no. 31, RSA.
35. Hamilton, *The Gorakhpur Report*, vol. 1, pt. 2, pp. 10-11.
36. David Scott, the Acting Magistrate of Gorakhpur, would note in 1812 that such disputes were 'of a most intricate nature and too numerous to be decided upon by the Magistrate as directed by the order of the *Nizamat Adalat* [Supreme Court at Calcutta] . . .' See Letter from David Scott, Acting Magistrate to M.H. Turnbull, Registrar to the Court of the *Nizamat Adalat*, Forth William, 11 March 1812, Letters Issued Register, GCR, *basta* 17, vol. 120, RSA.
37. This was noted by the Acting Magistrate of Gorakhpur in 1812. See Letter from David Scott, Acting Magistrate to M.H. Turnbull, Registrar to the Court of the *Nizamat Adalat*, Forth William, 11 March 1812, Letters Issued Register, GCR, *basta* 17, vol. 120, RSA.
38. Hamilton, *The Gorakhpur Report*, vol. 1, part 2, pp. 10-11.
39. See Report of W. Scott, Resident at Lucknow to Governor-General's office, FS Consl. December 1800, no. 43, NAI. A copy of the Bettiah

raja's petition to the Coll. of Gorakhpur was sent to Lal Ran Bahadur Sen, the brother of Raja Prithvipal Sen. See John Routledge, Coll. of Gorakhpur to Charles Boddam, Coll. of Saran dated 5 March 1803, Letters Issued Register, GCR, *basta* 16, vol. 2, RSA.

40. For a general history of the Palpali *raj* see Bishnu Prasad Ghimire, *Palpa Rajyako Itihas*, pt. 1 (Varanasi: Aryabhushan Press, 1988 [2055 BS]). See also Francis Buchanan-Hamilton, *The Gorakhpur Report*, vol. 1, ch. 4, pp. 166-8 and R.C. Majumdar, 'An Account of the Sena Kings of Nepal', *Proceedings of the Indian Historical Records Commission*, 12 (1929). For a genealogy of these kings see Shankarman Rajvamshi, *Sena Vamsavali* (Kathmandu: Bir Pustakalaya, 1963 [2020 BS]).
41. Hamilton, *The Gorakhpur Report*, vol. 2, p. 143; Hamilton, *An Account of the Kingdom of Nepal and of the Territories Annexed to this Dominion by the House of Gurkha* (Delhi: Manjushri, 1971 [1819]), p. 173. See also 'An account of the several powers bordering on ours and the Viziers Dominions', Add Mss 29210, pp. 89-95, especially p. 90, *Western Manuscripts*, BL; FS Procs, Consultation 43, December 1800, NAI.
42. See Hamilton, *The Gorakhpur Report*, vol. 1, pp. 331-3.
43. Hamilton, *The Gorakhpur Report*, vol. 3, pp. 32-3; D.R. Regmi, *Modern Nepal*, vol. 1, *Rise and Growth in the Eighteenth Century* (Calcutta: Firma K.L. Mukhopadhyaya), p. 313. See also Dinesh Raj Pant, 'Sri Panch Ran Bahadur Shahka Nauwota Aprakashit Lal Mohar', *Purnima*, vol. 58 (1983 [2040 BS]), pp. 37-9; Bishnu Prasad Ghimire, *Palpa Rajyako Itihas*, pt. 1 (Varanasi: Aryabhushan Press, 1988 [2055]), p. 96.
44. See Dinesh Raj Pant, 'Ashrit Rajyaupar', p. 52.
45. See W. Scott to Governor-General Lord Wellesley, December 1800, FS Consl. 30 December 1800, no. 39, NAI.
46. See letter from *kajis* Amar Singh Thapa and Dalabhanjan Pande to the Magistrate of Gorakhpur, received 2 September 1804, Boards Collections, F/4/185, vol. 2, pp. 87-90, IOR, APAC, BL.
47. See letter of John Adam, Secretary to Government to Lt.-Col. Paris Bradshaw, the Company's Agent on the Anglo-Gorkha frontier, 13 January 1813, FP Procs. 15 January 1813, no. 46, NAI.
48. The term *dhungo* is a metaphor for the state that finds frequent reference in Gorkhali documents. My use of the terms 'yams' and 'boulders' draw inspiration from the Gorkhali Raja Prithvinarayan Shah's statement comparing Nepal to a yam wedged between two boulders—India and China (*uparant yo raje dui dhungako tarul jasto rahe cha*). Cited in Yogi Naraharinatha and Baburam Acharya (eds.), *Rashtrapita Shri Panch Badamaharaja Prithvinarayan Shah Dev ko Dibya Upadesh* (Kathmandu: Prithvi Jayanti Samaroh Samiti, 1953 [2010 BS]), p. 15.
49. For locating the *pargana* and *tappa* divisions of the Gorakhpur-Butwal frontier, see Map 8.2 in this chapter.
50. See statement regarding the resources in the year 1212 *fasli* (CE 1804)

of each of the three divisions comprising the *pargana* of Binayakpur. Letters Issued Register, GCR, *basta*, 16, vol. 4, October 1805, pp. 294-6, RSA.

51. See *arzi* from the Lal Ran Bahadur Sen, *diwan* of the Butwal raja, in Extract from the Report of John Routledge, Coll. of Gorakhpur on the settlement of Butwal, 14 December 1802, in Letters Issued Register, *GCR, basta* 16, vol. 2, December 1802-February 1804, RSA.
52. See A. Ross, Acting Coll. to BOR, 10 March 1804, Letters Issued Register, GCR, *basta* 16, vol. 3, RSA.
53. While the raja of Gulmi claimed that Matka was made up of 19 *tappa*s, the Palpa raja asserted that the number was 18. See List of *Tuppeh*s of *Purgana* Mutka, Received 14 March 1804, in Claim of the Nepaul Government to possess the Zamindary of Butaul, Boards Collections, F/4/185, IOR, APAC, BL.
54. See Translation of an *arzi* from Shakti Prachand Shah, the minor raja of Gulmi, 17 January 1804, Boards Collection, F/4/185, pp. 46-50, IOR, APAC, BL. See also Translation of the deposition of Semnarayan, *qanungo* of Bansi on 24 April 1804, Boards Collection, F/4/185, pp. 256-60, IOR, APAC, BL.
55. See testimony of Kanak Niddhi Tiwari, *vakil* of the Palpa raja, in A. Ross, Coll. of Gorakhpur to BOR (with enclosures) 9 June 1804, Letters Issued Register, GCR, *basta* 16, vol. 3, pp. 100-23, RSA. See also translation of a letter from Gorkhali Kaji Amar Singh Thapa to Col. Ochterlony, 1807 confirming this in *Papers Respecting the Nepaul Wars* (*PRNW*), 2 vols. (London: J.L. Cox and Sons, 1824), vol. 1, p. 19.
56. Cited in *RRS*, vol. 3 (April 1971), p. 79.
57. See 'Translation of a petition (*arzi*) from Lal Ran Bahadur Sen, brother of Prithvipal Sen, the raja of Butwal to the Gorakhpur Coll.', in J. Fombelle to BOR, January 1804, Letters Received Register (30 June 1803-17 February 1804), GCR, *basta* 1, vol. 2, RSA.
58. Ibid. The Gulmi raja's sister Rajrajeshwari Devi was the eldest wife of the Gorkhali Raja Ran Bahadur Shah. In 1799, she had accompanied Ran Bahadur to Benares but later returned to Kathmandu in 1802 to assume the reins of government as the Regent. For details see, D.R. Regmi, *Modern Nepal*, vol. 2, Chs. 1-4.
59. The grant was made by Raja Girbana Juddha Bikram Shah, the infant king of Gorkha on 26 Bhadau 1860 BS (CE 1803). See translation of *sanad* from Raja Girbana Juddha Bikram Shah of Gorkha to Raja Shakti Prachand Shah of Gulmi, Boards Collections, F/4/185, pp. 234-6, IOR, APAC, BL. The grant stated that the terai belonging to Baldyang (*garhi*) has always belonged to Gulmi and it is today being taken from Palpa to be restored to Gulmi. Cited in Rajaram Subedi, *Gulmiko Ithashik Jhalak* (Gulmi: Kiran Pustakalaya, 1998 [2055 BS]), p. 127.
60. For details see translation of royal order (*sanad*) from Raja Girbana Juddha Bikram Shah of Gorkha to Raja Shakti Prachand Shah of Gulmi,

pp. 234-6 See also A. Ross, Coll. of Gorakhpur to BOR, 2 April 1804, Procs. of the Sadar BOR, Fort William, 17 April 1804, no. 4, *UPSA*; Petition (*arzi*) from Shakti Prachand Shah, the Raja of Gulmi, 14 March 1804, Boards Collections, F/4/185, pp. 226-32, IOR, APAC, BL.

61. See 'Translation of a petition from Lal Ran Bahadur Sen, brother of Prithvipal Sen, the Raja of Butwal to the Gorakhpur Collector', in J. Fombelle to BOR, January 1804, Letters Received Register (30 June 1803-17 February 1804), GCR, *basta* 1, vol. 2, RSA. See also Rajaram Subedi, *Gulmiko Itihasik Jhalak*, pp. 57-8. For more details about the Gulmi-Butwal-Gorkha nexus see the following: J. Ahmuty, Magistrate of Gorakhpur to J. Fombelle, Secy. to Govt. in the Department of the Ceded Provinces, 5 April 1804, Procs. of the Sadr Board of Revenue, Fort William, 10 April 1804, no. 25, UPSA; translation of an *arzi* from Shitab Rai, *vakil* of Lal Ran Bahadur Sen, manager for Prithvipal Sen, the Raja of Nepal, 26 December 1803, both in 'Claim of the Nepaul Government to possess the zamindarry of Butwal', Boards Collection, F/4/185, pp. 36-38, IOR, APAC, BL.
62. At least this was what the Palpali raja claimed. Nothing specific was mentioned regarding the extent of the lands in the plains that the Palpali raja was entitled to. See, translation of an *arzi* from Shakti Prachand Shah, Raja of Gulmi, 25 May 1804, Boards Collections, F/4/185, pp. 191-6, IOR, APAC, BL.
63. See BOR to A. Ross, Coll. of Gorakhpur, 20 August 1804, Letters Received Register (10 August 1804-2 April 1805), GCR, *basta* 1, vol. 4, RSA.
64. For a brief history of the Pyuthana kingdom, see Gitu Giri, *Pyuthana Rajyako Itihasik Jhalak* (Pyuthana: Jilla Vikas Samiti, 1995 [2052 BS]).
65. D.R. Regmi, *Modern Nepal*, vol. 1, p. 324; Gitu Giri, *Pyuthana Rajyako*, p. 20.
66. The officials deputed were Jasoghar Pant, Dashrath Tiyari and Dalel Singh. They were deputed to survey all the lands in Pyuthana. See Gitu Giri, *Pyuthana Rajyako*, p. 49.
67. D.R. Regmi, *Modern Nepal*, vol. 1, p. 11.
68. See *Calendar of Persian Correspondence* (New Delhi: National Archives of India, 1949), vol. 9, nos. 653, 1737.
69. Ibid.
70. See Governor-General-in-Council's Political Letter to the Court of Directors, 4 August 1791, paragraphs 68-79, in *Fort William India-House Correspondence, 1787-1791*, vol. 16 (Delhi: National Archives of India), p. 397.
71. For details see Boards Collections, F/4/422, no. 10381, p. 426, IOR, APAC, BL; BOR to A. Ross, Coll. of Gorakhpur, 20 August 1804, Letters Received Register (10 August 1804-2 April 1805), GCR, *basta* 1, vol. 4, RSA. The outstanding balances due from Sheoraj rose from Rs. 2,656 in 1801-2 to Rs. 7,650 in 1803. For details on the above see the following,

A. Ross, Coll. of Gorakhpur to BOR (with enclosures), 9 June 1804, Letters Issued Register, GCR, *basta* 16, vol. 3, pp. 100-23, RSA; A. Ross, Coll. Gorakhpur to BOR, 8 November 1804, in *Procs. of the BOR*, Forth William, 20 November 1804, no. 10, *UPSA*; BOR to Coll. of Gorakhpur (Extract from the Procs. of the Secret Department), 23 October 1806, GCR, *basta* 1, vol. 7, RSA.

72. See for instance the following. Order to local functionaries in Sheoraj regarding reclamation of waste lands, AD 1797, RRC, vol. 25, p. 346; Jhungha Chaudhari, Mansukha Chaudhari and Santhokhi Chaudhari asked to make the *bandobast* for the *raiyat* to cultivate in *pargana* Sheoraj. One Jhungha Chaudhari was also given the *kalabanjar mauza* of Maladeva (?) on a *jagir* basis, AD 1800, RRC, vol. 24, p. 32; Tax assessment rate applicable to the four *varnas* and 36 *jat* in *pargana* Sheoraj, *zilla* Pyuthana, AD 1800, RRC, vol. 24, p. 31; Order regarding new villages settled by Kaptan (Captain) Chandra Bir Kunwar on *bekh buniyad* lands in Sheoraj, AD 1812, *RRC*, vol. 40, p. 341; Chaudhari and Laskari Chaudhari are appointed *chaudharis* for the whole of Sheoraj. They were granted one *kalabanjar mauza* Shankarpur (?) in *praganna* Sheoraj as *jagir*, AD 1800, *RRC*, vol. 24, p. 32; Birya Rokaya deputed to Sheoraj along with troops under his jurisdiction, AD 1797, RRC, vol. 25, p. 639; *subedar* Parsu Ram transferred from Sheoraj to Gorkha to construct embankments on *sera* (crown) lands, AD 1797, RRC, vol. 25, p. 639.
73. *Chaudhrai* grant made to Jas Raj Chaudhari, from *pargana* Deokhori-Sheoraj, AD 1796, *poka* 9, no. 8, sr. no. 919, *Lagat Phant* (Archives of the Department of Land Revenue in the Ministry of Finance, Government of Nepal), Kathmandu.
74. See letter of J. Ahmuty, Coll. of Gorakhpur to N.B. Edmonstone, Secy. to Govt., FS Consl. 16 January 1806, no. 106, para 2, NAI.
75. See letter of Alexander Ross, Coll. of Gorakhpur to Charles Buller Secy., BOR, 12 February 1805 in Letters Issued Register, GCR, *basta* 16, vol. 3, RSA; and ibid. in *Proc. BOR for Ceded and Conquered Provinces*, Procs. 31 July 1810, no. 31A, *UPSA*.
76. Jurawan Chaudhari held 39 villages. The Gorkhali officials were Faujdars Maniraj Bhaju, Tarapeet Upadhyaya, Jemadar Lashkari Chaudhari, and Mutsaddi Shiv Baksh. At least two Rajput zamindars were also involved in these disputes, one Hanumant Singh and another Pahalwan Singh. My information is derived from the following sources: J. Grant, Coll. of Gorakhpur to Board of Commissioners, 22 September 1811, Revenue Letters Issued Register, GCR, *basta* 17, vol. 18, RSA; D. Scott, Acting Magistrate of Gorakhpur to G. Dowdeswell, Secretary to Government, 19 November 1811, Letters Issued Register, GCR, *basta* 25, vol. 164, pp. 100-6, RSA; FP Consl. 17 January 1812, no. 46, NAI; See also Letter from J. Carter, acting coll. to Board of Commissioners,

22 January 1819, Letters Issued Register, GCR, *basta* 17, vol. 122, pp. 144-8, RSA.

77. Maniraj is also known to have possessed certain villages in the Butwal *terai* on a *sir* tenure (those lands cultivated by a landholder or village zamindar directly or with hired labour). See translation of a report from Mirza Hussain Ali Beg, tehsildar of pargana Tilpur, Benaiyakpur, 30 October 1814, cited in *PRNW*, vol. 1, p. 177.
78. D. Scott, acting magistrate to George Dowdeswell, 16 January 1812, Letters Issued Register, GCR, *basta* 25, vol. 164, pp. 149-51, RSA.
79. Some lands in Sheoraj were given to one Kunwar *sardar* (Chandra Bir Kunwar? Bal Narsingh Kunwar? But probably the former). See Ganapat Upadhyaya to Ujir Singh Thapa, Phalgun 9, no year, in Shankarman Rajbamshi (ed.), *Shahkalin Chittipatra Samgrah*, vol. 2, pp. 79-85.
80. Translation of a letter from *budakaji* Amvar Simha Thapa to the raja of Gorkha, 12 April 1815, FS Procs. 16 May 1815, no. 61, NAI.
81. Kaji Amar Singh Thapa to Bhim Sen Thapa and Ran Dhwaj Thapa, April 1813, in Shankarman Rajbamshi (ed.), *Shahkalin Chittipatra*, pp. 6-8.
82. The actual term used is *mamilat ma yeti yeti paisa diyako ho.* Ibid., p. 8. It probably means *muamalaat* or pertaining to official affairs. The sense of the statement probably is that Sheoraj was included in the management of government affairs.
83. For similar confusion concerning the exchange of gifts along the Anglo-Gorkha frontier see Bernard A. Michael, 'When Soldiers and Statesmen Meet: "Ethnographic Moments" on the Frontiers of Empire', in Stewart Gordon (ed.), *Robes of Honour: Khil'at in Pre-Colonial and Colonial India* (New Delhi: Oxford University Press, 2003), pp. 80-94.
84. *Bir Pustakalaya Ithihasik Chittipatra Samgraha*, no. 400, NAN.
85. See Raja of Nepaul to Governor-General, 4 May 1814 & 3 June 1814 in FS Consl. 23 June 1814, nos. 22 and 23, NAI.
86. After the Anglo-Gorkha War, in 1816, a claim to the *tappa* of Sheoraj was put forward by one Soobah Lal Sahye, a relative of the raja of Pyuthana. His claims were denied after it was decided to return Sheoraj to the Gorkhalis. The Palpali royal family headed by Rattan Sen, and now resident in Gorakhpur, protested this handover. See Petition of Soobah Lal Sahye, 9 January 1818, Letters Issued Register, GCR, *basta* 17, vol. 122, pp. 189-90, RSA; and Boards Collections, F/4/550, no. 13378, 21 pp, IOR, APAC, BL.

CHAPTER 9

The Utility of Military Power in British Indian Imperialism, 1838-1842

FRANK H. WALLIS

INTRODUCTION

The question of who is fit to rule an empire is less important than the question of how to acquire, conquer, or build one. In the nineteenth century, achievement of the latter absolutely required the use of military power. This brief essay outlines the way in which military deployment sustained Britain's imperial policy in South Asia, and suggests how the military was a component of imperialism.[1]

British imperial expansion to the north-west of India in 1838 followed a dream of potential conflict with Russia beyond the periphery of the extant empire in the subcontinent. It did not matter if the threat from Russia was actual or fantastical: British proconsuls behaved as though it was real.[2] The odd result was an enormous increase in the size of British holdings in the subcontinent from 1838 to 1849, involving Afghanistan, Sind, and Punjab. The immediate pretext for British invasion was the restoration of Shah Shuja, the deposed ruler of Afghanistan, who obtained his decades-old goal of regaining the throne of Kabul by the summer of 1839, courtesy of a British army.

Lord Auckland's expedition westward into Afghanistan beginning in 1838 was supposed to be a response to Russian ambitions in Central Asia. It had the effect of drawing British power far to the west of extant political boundaries of India. The Great Game in Central Asia between the British and Russian empires began after Waterloo in 1815 when the British tried to build a political frontier, or containment zone, reaching from Constantinople to Kabul. This Game melded into the Eastern

Question, or how the corrupt and decaying empire of the Turks could be propped up as a bulwark against the Russians. The connection between these two issues was the notion of preventing Central Asia from being converted into a lever with which to pry Britain out of power in Europe.[3]

A price would have to be paid for extending British power into Central Asia. Despite the Envoy William Macnaghten's cheerfully optimistic dispatches from Kabul even in October 1841, a general insurrection against the British puppet rulership of Shuja was underway. In November, the British garrison and political men were attacked and besieged in Kabul. Before the end of the year Macnaghten would be dead, killed by Akbar Khan. The Army of the Indus would be forced to retreat back to Peshawar in the first week of January 1842, harried at all points and extirpated at the hands of Afghan tribesmen.[4]

Lord Ellenborough, who replaced Auckland, tried to remedy the Kabul disaster with a punitive expedition in 1842, successful in its intent of massive destruction but productive of no lasting power or political achievement, symbolized by the capture of the gates of Somnath as a war trophy (whose provenance proved to be counterfeit). The military option intended by Auckland in 1838 to achieve a preemptive object against Russia had been a disaster without parallel in British history, but the loss proved something more important: that the alleged Russian threat was a fraud at best. For a generation after 1842 the Government of India (GOI) forgot about the Russian 'threat' on the distant periphery of the empire and concentrated instead on building up the British Empire on the subcontinent of Asia, which meant annexing Sind, Gwalior, Punjab, and Kashmir. Thus, in another sense, the expedition to the West was not so much a failure as an opportunity for territorial aggrandizement within India, a chance or excuse to consolidate and secure the 'natural' boundary on the Indus for the British Empire.[5] Meanwhile, the deposed Dost Mohammed returned to Kabul to rule Afghanistan.

THE CONQUEST OF SIND

The task of conquering Indus required armies and application of military power. Forcing the *amirs* of Sind to capitulate and suffer their country to be used as a launching pad for the

Afghanistan invasion became an obsession for Auckland as his Army of the Indus marched toward Central Asia. The British annexation of Sind is an example of the imperial trait of expansion based on presence. Auckland placed troops in Sind as part of his expedition to the westward, and Ellenborough expanded the concept relentlessly until the lower Indus came under the military control of the British government.[6] The 'native' rulers, the *amirs* of Upper and Lower Sind, knew what was happening and tried to make an adjustment in the face of foreign aggression, presence, and expansion, dealing with the British and meeting their incessant and growing demands until at last they stood up as men and resisted imperial encroachment.

Auckland's great plan for the expedition to the westward in the spring of 1838 had momentous consequences for the disposition of Sind under British control and eventual annexation into the empire.[7] The peripheral imperative of gaining stability and security on the Afghan frontier of India required two things of Auckland: (i) securing Ranjit Singh (ruler of the Punjab) to the GOI, and (ii) subordinating Sind to the GOI. The difference between these two policies was based on a prejudice of the British in favour of an old ally, the Maharaja of Lahore, over the congeries of 'native' princes inhabiting the Indus valley of Sind. In 1838, the GOI finally extracted at gunpoint a political treaty, meaning a British political agent resident in Hyderabad (not to be confused with the capital of Nizam in Deccan) would be allowed to represent British government interests directly. Treaties of commerce, perpetual friendship and cooperation were the operative British techniques for thirty years. Auckland extorted further concessions from the *amirs*, such as making a tribute payment to support Shuja. Despite their cooperation in Auckland's expedition to the westward, the *amirs* were accused of plotting against British interests and of writing vague letters to the Persian Shah that may have expressed support for his cause in Herat (Afghanistan).[8] This was an absurdly thin pretext upon which permanent occupation of Sind was based. General Sir Charles Napier was detailed to disarm and subdue any resistance to British interests in Sind, and this intentionally ignorant man carried out orders to perfection, forcing the *amirs* into an aggressive stance until they opposed the foreigner with an army, inevitably destroyed by the troops of Napier in 1843.

The Battle of Miani on 18 February 1843 put an end to the political independence of Sind. Napier's forces lost 62 officers and enlisted men, while Sindian casualties were estimated at 5,000. Lord Ellenborough, who assumed the proconsulship in early 1842, declared Napier to be Governor of the Province of Sind, with Hyderabad as its capital. Ellenborough, whose annexationist vision prompted the posting of Napier to Sind, knew the general would succeed in subduing the Indus valley and bring it under British control without having to endure years of negotiations at the hands of political officers who had botched imperialism in Afghanistan. It should be clear that Ellenborough's concept of imperial government was heavily weighted towards punishment. He wrote to the Secret Committee that the British position was now more advantageous because of bloodshed than due to pacific measures. War had been inevitable, so the GOI was fortunate in the timely treachery of the *amirs*, forcing the British-Indian Army to destroy them forever. British forbearance was tried for three months under Napier, and the general did his best to control the 'barbarians' short of making war, and the events of that period gave a 'peculiar glory' to Napier, a fame which always attended success in war, '. . . obtained in the prosecution of measures which had for their object the preservation of peace'.[9]

For James Outram, his adjutant in Sind, Napier began his ascendancy in Sind with a peremptory tone toward the *amirs* at their first meeting in September 1842, reading aloud to them a prepared statement which omitted the traditional complimentary preamble so much desired in traditional South Asian communications, and simply spoke to them in an offensive dictatorial tone. Second, Napier's menacing attitude and constant military parades and preparations caused anxiety among the *amirs*, impelling them to adopt self-defense, which by itself was a ground for invasion, at least in the mind of Napier. This would have been a violation of Ellenborough's policy of mollifying the *amirs*, and in fact the governor-general took a softer line toward them than Napier. But the general often hid Ellenborough's moderate policy recommendations and replaced them with more severe ones. Third, Napier invaded Sind in December 1842 and removed Rustam Khan and replaced him with Ali Murad in Upper Sind. Fourth, Napier withheld critical documents from Ellenborough, failing to transmit copies of Outram's notes on conferences with the Hyderabad *amirs* (8-12 February 1843), letters from the *amirs* to Napier protesting their treatment, and Outram's views in their favour, to the governor-general, thus securing a decision against them on *ex parte* evidence. Even when news of their existence surfaced in London, Napier stalled and failed to supply all of them. Fifth, Napier accused all the *amirs* of attacking Outram, when it was only one, Mir Shadad of Hyderabad. Sixth, Napier induced the *amirs* at the Battle of Miani to surrender on the condition that those leaders not at the scene of combat would be well treated. But these *amirs*, along with their families and male relatives, were arrested and transported to India along with those *amirs* who actively resisted the British. Seventh, Napier changed the way of British governance in the empire, from moral force to physical force. Outram thought British interests in the East required consolidation and not extension. Napier's 'unjust war of aggression' was a crime that had unfortunately been attributed in the public mind to Ellenborough.[10]

Outram was too kind to Ellenborough, whose annexationist impulse prompted the posting of Napier to Sind. He knew the general would succeed in subduing the Indus valley and bring it within the British orbit, without having to endure years of negotiations at the hands of the politicals. Ellenborough was just as paranoid about the intent of the *amirs* as Napier, who wrote in his journal in October 1842 that 'We have no right to seize Sinde; yet we shall do so, and a very advantageous, useful, humane piece of rascality it will be.'[11]

Despite fulminations in the imperial metropole about Ellenborough's territorial aggrandizement and for exceeding instructions, the annexation of Sind was allowed to stand.[12] For Ellenborough, the annexation of Sind was an acquisition of the highest value, shortening the lines of communication between England and the main body of the army in the North-West Frontier, cutting off the Punjab from European succour, and affording a large surplus land revenue for the GOI.

THE ABSORPTION OF PUNJAB

The Tripartite Treaty of 1838 brought the Punjab and GOI together on the side of Shuja's restoration in Kabul, and the Lahore state followed the decision of Ranjit Singh's to support the British in their invasion of Afghanistan. Auckland displayed unusual solicitation for Ranjit Singh's position when he agreed to take an invasion route *around* Punjab and not through it, leaving Sind to pick-up the tab for that honour. The British were fortunate in dealing with Ranjit Singh, because when he died in 1839 the successor Lahore government experienced continual upheaval as rival factions fought for supremacy, resulting in a string of assassinations. To the British the state of affairs in Punjab appeared as 'near anarchy'.

Within days of his nomination as governor-general, Ellenborough called upon Lieutenant Henry Marion Durand in October 1841 to develop a plan for the invasion of Punjab and consulted with Wellington about it, who advised him in April 1842 the best method for transporting troops across the Sutlej would be with pontoon bridges. The Secret Committee advised Ellenborough in April 1843 not to conduct military operations beyond the Sutlej River but if Lahore was hostile then the new frontier of the British Empire would have to be the summit of the Himalayas. In India, Ellenborough's first responsibility was to shore up the British position in the north-west, and we have already seen his policy and measures enacted with regard to Afghanistan and Sind, the former being evacuated and the latter being annexed partly as an invasion base for the conquest of Punjab. After the September 1843 assassinations, Ellenborough informed London that he could detect no Sikh hostility against the GOI, and in October informed Wellington that Punjab would inevitably fall to the British, and that the new prime minister, Hira Singh, the son of Dhian Singh, seemed to be pro-British. The governor-general had a notion that developed into policy—the lowlands of Punjab must be under the control of Hira and the British, while the hills and mountains of Kashmir and Jammu far to the north would be bestowed on the Dogra chief Gulab Singh, in a sort of dyarchy splitting up the power of the Khalsa kingdom in Punjab.[13]

Ellenborough confided to the Duke of Wellington and to the Secret Committee that an invasion of Punjab was to take place in late 1845, and that he had taken measures to prepare for this event. The imperial authorities in India and London assumed that Punjab would be absorbed into the empire at some point, and 1845 seemed the opportune time, based on the convenient pretext of incessant political assassinations in Lahore, and the Khalsa army showing signs of republican self-rule. Sir Henry Hardinge, who replaced Ellenborough in July 1844, was not averse to the conquest of Punjab, but neither was he very eager to annex it. He preferred the frugal technique of making the 'natives' manage their own with partial submission to the *imperium.* He wanted Punjab to operate as a convenient buffer state against putative Muslim invasion from the west. For Hardinge, the trigger for conquest was British anxiety about the behaviour of the Sikh army. The example of the Khalsa troops extorting concessions from the Lahore *Durbar* must not be repeated in British territory. It was better to preempt the Sikhs and thereby prevent mutiny in British-India. Yet, one wonders whether the British would have been satisfied with a strong Punjab in control of a well-disciplined army. Some other pretext would then have to be invented. Similarly, a weak state with a weak army would do the British Empire no good as a buffer state. It would have to be absorbed.[14]

By December 1845, on schedule, Hardinge had amassed a powerful army on the Sutlej, waiting for a signal to invade Punjab. Alarmed, the Sikh state moved an army into its own territory on 8 December on the left bank of the Sutlej, not on British controlled land, but Hardinge ordered his own troops to advance on 11 December, despite having no intelligence of Sikh troops advancing upon British positions in British territory.[15] According to British sources, Rani Jindan embarrassed her generals into marching against the British partly as a question of honour, but more likely as a tactic agreed among her coterie to rid themselves of a competitor for state power, the Sikh Army, by sending it towards an inevitable defeat. At the battles of Mudki and Ferozeshah in late December, and at the Battle of Sobraon in February 1846, the British defeated the Sikhs and imposed terms. Hardinge became protector of the boy Maharaja Dulip Singh; a British garrison was established in Lahore; the Sikhs had to pay a large fine; the army was reduced by two-thirds; British troops were allowed to go anywhere in Punjab at will, and occupy any place; one-third of the Khalsa Kingdom was annexed to the British Indian Empire; the Rani was removed from political power. The British resident became *de facto* ruler of Punjab and guardian of the young Dulip Singh. Collaborators such as Gulab

Singh were richly rewarded. The British spoke of restoring a strong government in Lahore, but their actions proved contradictory and left Punjab crippled and dismembered. The British called this the restoration of friendship.[16]

The home authorities, including Sir Robert Peel, the prime minister, were amazed at British forbearance in the Punjab.[17] In August 1847 Hardinge reported that the reform of 'barbarism' was well under way, and that the land tax system was being reformed (changed). But, spies and British agents reported growing signs of disaffection. With the arrival of Lord Dalhousie in January 1848, the GOI became even more aggressive toward Punjab, and the new proconsul was a vigorous and committed zealot for imperial expansion. One would be hard pressed to name a governor-general more enthusiastic for imperialism. In April 1848, an incident of anti-British resistance at Multan—the killing of two British agents—fuelled events which resulted in the complete annexation of Punjab. From a nation reconciled to British domination, Punjab was now viewed as a vast conspiracy against the English, in the space of a week's time. Both the agent in Lahore, Frederick Currie, and Lord Dalhousie in Calcutta believed this was the symptom and beginning of a general rebellion, and they decided to wait several months before launching a mission to retake Multan. Their objective was to give enough time to the rebels to let the insurrection spread throughout Punjab.

Swiftly, Currie set about discovering anti-British plots, executing Sikh leaders, banishing the Rani to Hindustan, while Dalhousie assembled an army of punishment to teach all of India about British forbearance.[18] He believed a strong and friendly Punjab on the western periphery of the British Indian Empire was a hopeless prospect: Sikh perfidy was inbred, treachery part of their national character. By August 1848, Dalhousie had decided that annexation was the only option to preserve imperial stability on the western frontier. He expressed joy and elation at the news of rebellion in Hazara province far to the north, because a distinct anti-British leadership was presenting itself in the open, ready to be exterminated. The president of the India Board, John Cam Hobhouse wrote on 24 November that HMG decided Punjab must be reconquered and occupied, but assumed no final decision about permanent administration would be made until the rebellion was crushed. Annexation was regarded as an extreme measure, but they would endorse any choice Dalhousie deemed expedient.[19] The irony of the situation was apparent to Currie, who told Dalhousie that if the GOI was now at war with the Lahore state, secretly of course, then it put him in an odd and anomalous position, because Currie was the head of that state.[20] Hobhouse wrote to the proconsul that the Lahore government was being punished for the mistakes of the British. The *sirdars* were by treaty subordinate to the GOI, and it would be wrong to punish them as though they had been masters of their own country. By treaty they were not.[21] Technically speaking, British functionaries were as accountable as the Sikh *Durbar* for any mistakes that may have led to the rebellion. Still, the situation called for military solution, and Dalhousie was not one to let ironical arguments dissuade him from imperial conquest.

Sikh independence was finally extinguished at the battles of Chillianwalla and Gujerat in January and February 1849. The terms given to Dulip to sign on 29 March were not really necessary. Dalhousie was going to enact British sovereignty over Punjab whether Dulip and key *sirdars* affixed their signatures or not.[22] The annexation was irrevocable. London expressed surprise over this news, but chose to ignore Dalhousie's peremptory decision and heartily endorsed the policy, and the proconsul assured his home audience that annexation was inevitable and necessary: reversal would accomplish nothing, for in a few years the GOI would have to annex Punjab anyway.

THE DYNAMICS OF BRITISH IMPERIALISM

Perceptions of lack of stability on the periphery allowed the imperial mind to accept an aggressive policy, to advance imperial presence beyond current borders into what used to be considered frontier lands. The imperial frontier obtained an almost mysterious character in the estimation of imperial decision makers, advancing as if by inevitable momentum. In Afghanistan, Auckland focused his policy on setting up his puppet ruler Shuja in Kabul to steady the imperial periphery and bring Maharaja Ranjit Singh of Punjab into the imperial orbit. At the same time he sought to complete the isolation of Punjab with a project of interfering in Sind, establishing a British resident and basing thousands of troops there for the additional purpose of maintaining secure lines of communication for a vastly expanded imperial boundary. The Indus beckoned to British imperialists. They were already in central India. Why not advance the boundary a bit, say 1,200 miles west and north? With new responsibilities in Kabul, Punjab was too dangerous and restive to leave unmolested in the rear of British forward positions on the extreme edge of the periphery. Presence implied expansion, especially if there was a threat on the periphery, and the expedition westward was nothing if not an elucidation of the principle.

The irony is that events proved the British position to be based on fraud. Foreign Minister Palmerston had been in communication with his Russian counterpart in late October 1838 and informed the Cabinet that he was satisfied Russia did not harbour any hostile designs to the interests of the British in India.[23] It was in the same month that Auckland publicly announced his invasion of Afghanistan, a course he refused to alter or rescind even when news came that the Persians had abandoned their attack upon Herat. Before the disaster in 1841-2, the imperialists were discussing ways and means of expanding further into Central Asia, perhaps to the Oxus and an imaginary border with the Russian Empire.

The presence of British troops in Sind implied submission to the foreign imperialist. It was deemed a waste of time to negotiate among the various *amirs* who ruled the country without the presence of British troops nearby. A reserve force would need to be diverted to Lower Sind, as far as the port of Karachi, because it was believed absolutely necessary for the British forces to have secure communications from the sea.[24] But with the partial occupation of Sind in 1838, the path had been cleared for annexation under Napier, a self-confessed rascal. Despite Auckland's proclamation at Simla in October disavowing territorial aggrandizement, the imperative of presence/expansion worked an amazing revolution in policy. British presence annoyed the *amirs*, and the British used this to provoke hostile reactions that would supply a pretext for outright annexation of Sind in 1843.

The British annexation of Sind is an example of the imperial trait of expansion based on presence. Auckland placed GOI's troops in Sind as part of his expedition to the westward. Initially, the Lower Indus came under the military control of the GOI.

First, came the treaty of protection, submission, and subordinate cooperation, yielding up partial sovereignty to the British. The capitulation of the leading *amir* of Upper Sind, Rustam Khan, took place on 25 December 1838 as he ceded Bukkur to the GOI and placed Upper Sind under British protection.[25] In early 1839 the *amirs* of Hyderabad (Lower Sind) were forced to cede the entire right bank of the Indus all the way to Baluchistan, along with Karachi, Sukkur, and Tatta, giving the British 'perfect supremacy' over Sind, as if the region was going to be ruled directly by the British. The British view was that Sind formed a natural and integral portion of India, and the *amirs* had no choice but to bind their interests permanently with those of the GOI. In Auckland's view the *amirs* accepted British administration of Sind from the moment British troops entered their country, an instantaneous expansion of British imperialism through the mere presence of British soldiers and sepoys.[26] The *amirs* were made to understand that their own behaviour had caused the GOI to station troops at fixed points throughout their country. Petty grievances such as an *amir*'s vague letters passed to the Shah of Persia would harden into causes of war later on, made convenient pretexts for military aggression against a harmless power on the periphery of the British Indian Empire.

As London devolved upon Ellenborough the authority to make an empire in India with minimal interference from the distant home authorities, so did Ellenborough devolve power to Napier to crush an independent Asian state far to the westward, bringing it within the imperial orbit.[27] The British government had scarcely obtained a footing in Sind before an obsolete claim of 21 lakh on behalf of Shuja was pressed on the *amirs*, in addition to free trade on the Indus. Napier supported the venal and treacherous Ali Murad, almost from the time the general entered Sind in September 1842 at the head of an army, replacing the British resident, ostensibly to negotiate more land and economic concessions from the *amirs*, some of whom knew what this intrusion meant, and who then offered almost symbolic resistance to foreign interference.[28] The Battle of Miani on 18 February 1843 put an end to the political independence of Sind, and Napier was named governor, with Hyderabad as the capital. Ellenborough was denounced in the metropole for annexing Sind, but this latest expansion of empire was never countermanded.[29] Napier's presence on the Indus implied expansion of the British-Indian Empire beyond the periphery to all of Sind.

Annexation of the Punjab was the crowning glory of the expedition to the westward, symbolized forcefully in the purloined Koh-i-Nur diamond being handed over to the Queen as a fitting addition to the Crown jewellery collection. Expansion of the British into Punjab was implied through the magical growth of army cantonments moving steadily westward along the Sutlej River beginning in the 1830s. British presence on the frontier with Punjab acquired a self-fulfilling rationale for continual advancement, and Punjabi nationalism would be stifled in the embrace of British 'friendship' under Dalhousie. But first, Lord Hardinge, who wrote that 'A military rule is congenial to barbarians . . . ,'[30] presented himself at the head of an invasion army at Ferozepore in 1845, a recent British acquisition on the Sutlej. The proconsul would have liked to take his army all the way to Peshawar but disavowed complete territorial aggrandizement in favour of the more financially appealing solution to the problem of reconciling Sikh government and British imperial frontier imperatives through self-managed submission.[31] Hence, three treaties of 1846 extorted via conquest gradually but deliberately placed the heavy stamp of British legality on Punjabi submission to foreign presence and domination. London and Calcutta congratulated themselves on their purported lack of zeal for expanding just for the sake of gaining new territory for the British Indian Empire, but nevertheless took one-third of the Punjab for themselves (the Jullundar *doab*), and gave another third (Kashmir) to Gulab Singh.[32] Hardinge's halfway annexation was insufficient for the purpose of British imperialism, which demanded a perfect periphery, a situation unobtainable in the real world.

The British presence in Punjab following the invasion of 1845-6 implicated the inevitability in expansion two years later under Lord Dalhousie as that governor-general waited with scientific forbearance for his opportunity to stifle remaining Sikh independence.[33] Resistance to British presence and domination was something the British agents knew to be tangible but not yet manifest in concrete action until an incident at Multan in 1848. Official reports were uniform in expounding the wonder-working influence of the new British residency system in Lahore upon all Punjabis, bringing happiness to a benighted people, but the underlying cynicism of the imperialist concerning the fidelity of 'natives' remained the primary belief at the top of the imperial chain of command. Dalhousie believed the Sikhs hated British presence and both sides would eventually reveal their true intentions. With the killing of two British agents in Multan in 1848, the governor-general had his pretext for ending the Lahore state—paradoxically a British controlled state since the famous victories of Hardinge—and pushing the official political border of British India to Afghanistan. Dalhousie thought he had *carte blanche* to do as he pleased with Punjab. He believed Sikhs could never really entertain feelings of friendship for the British, not after their humiliation in 1845. It was pointless to continue funding this experiment in a British resident operating in Lahore through a putative Sikh bureaucracy.[34] Dalhousie believed the Sikhs had to be saved from their own destructive impulses.[35] There was some resistance in London, but John Hobhouse, president of the India Board, assured Dalhousie that if he felt compelled by the urgency of the case to annex without instruction from the home authorities, then HMG would put the most favourable construction on his actions.[36] Hobhouse confided to Dalhousie that the fidelity of barbarians depended on the character of their civilized masters: 'I have not as yet seen any distinct proof that the white man has lost any of his superiority over the dark race . . . these Asiatics expect to be led. . . .'[37] The military humiliation of the Sikhs was completed with the battles of Chillianwalla on 13 January 1849, and Gujerat on 21 February 1849. For Dalhousie, the British arms needed restoration in the public eye, to reverse the 'sinister rumours that our star was waning'. Therefore, it would have been both cowardly and mad not to annex the Punjab.[38]

But, any government puts a spin on aggrandizement with the excuse of national defense and the result is the same: an independent nation is conquered and their land taken over by foreigners. The British imperialists in London and Calcutta believed they had accomplished a great thing in at long last settling the buffer area between India and Central Asia, but all they had achieved was moving the goal posts further west, with the intervening playing field now under direct British management instead of under client operation on their behalf, as happened in the time of Ranjit Singh.

CONCLUSION

The British Indian empire did not happen by accident or inadvertence. Enough has been said about successive governors-general to illustrate a policy truth on empire building at this time: precise measures may have been contingent, but the intent was always the same, and both the metropolitan and proconsular agents knew what the end game was. The presence of an aggressive set of white foreigners in South Asia implied most forcefully the intent to dominate the 'native' population, for pecuniary and political objectives. Perceived threats on the frontier acted as a magnet, sucking out British power from the centre to the periphery, inciting policy makers to advance thousands of troops to push the sway of domination further west. Hobhouse, Ripon, Ellenborough, and Dalhousie were the most forceful exponents for moving the official boundary of British Indian empire further west and north in the 1830s and 1840s, establishing not just a strong buffer on the periphery, but bringing this vast territory under direct British control. This move was not the product of central planning, but of central imperial thinking about what was right for the British Indian empire.

Leaders sometimes act on ideology and attitudes, i.e. beliefs.[39] However, both human agency and structure must be accounted for in explanations of imperialism, because there must be a reason for military action. Ideas, beliefs, values, interests, norms all need to be analysed, because they are part of the decision-making system. The British expedition westward from 1838 to 1849 was the result of human decision-making based on the perception of danger to the periphery. The policy makers were a microscopic minority stationed at London and Calcutta. These men worked very hard to achieve domination of South Asia, from Calcutta to Kabul and from Ceylon to Karachi. The factors that motivated their policy options involved: (i) the notion that presence implied expansion; (ii) a racist attitude of contempt for the 'native' peoples; (iii) a belief in the mission of civilization against barbarism; (iv) showing resolution and toughness to the 'natives'; (v) punishing on the periphery for those who failed to comply with collaborationist demands; (vi) military aggression masked as self-defense; and (vii) the assumption that 'natives' were incapable of self-rule.[40]

A triad of necessary conditions for expansion of empire has been postulated by Alexander Motyl, and this essay has suggested some examples from the case of British India in the early years of Queen Victoria. First, without an expanding economic base an empire cannot pay for its administration and defense, hence the need for continual resource flows within imperial channels. The stability of an empire depends partially on the efficiency of the core/periphery channel, vital in conveying resource flows.[41] This process of resource allocation and acquisition has been noted in the British diplomatic missions to the 'native' princes along the Indus, whose apparent goal was an increase in British commerce, and the governor-generals, especially Dalhousie, and their obsession with securing land revenue in newly acquired territory, to pay for military occupation and future operations and campaigns. In addition, to make sure that peripheral elites cannot band together to resist the core and disrupt resource flows to the metropole, was another factor that troubled the imperialists. This condition was of paramount concern to all British rulers in India, achieved through bribes, co-opting, and punishment of 'native' elites. Finally, the peripheral elites competing with each other for the favour of the metropole ensured their continued dependence on and submission to the imperial power. This condition was achieved via military pressure forcing numerous treaties of friendship and subordinate cooperation on the leaders of Sind and Punjab, spelling out the degree of their submission and collaboration.[42]

NOTES

1. My ideas on imperialism have come from political science. The most influential on my thinking begins with Alexander J. Motyl, *Imperial Ends: The Decay, Collapse, and Revival of Empires* (New York: Columbia University Press, 2001), who explores core/periphery issues in depth. Motyl based his theory on John Galtung's, 'A Structural Theory of Imperialism', *Journal of Peace Research,* vol. 8 (1971), pp. 81-117. He also saw a similarity between imperial decay and totalitarian decay, for

example in Karl Deutsche, 'Cracks in the Monolith', in H. Eckstein and David Alper (eds.), *Comparative Politics* (New York: Free Press, 1963), pp. 497-508. I also recommend Jack Snyder, *Myths of Empire: Domestic Politics and International Ambitions* (Ithaca, NY: Cornell University Press, 1991) and Charles A. Kupchan, *The Vulnerability of Empire* (Ithaca, NY: Cornell University Press, 1994). Tony Ballantyne in *Orientalism and Race: Aryanism in the British Empire* (New York: Palgrave, 2002), took issue with the concept of core/periphery because it 'privileged' the metropole over the 'subaltern' (natives). He preferred to see empire as a web of interrelated indigenous cultures which had little to do with London (pp. 11, 14), but without the metropole, how can there have been a British empire? This interpretation is close to saying there was no such thing as a British empire. Of special interest, see Anthony Pagden, *Lords of all the World: Ideologies of Empire in Spain, Britain and France c. 1500-c.1800* (New Haven, CT: Yale University Press, 1995); George Lichtheim, *Imperialism* (New York: Praeger, 1971); Wolfgang J. Mommsen, *Theories of Imperialism*, pb. edn., originally published as *Imperialismus theorien* [Göttingen, 1977] (Chicago: University of Chicago Press, 1980; 1982); Michael W. Doyle, *Empires* (Ithaca: Cornell University Press, 1986); Charles S. Maier, 'Imperial Limits', in Andrew J. Bacevich (ed.), *The Imperial Tense: Prospects and Problems of American Empire* (Chicago: Ivan R. Dee, 2003).

2. See, for example, Edward Ingram, *The Beginning of the Great Game in Asia, 1828-1834* (New York: Oxford University Press, 1979), pp. x, 7, 328-30; Malcom E. Yapp, *Strategies of British India: Britain, Iran, and Afghanistan, 1798-1850* (Oxford: Oxford University Press, 1980); Peter Hopkirk, *The Great Game* (New York: Kodansha, 1990), p. 2.
3. Some classic postulates begin with Alfred Lyall, *The Rise and Expansion of the British Dominon in India*, 3rd edn. (London: John Murray, 1894). Also, Michael H. Fisher (ed.), *Politics of the British Annexation of India, 1757-1857* (New Delhi: Oxford University Press, 1993), pp. 21, 23.
4. J.A. Norris, *The First Afghan War, 1838-1842* (Cambridge: Cambridge University Press, 1967); Donald Sydney Richards, *The Savage Frontier: A History of the Anglo-Afghan Wars* (London: Macmillan, 1990).
5. Barbara N. Ramusack, *The Indian Princes and Their States [The New Cambridge History of India*, III.6] (Cambridge: Cambridge University Press, 2004), p. 62. She rightly sees British imperialism as a racist gambit. Ramusack also disagreed with the periodization of William Lee-Warner, *Protected Princes of India* (1899), who posited three distinct phases of intervention in native states from 1790 to 1890, in favour of a dynamic tension in which the GOI adapted intervention or non-intervention as the case merited, never entirely abandoning one or the other (pp. 56-8). Ramusack follows an Indian historian on this, K.M. Panikkar, *The Evolution of British Policy Towards Indian States, 1774-1858* (Calcutta:

S.K. Lahiri, 1929), who also pointed out that the GOI could and did idly watch 'native' princely misrule for years or decades, while other states were subject to interference almost on a whim (pp. 74-5, 88-9).

6. 'Copy of the Treaty with Ranjit Singh and Shah Shuja, Lahore (26 June 1838); and of the Treaty between Ranjit Singh and Shah Shuja (12 March 1834) . . .' [India Board, 1838], MSS EUR F213/97, 1-2. Hereafter, Tripartite Treaty, MSS EUR F213/97, India Office Records (IOR), British Library (BL), London.
7. Of the numerous works on the annexation of Sind, see Robert A. Huttenback, *British Relations with Sind, 1799-1843: Anatomy of Imperialism* (Berkeley, CA: University of California, 1962); Robert A. Huttenback, 'The Annexation of Sind', in Michael H. Fisher (ed.), *Politics of the British Annexation of India, 1757-1857* (New Delhi: Oxford University Press, 1993), pp. 224-48; H.T. Lambrick, *Sir Charles Napier and Sind* (Oxford: Clarendon Press, 1952); Kala Thairani, *British Political Missions to Sind . . . 1799 to 1843 . . .* (New Delhi: Orient Longman, 1973). Of note are books by the brother of the general who subjugated Sind, William F.P. Napier, *The Conquest of Scinde* (London: T&W Boone, 1845); and *The Life and Opinions of General Sir Charles James Napier*, 4 vols. (London: John Murray, 1857); and a ferocious rejoinder by his nemesis James Outram, *The Conquest of Scinde: A Commentary* (Edinburgh: Wm. Blackwood and Sons, 1846).
8. William Macnaghten to James Outram, 26 July 1838, 'Instructions', MSS EUR F213/46, nos. 3, 4, 8.
9. Napier to Ellenborough, 18 February 1843; Ellenborough's Notification of Victory, 5 March 1843; Ellenborough to S. Comm., 13 March 1843, *Parliamentary Papers* (*PP*), 1843, vol. 39, pp. 567, 570, 569; 571-2. Modern scholars may be startled at the disparity in casualties for this battle and several others in which British armies fought Indians. At least one social historian has noted that this was due less to superior technology than to more efficient social organization: Stephen Peter Rosen in *Societies and Military Power in India and its Armies* (Ithaca, NY: Cornell University Press, 1996), pp. 258-61, claims that the European armies had better discipline than the 'natives', due to the latter's closer connection with local society, reflecting ethnic and caste preoccupations and divisions.
10. James Outram, *The Conquest of Sinde: A Commentary* (Edinburgh: William Blackwood and Sons, 1846), pp. 535-47.
11. Napier, *Life of Napier*, vol. 2, p. 203.
12. 4 February 1843; 3 June 1843, MSS EUR F213/31, nos. 254, 261.
13. S. Comm. to Ellenborough, 3 April 1843, MSS EUR F213/31, no. 257; Ellenborough to Wellington, 20 September 1843, in Gonda Singh, *Private Correspondence Relating to the Anglo-Sikh Wars* (Amritsar: Sikh History Society, 1955), pp. 54, 57. Ellenborough to Wellington,

20 October 1843, in Jagmohan Mahajan, *Annexation of the Punjab: A Historical Revision* (1949, rpt., New Delhi: Spantech Publishers, 1990), p. 7. Gulab had been a supporter of Sher Singh who was assassinated, assisting in his accession to power in 1841 with thousands of hill tribesmen sent to overawe the Sikh Army in Lahore. When he withdrew from the capital city Gulab absconded with 16 wagons laden with Ranjit's treasure, and 500 horsemen (each of them carried away a bag of gold coins). Part of this loot would be handed over to the British in 1846 in return for Kashmir and Peshawar (p. 56).

14. Hardinge to Ripon, 23 January 1845, Ripon Papers, Add MSS 40871, 88-94, BL.
15. G. Singh, *Correspondence*, pp. 90-100.
16. Treaty of Lahore (9 March 1846); Treaty of Bhyrowal (16 December 1846).
17. Peel to Hardinge, 24 February 1846; 4 March 1846, Peel Papers, Add MSS 40475, 154; 160; Peel to Hardinge, 4 April 1846, Peel Papers, Add MSS 40475, 198, BL.
18. Dalhousie to Currie, 21; 25 December 1848, G. Singh, *Correspondence*, pp. 141, 142.
19. Hobhouse to Dalhousie, 24 November 1848, F213/27, 77, 78; Russell to Hobhouse, 20 November 1848, F213/12, 40, BL. When Dalhousie was penning letters to London about new victories in Punjab, Russell wrote '. . . that our Empire of India can be delivered from dangers of no ordinary kind' only through an application of Dalhousie's abilities, by a bold and fearless execution of powers entrusted to him by HMG. (See Russell to Dalhousie, 8 March 1849, F213/12, 85.) At the same time Wellington had no notion as to Dalhousie's plans for Punjab, but wrote him that no matter the principle on which he formed a settlement, it was the army which would sustain it. (See Wellington to Dalhousie, 5 March 1849, F213/12, 88, 89.)
20. Currie to Dalhousie, 12 October 1848, G. Singh, *Correspondence*, pp. 102, 104, 105, 107.
21. Hobhouse to Dalhousie, 24 November 1848, F213/27, 77, 78; Russell to Hobhouse, 20 November 1848, F213/12, 40.
22. Dalhousie to Hobhouse, 8 October 1848, BL, IOR, Broughton Collection, F213/23, 229-32.
23. Palmerston to Count Pozzo di Borgo, 20 December 1838, Broughton Papers, Add MSS 36469, 302-4, BL.
24. John Cam Hobhouse to Auckland, 4 February 1841, MSS EUR F213/8, 24; 48.
25. Alexander Burnes to Henry Torrens, 25 and 28 December 1838, *PP*, 1843, vol. 39, pp. 162-3, 167.
26. Auckland to Secret Committee, 13 March 1839, 10 February 1839, BL, IOR, MSS EUR F213/35, nos. 1 and 12.

27. Ellenborough to Secret Committee, 8 June 1842, *PP*, 1843, vol. XXXIX, p. 394.
28. Ellenborough to Napier, 4 November 1842, 14 November 1842; Napier to Ellenborough, 5 November 1842, *PP*, 1843, vol. XXXIX, pp. 495-6, 500, 507, 509.
29. Ellenborough to the Earl of Clare, 26 March 1843, quoted in Algernon Law, *India under Lord Ellenborough, March 1842 - June 1844* (London: John Murray, 1926), p. 64.
30. Hardinge to Currie, 5 May 1850, quoted in Gonda Singh (ed.), *Private Correspondence Relating to the Anglo-Sikh Wars*, p. 50.
31. Hardinge to Lord Ripon, 23 January 1845, Ripon Papers, Add MSS 40871, 88-94, BL.
32. Ripon to Peel, 10 March 1846; Ripon to Queen Victoria, 23 March 1846; Ripon to Hardinge, 23 March 1846, Ripon Papers, Add MSS 40875, 252; 324; 328.
33. Dalhousie to Couper, 10 May 1848, in J.G.A. Baird (ed.), *Private Letters of the Marquess of Dalhousie* (London: Blackwood and Sons, 1910), pp. 25-7; Dalhousie to Hobhouse, 4 May 1848; Dalhousie to Hogg, 29 May 1848, BL, IOR, F213/23, 45; F342/11, 8.
34. Dalhousie to Hobhouse, 15 August 1848, BL, IOR, Broughton Collection, F213/23, 170-1, 173, 174, 176.
35. Ibid., 182-3.
36. Hobhouse to Dalhousie, 23 October 1848, F213/27, 60-2.
37. Ibid., 64.
38. Dalhousie to Hobhouse, 15 June 1849; 14 July 1849, F213/24, 103-4; 21.
39. Motyl, *Imperial Ends*, pp. 31, 33, 36.
40. Forthcoming monograph from the author, Mellen, 2009.
41. Motyl, *Imperial Ends*, pp. 12, 48, 52.
42. Compare with Motyl, *Imperial Ends*, pp. 23, 89.

CHAPTER 10

'Crusade against Arms': 'Murder' and 'Disarmament' in Trans-Indus Districts of the Punjab, 1890-1900

GAGAN PREET SINGH

INTRODUCTION

The factors responsible for the successful colonial conquest by imperial powers are still widely debated. The emphasis has been on the role played by the technological superiority of colonial powers particularly in relation to their superior arms.[1] However, the majority of the works on the colonial conquest have failed to discuss the consolidation of colonial conquest which made colonial rule viable. This important end stage was the disarmament of the 'natives' particularly in those societies where arms were in wide circulation. It was not possible for colonial regimes to successfully control, rule and pacify such subject populations until they were denied access to their arms. In the beginning of British rule, the disarmament of the 'natives' was a strategic question. But, later on the possession of arms in the hands of the 'natives' was directly linked to the indices of violent crime by the colonial state. As a result of which disarmament was described as another measure of crime control by the colonial officials. Unfortunately, there has been a lack of studies on the colonial disarmament policies. The classic case of colonial disarmament is popularly understood as the disarmament of Hindustan after the upheaval of 1857-8. However, the first case of disarmament took place after the annexation of the province of Punjab by the British in 1849.

The colonial societies differed from the pre-colonial successor states in a very important manner, that is the circulation and possession of arms was closely regulated and controlled by the colonial state. This important aspect of colonial state has been accepted by generations of historians so naturally that it does not enter into dominant historiography as a colonial innovation to subjugate the colonized and maintain public order. But, no reference is made to the Arms Act and regulation of arms in society. The regulation of arms in society is taken for granted and as a result there is a general lack of studies around Arms Act as such. One of the objectives of this paper is to draw attention towards the importance of arms in society in general. A study of a particular region and its relationship with arms may help us in this regard. Such an important region was the province of Punjab during the nineteenth century. However, in this paper we are concerned only with six trans-Indus districts of the province of Punjab at the end of the nineteenth century.

The province of Punjab even after annexation had a very peculiar relationship with arms in general. As a result, a study about disarmament and of the Arms Act in relation to Punjab is required. This may be argued on the basis of the following factors. First, it was perhaps the province of Punjab where for the first time a massive disarmament was carried out after annexation. Considerable difficulties were faced by the colonial state while disarming the province of Punjab. For instance, in the Manjha region (Central Punjab) alone it was estimated that 30,000 arms were in the hands of the 'natives'. Sirdar Lehna Singh was appointed to carry out this disarmament by the board of control. He reported that 'many had gone into the British Provinces in hopes of procuring service there and some had gone to Jummo, others had burned their pieces to avoid detection'.[2]

Second, the exemptions under the Arms Act of 1878 for army personnel were unique to the province of Punjab. It was solely in the case of Punjab that such exemptions were provided and extended to all commissioned officers of the Native Army, pensioned or on active service, and all non-commissioned officers and men of the Native Army. They were exempted from having a license for their personal weapons and a 4 *anna* annual fee. However, the other 'martial-races' including even the Gurkhas were not exempted. For instance, the Gurkha officers and soldiers stationed in Dharamsala were required personal licenses and payment of a fee of 4 *anna*s annually to use a gun for sporting purposes.[3]

Third, the other state institution, i.e. the police, had a unique armed character. The whole of the force was drilled, disciplined and instructed in the use of fire-arms, and there was no distinction between an armed reserve and an unarmed contingent. This was not the practice in the other provinces where armed and unarmed units were separate.[4] Before 1875, policemen were enlisted for provincial, municipal and cantonment police separately and the provincial armed police only were drilled in the use of firearms. After 1875, the several bodies of police were merged into one force and every man enlisted for general duty was thoroughly instructed in the use of firearms and in company drill.[5] In Punjab, each policeman was required to fire 12 shots a year only at target practice with ball ammunition.[6] In contrast, police in other provinces was poorly trained in the use of firearms. For example, in the case of North-West Provinces, in some of the encounters with outlaws and dacoits, hundreds of rounds of ball cartridges were fired by the rank and file of the police and sometimes not even a single one was hit due to lack of training in firearms.[7]

Fourth, some of the areas were never disarmed like the trans-Indus frontier districts of the province of Punjab. It was chiefly with the disarmament of the trans-Indus districts of the province of Punjab that it was later organized as a separate province called the North-West Frontier Province. The disarmament of these districts was carried out in a partial manner at the end of the nineteenth century. While carrying out the disarmament of central Punjab and Hindustan, home-to-home searches were made and force was used. In the case of disarmament of these districts the situation was completely different. As discussed below, the state was so weak and helpless that it could not carry out general disarmament even in a one single village. One crucial difference was that the disarmament in Punjab and Hindustan was carried out after massive upheavals while in the case of these frontier districts the situation was completely different. As a result an altogether different strategy was adopted by the colonial state.

T.R. Moreman in one of his contributions has linked the tightening of regulation of arms in the frontiers during this period with uprisings and illicit arms trade across the frontier. During these risings in 1897-8, imperial losses exceeded those suffered during the Second Afghan War.[8] According to Moreman, the use of rifled 'arms of precision' against British expeditions during the

1897-8 campaigns gave the Indian Army serious cause for concern. These arms were secured by the enemy from British territory through illegal trade. The disarmament of these districts was an effort to stop the supply of 'arms of precision' to the enemy.[9]

This study will reveal that the actual cause behind the disarmament was not the unwanted supply of 'arms of precision' to the independent tribes through illegal trade. Rather, it was due to the 'murder of White military and civil officials' by weapons of assassins—pistols and knives. This is not to deny the fact that during hostilities with the independent tribes, possession of 'arms of precision' by the enemy were a serious concern of the British but during the peace time 'knives and pistols' were a more serious threat and a serious challenge to White presence in this region. Moreover, Moreman's work suffers from his over-reliance on official sources, particularly military ones. William K. Storey in an article notes that the history of technology cannot be written without mentioning the history of technological skill which exists at the intersection of the human and the material.[10]

This paper is divided into four sections. The first section will discuss the history of the Arms Act in general in relation to these districts. The second section discusses the history of murders in the frontier districts and criminality of the Pathans in general. In the third section an effort has been made to study the particular circumstances which appeared at the end of the nineteenth century and the final section deals with the actual disarmament measures carried out in these districts.

I

Punjab was annexed on 29 May 1849. Immediately after annexation, a Board of Administration was constituted having wide powers and unrestricted control over all matters pertaining to the Punjab. Apart from dealing with the Khalsa Army, 'native' aristocracy and law and order, the board undertook fairly extensive public works. The possession of gunpowder and warlike weapons of every description such as *zamburaks* (camel swivel guns), muskets, matchlocks, swords, spears, daggers and the like was prohibited to all classes within the Punjab, other than European subjects and government servants.[11] The district authorities were empowered to grant licenses after due verifications.[12] With the intention of recovering the arms of the disbanded soldiers of the Khalsa Army, Sirdar Lehna Singh was given the charge of Manjha, Sirdar Khan Singh Man was placed at Lahore, Sirdar Ram Singh Jelawallah in the *doab* between Ravi and Chenab and the Jhelum, and Sirdar Chutter Singh between Jhelum and the Attock.[13]

General disarmament was never carried out in the trans-Indus districts. As early as April 1849, the secretary of the Board of Administration argued that it was not expedient to disarm the population in Hazara and the trans-Indus districts though measures might 'prudently be adopted to deprive of their weapons, parties notoriously disaffected or likely to give trouble'.[14] Accordingly, in the beginning, the police force of the province was divided into two distinct bodies, the police of the Peshawar and Derajat Divisions, commonly called the trans-Indus Police (including the six districts of Hazara, Peshawar, Kohat, Bannu, Dera Ismail Khan and Dera Ghazi Khan) and the Cis-Indus Police, comprising the remaining 26 districts of the province of the Punjab.[15] The Peshawar Division contained the frontier districts of Hazara, Peshawar and Kohat. The Derajat Division contained Bannu, Dera Ismail Khan, Dera Ghazi Khan and Muzaffargarh. All the districts in the Peshawar and Derajat Divisions comprised the trans-Indus districts of the British Indian Empire. Within the trans-Indus districts, large towns were few. At the end of the nineteenth century those with more than 50,000 inhabitants included only Peshawar. Towns with less than 50,000 but more than 20,000 inhabitants were Kohat, Dera Ismail Khan and Dera Ghazi Khan.[16] Formerly a part of the province of Punjab these districts were converted into a separate province in November 1901, called the North-West Frontier Province with an area of 42,647 sq km.[17]

It was only in March 1855 that for the first time the interdiction of the public carrying of arms was sanctioned in these trans-Indus districts at the instance of the chief commissioner of the Punjab. However, no prohibition on the possession of arms was introduced at this time. Even prohibition on carrying of arms was first introduced as an experimental measure for six months. Proclamations were issued by the chief commissioner prohibiting the wearing of arms on ordinary occasions except by certain specified classes of persons holding a license from the district officer.[18]

At the end of the six months of the above experimental measures the deputy commissioner of these districts noted that the regulations were more or less successful. Captain James, the deputy commissioner of Peshawar, considered that it had been attended with great advantage. However, the measure was not carried out in certain border tracts of the Peshawar district. Captain Becher, the deputy commissioner of Bannu, also wrote that the proclamations had been complied with except in the frontier villages of Agor, Konch, Bogarmang and Kagan, and all along the bank of the Indus opposite to independent territory. From Kohat, the Deputy Commissioner Captain Coke reported that the orders had not been enforced owing to the physical configuration of the country and the hostile relations of the people with the neighbouring trans-border tribes. These reports were corroborated with reports from the Derajat Division. The Government of India (GOI) in the Foreign Department letter dated 26 June 1856 maintained this prohibition. It remained in force (though in Kohat and the border tracts of Peshawar and Hazara, it was not always acted upon) till the introduction of the Arms Act XI of 1878. A committee was appointed in 1878, few months before the promulgation of Act XI to consider measures of border defence for the Peshawar Division. The members of the committee were unanimously of the opinion that the carrying of arms should continue to be prohibited except under licenses granted to persons of position and the prohibition should be more rigidly enforced than appeared at the time to be the case.[19]

It was pointed out by two members of the committee, the commissioner of Peshawar, F.R. Pollack and the inspector-general of Police, H.N. Miller, that the large number of murders which filled the criminal records of the Peshawar district was, in great part, due to the unlimited possession of arms by the whole population, The chosen weapons of assassination were the knife and pistol-weapons not required in the ordinary pursuit of life, nor required for frontier defence. If possession of these were prohibited, great decrease would at once become discernible in the number of murders annually committed.[20]

Even then the Griffin Committee recommended only prohibition on the carrying of arms and no general disarmament. The final recommendations of the Griffin Committee were that it was inexpedient to prohibit the possession of arms in the Peshawar and Kohat districts; that the carrying of arms should be, as now, prohibited, except under license; that these licenses should only be granted to persons of position and their followers; and a large discretion should be allowed to the district officer in the matter of individual disarmament. The number of arms in the district should be registered, which had already been done in Peshawar without any difficulty, and further the deputy commissioner should have full power to forbid the possession of arms by persons of known bad character.[21]

In the case of Hazara district, the committee recommended that with the exception of the villages immediately on the border, all others should be disarmed and placed under the general provisions of the Arms Act in the same manner as the other Cis-Indus district, and that, under it, the possession of arms without a license should be considered to be an offence. The committee argued that there was nothing in the character of the Hazara district to separate it from the system in force in the remainder of the Punjab.[22] At this point of time F.W.R. Fryer, divisional commissioner of Dera Ghazi Khan argued that Baluchis should be exempted from the Arms Act. He argued that the latter themselves were the most efficient guards in that part of the border. They provided considerable aid to the government during the raid on Asni in 1857 and the raid on Harrand in 1867. In fact, armed Baluchis were often used by the government. Fryer narrates a personal experience: 'I start, as I have said, today for the Khetran Valley, accompanying the Governor-General Agent for Baluchistan and with me I am taking 250 armed Baluchis.' It was expected by Fryer that licenses will cause considerable irritation. As he said: 'the Baluchis are very proud of being allowed to go armed, and they will urge that their arms have always been used on the side of the Government, and will consider restrictions placed on their possession of arms as reflection on their loyalty'.[23]

The recommendations of the Griffin Committee were accepted by the lieutenant-governor of Punjab, Sir R. Egerton in March 1878.[24] But, the legal aspect of the case completely changed with the issue of Government of India Notification No. 515 of 6 March 1879, which in Rule III(c) exempted the districts of the Peshawar and Derajat Divisions (with the exception of so much of the Hazara district) from the operation of Section 13 of Act XI of 1878. Earlier the Government of Punjab had already issued a notice in this regard in a notification dated 16 May 1878.[25] Interestingly, the old prohibition against the carrying of arms was popularly supposed to be still in force and was acted upon in practice till late 1890s. The result of above notification was to remove all legal restrictions to the possession of arms in the trans-Indus region.[26]

The question of prohibiting the carrying of weapons in these districts was again raised by Colonel Ommanney, the commissioner of Derajat Division, in April 1897. Among the local officials, the views of the district superintendent of Peshawar, Mr. Hastings, were that knives and pistols should be confiscated throughout the district except in a very few and special instances where licenses should be granted for keeping or carrying them; that other arms, viz., guns, rifles and swords should only be carried with a license; that all guns, rifles and swords should be registered and numbered.[27] The deputy commissioners of two districts agreed with Colonel Ommanney that the carrying of arms without a license should be absolutely forbidden by withdrawal of the exemption from the provisions of Section 13 of the Arms Act. In fact, Peshawar district authorities went a step further by proposing that daggers and pistols should not even be possessed; these weapons should be in general confiscated, while other arms should be 'registered and numbered'.[28] However, no prohibition on the possession of weapons took place.

II

The inhabitants of Dera Ghazi Khan district were predominantly Pathans and Baluchis, the latter being in majority. The colonial ethnographies insisted that each caste and community had behavioural as well as physical and cultural characteristics common to all its members.[29] It was argued that Pathans possessed some of the worst traits of human character. Treachery, tyranny, revenge, bloodthirstiness, pride, superstition, boastfulness and cupidity were the general terms used to describe their attributes.

They were defined as avaricious, treacherous, and entirely unscrupulous in character. For instance Denzil Ibbetson in his *Panjab Castes* (1916) provides us with such colonial aggregates, particularly in relation to Pathans and Baluchis of the trans-Indus districts of Punjab. While giving the description of Pathan as an aggregate, he mentioned that the true Pathan is the most barbaric of all the races with which we are brought into contact in Punjab. He was bloodthirsty, cruel, and vindictive in the highest degree and he does not know what truth or faith is, insomuch that the saying *Afghan baiman* had become into a proverb among his neighbours; and though he was not without courage of a sort and was often curiously reckless of his life, he would scorn to face an enemy whom he could stab from behind, or to meet him on equal terms if it were possible to take advantage of him, however, meanly. The colonial ethnographer often substantiated his arguments with the help of proverbs. The proverbs cited in the case of Pathans are: '. . . the bigot of most fanatical type, exceedingly proud, and extraordinarily superstitious.'[30] The Pathans were considered as a favourite class for enlistment in the combatant branches of the Indian Army, especially after the introduction of 'class' system of recruitment by Lord Roberts in 1893.[31]

On the other hand, the 'Baluch' was described by Ibbetson in relation to Pathans as less turbulent, less treacherous, less bloodthirsty, and less fanatical and he has less of devil in his nature. Still, the criminality of Baluch as a tribe was emphasized by Ibbetson. He argued that a Baluch was a thief by tradition and descent. To corroborate his argument he mentioned some of the Baluch proverbs, such as, 'God will not favour a Baluch who does not steal and rob' and 'the Baluch who steals secures heaven for seven generations of his ancestors'.[32]

The Baluch and Pathan inhabitants of these districts usually carried arms with them as part of their culture. They were also known for production of arms in the pre-colonial era. However, the indigenous tradition of arms production was controlled, subjugated and replaced by the colonial state. Rifles used to be manufactured in the suburbs of Kohat, but the industry was nearly destroyed by the introduction of British rifles. Even then arms were a part of Pathan or Baluch dress code. For example, while describing the dress of a 'Baluch' it was argued that 'he usually carried a sword, knife and shield'.[33] Again, while describing the dress code of the 'Pathan' was it included 'the long heavy Afghan knife and the matchlock or *jezail*'.[34]

The one perennial problem at the frontier was the practice of feuding and raiding among the tribes. The tribes were under constant threat of being attacked. As one contemporary observer narrated:

> The Pathans beyond and upon our frontier live in frontier villages, to which are attached stone towers in commanding positions, which serve as watch-towers and places of refuge for the inhabitants. Small raids from the hills into the plains below are still common; and beyond the Indus the people, even in British territory, seldom sleep far from the walls of the village.[35]

More important is the fact that the British, particularly after 1878, depended on the 'natives' for the defence of these districts.

In 1878, the Border Defence Committee was appointed to deal with problems of border defence for the Peshawar Division. It was for the first time on the recommendations of this committee that the Border Police Force was constituted. The committee had recommended that at the more exposed positions of the Peshawar and Kohat districts, posts should be established at which a mixed body of frontier police and border militia should be stationed. The border police was to be recruited from British territory, cis- and trans-Indus regions—the greater proportion, however, was expected from Peshawar district. It was recommended that a large proportion of them should be nominated from among the villages immediately contiguous to independent territory, and to give service, as officers of this corps, to relatives of the more important *khans* and *maliks* of the district.[36] On the other hand, the frontier militia was supposed to be an attempt to enlist the tribes beyond the border in its defence. It was argued that there was every reason to hope that by enlisting a considerable number of the members of the more important tribes, along the whole length of the border, in a frontier militia and assigning to them allowances for services therein performed, the peace at the border may be more generally preserved.[37]

At the same time all the villages at the border were supposed to maintain a certain number of adult males in readiness for defence. All those villages in the Peshawar district were under the terms of the land revenue settlement and bound to render such services. In return, a certain proportion of their assessment was suspended while they performed the police services.[38] At the same time at the recommendation of the Border Defence Committee, for the first time the policy of providing arms to the villages on the British side of the frontier for defence was initiated. It was argued by the committee that serviceable arms should be provided to border villages by the government. Each person receiving a government weapon was supposed to produce a security equal to the value of the weapon. These arms were to remain the property of government, subject to periodical inspection.[39]

To deal with crime and other problems related to the frontier, special regulations were passed by the government to administer these areas, and the most important was the Frontier Crimes Regulation of 1887.[40] Even then the crime in these districts remained high, particularly the number of murders committed annually. The *District Gazetteer of Rawalpindi* informs us the cause of almost every murder committed in the district was a woman. In a large percentage of these cases there was practically no evidence, though there was generally strong circumstantial evidence against the murderer, regarding whose guilt there was no doubt. The findings of the *jirgas* enabled the deputy commissioner to punish the offender and, by enforcing the executive measures authorized by the regulation, prevented the matter growing into a blood feud.[41] The colonial officials pointed out the cold bloodedness with which these crimes were committed.[42]

The number of murders in Punjab in relation to its population was much higher in comparison to any other province of the British Indian Empire. As early as 1861, it was argued that compared with other provinces of India and with England, the number of murders in the Punjab was great in proportion to population. Most of the murders occurred in the frontier districts, where 'the people' were 'armed, blood feuds rife and evasion of justice by flight beyond the border a matter of the greatest facility.[43] For instance, during the year 1897, the ratio of murder was 1 to 150,765 persons in Bengal, in North-West Provinces it was 1 to 68,978, while for the Punjab the ratio was 1 to 28,933.[44]

Although having a very scanty population, the trans-Indus districts of Punjab recorded a large number of murders in relation to their population in contrast to the cis-Indus districts of the Punjab as a whole. For instance, from the very beginning, in the year 1861, the cis-Indus districts recorded 145 murder cases while the trans-Indus districts alone recorded 108. In fact, from 1861 to 1899, during the years 1866, 1867, 1870, 1871, 1872, 1874, 1879, 1880 and 1883, the trans-Indus districts reported more cases of murder in comparison to the rest of the Punjab.[45] Further, within the trans-Indus districts, the Peshawar Division was responsible for majority of the cases. The rise in the number of murders was also due to the introduction of new method of counting murders in 1897. While earlier more than one murder in a single case was counted as a single case, now one person killed meant one murder case. Thus, ten murders occurring in a single raid would formerly have been shown as one case, whereas now they were entered as ten cases. But, as the Report for the year 1899 argued, 'it is safe to say that the effects of the new method on the statistics of the Province as a whole can hardly be great'.[46]

In 1900, the four trans-Indus districts of Peshawar, Kohat, Bannu and Dera Ismail Khan were responsible for 275 out of the 799 murders in the province of Punjab. In Rawalpindi, 71 cases were reported, in Lahore 47 and in Amritsar, 33 cases. The remaining 24 districts had 270 cases amongst them.[47] In fact, the number of murders remained high even afterwards. From 1902 to 1903 the figures of murders reported fluctuated from 19 to 20 for Hazara district, 82 to 90 for Peshawar, 40 to 43 for Kohat, 20 to 36 for Bannu and 9 to 11 for Dera Ismail Khan.[48] The total number of murders committed in these districts was 192 for 1902, 180 for 1903, 152 for 1904 and 196 for the year 1905.[49] In the trans-Indus districts, the situation of Peshawar district was most vulnerable. During the year 1898, the number of murders reported in Peshawar district reached the high figure of 144 cases. Up to 1895, the average of murders did not reach the figure of 90, while the average of three years from 1896 to 1898 was around 124. As compared to the neighbouring Hazara district, in Peshawar one person in every 4,850 persons was murdered, while in Hazara the ratio was 1 to 47,000.[50] In Bannu district, where the number of murders in 1898 was 69, the ratio was 1 to 6,004 of the population.[51] Tables 10.1 and 10.2 give an idea of

TABLE 10.1: MURDER CASES REPORTED IN THE DISTRICTS OF HAZARA, PESHAWAR, KOHAT AND BANNU DURING 1875-99

Year	Number of murder cases reported in district			
	Hazara	Peshawar	Kohat	Bannu
1875	13	81	11	11
1876	8	71	18	8
1877	17	66	25	11
1878	16	52	30	14
1879	32	55	47	17
1880	24	60	39	25
1881	12	56	39	17
1882	20	69	31	20
1883	21	75	25	11
1884	22	81	25	21
1885	27	101	35	33
1886	13	109	33	27
1887	41	96	38	38
1888	23	79	40	35
1889	19	43	36	23
1890	24	56	47	29
1891	22	52	35	32
1892	24	77	28	26
1893	17	66	26	32
1894	18	89	33	43
1895	29	87	28	53
1896	28	106	35	69
1897	27	141	38	54
1898	13	149	31	72
1899	19	135	55	82

Source: 'Disarmament of the Frontier Districts of the Punjab', Home (Public), July 1901/209-213, p. 7, NAI.

TABLE 10.2: STATEMENT SHOWING THE NUMBER OF MURDER CASES REPORTED IN VARIOUS DISTRICTS OF PUNJAB DURING 1893-9

District	1893	1894	1895	1896	1897	1898	1899
Hisar	11	10	28	13	23	18	19
Delhi	13	8	10	14	16	11	20
Ludhiana	26	26	11	19	12	15	12
Ferozepur	24	24	41	46	29	37	42
Lahore	43	38	31	39	38	42	52
Amritsar	15	25	23	26	25	31	29
Rawalpindi	62	39	68	60	72	55	60
Hazara	19	18	29	28	27	13	19
Peshawar	67	89	87	106	141	149	135
Kohat	32	33	38	35	38	31	55
Bannu	33	43	53	69	54	72	82
Dera Ghazi Khan	18	35	26	29	36	33	49

Source: The sources are Annual Police Administration Reports of Punjab (PARP) for respective years. See, PARP, 1994, pp. 8-9, PARP, 1895, pp. 8-9, PARP, 1896, p. 8, PARP, 1897, p. 8, PARP, 1899, pp. 12-13.

such crimes committed in selected districts of North-West Frontier Province and Punjab.

The colonial strategy to deal with this crime relied on the disarming of these districts. But the emphasis at least initially was not at all on the 'arms of precision', it was rather on what was classified by colonial officials as 'arms of assassin'. Most of the murders were committed with the help of knives and pistols. During the year 1894, out of 102 murder cases, 36 were committed with the help of knives while 33 with the help of firearms, mostly pistols.[52] In the year 1897, out of 133 murders reported in Peshawar district, 60 were committed with the help of knives while 40 were committed with pistols. During the year 1898, out of total 149 murders committed in Peshawar district, 39 were committed by pistols while 63 with the help of knives, and 16 with guns. Similarly, during the year 1899, 51 murders were committed by pistols and 51 by knives. In fact, as late as 1903, out of total 180 murders reported during the year in North-West Frontier Province, 19 were committed with pistols and 79 with the help of daggers and knives. However, there were some exceptions. For instance, in Bannu district during the year 1898, out of the total murder cases reported, 38 per cent were committed with guns, while 30 per cent with the help of knives and only 3 per cent with the help of pistols. It was recognized that guns were increasingly used for murder.[53] However, a general understanding remained that small weapons like pistols, daggers and knives were responsible for the murder committed and it was necessary to confiscate these weapons from the hands of the 'natives' to control the resurgence of violent crimes. As discussed below a moral language was used against these weapons such as a 'crusade' or a 'war must be waged' against these weapons.

At the same time not denying the fact that the impact of the frontier was significant in these districts, but in the case of murders, the contribution of trans-border men was not much important. For instance during the year 1900, out of 106 murders committed in Peshawar district (excluding those committed by poison), only 16 were committed by the trans-border men. Similarly, in 1898 out of the 13 murders committed in Hazara district, only one was committed by trans-frontier men. While in Kohat, out of the 46 murders committed in 1900 (excluding those committed by poison), only 8 were committed by trans-border men.[54]

The question arises, as it is very much evident from the above description, the number of murders committed in the trans-Indus districts remained high from the very beginning of the colonial occupation. Then, why was disarmament carried out at the end of the nineteenth century? What was so important about murders committed during the end of the nineteenth century in these districts?

III

The North-West Frontier was strategically very important in the British imperial policy. A large part of the army was concentrated along the North-West Frontier. Over time, a heavy military presence in this region led to the emergence of cantonments and other army-dominated settlements. The cantonments as urban centres were a typical colonial innovation. They evolved into a 'peculiar' institution.[55] Cities came up around these cantonments gradually due to their commercial importance. This process was further aided by colonial frontier policy. For example, due to the strategic importance of frontier, all important centres on the frontier were linked with road and railway network with the interior of British India. Among all important centres on the frontier the most important was Peshawar, the largest military base on the frontier.[56] The cantonments often promoted commercial activity. They became markets that often blossomed into urban centres, particularly if the *bazaar* had developed into a large town, consuming much of the produce of the nearby countryside. Their existence was purely on account of military needs.[57]

But this was only one aspect of the picture. On the other hand, the transformation of cantonments from a purely military base to markets and over time into commercial centres led to an increasing interaction between the 'natives' outside and the personnel in the cantonments. The increasing scale of movements of the 'natives' into the cantonments was a serious concern for the military officials. Nowhere was this concern more serious than in Peshawar and other cantonments and military camps in the region. The problem reached heightened proportions at the end of the nineteenth century when cantonments and other municipalities became targets of crime by the 'natives'. A large number of thefts, burglaries and arms thefts were becoming increasingly alarming for the military officials. As Robert Warburton argued 'the audacity with which thieves in Peshawar Valley broke into houses and barracks in Cantonments, and carried off property, chiefly rifles, baffles all description'.[58] But, perhaps two forms of crime were completely new and became a serious challenge to the colonial state. These included the murder of the white military and civil officials and 'native' soldiers at the end of the nineteenth century and secondly the increasing number of arms thefts from the cantonments and the military camps. The arms thefts were often perpetrated with violence.

It seems that at the end of the nineteenth century there was an increasing threat to the lives of 'white' men in this region. The presence of Pathans in 'white' settlements was a serious concern for the colonial officials. As one contemporary police official argued,

> Large number of Pathans, desperadoes whose residences and even names are unknown, ostensibly working as labourers, reside in Simla and other Hill sanatorium, the seats of Government in India, where their presence, considering their known reckless and ruffianly [*sic.* ruffian] character [was] . . . undesirable.[59]

Apart from White military officials and personnel, threat was also perceived by the other 'white' men residing in the frontier region. For instance, the Public Works Department of the Government of India proposed to arm the plate-layers and station staff in the Sind-Pishin section of the railway with swords as the means for protecting themselves against the attacks of 'fanatics'. However, it was sanctioned that Europeans and Eurasian employees, who were members of the Railway Volunteer Corps and who were trained in the use of firearms, should be provided revolvers while in the case of 'natives' and thou who were not volunteers, swords were considered sufficient.[60] In 1900, the adjutant-general in India argued:

> The safety of European life in Peshawar is probably less than in the first days of the occupation. Such a blot on our administration and the inability to secure the lives of the government servants can only be regarded by the people as a sign of weakness.[61]

The other form of violent crime was arms thefts. The cases of arms theft by the frontier men were reported from as far as districts like Gurgaon, Ludhiana, Ambala and Karachi.[62] Most of the cases of arms theft took place in the trans-Indus districts.[63] In fact, cases of arms theft in a cantonment can suggest that how serious the challenge was.[64] More important is to recognize the fact that in most cases, particularly arms thefts, which reached an alarming proportions during the end of the nineteenth century,[65] were committed with the help of arms and hence violent.

During June 1895, Lieutenant-General William Lockhart, commanding the forces in Punjab, argued for prohibitions on carrying of arms. The context for these proposals originated from the murder of Lieutenant Limond of 6th Punjab Infantry. Lockhart justified death penalty for carrying of arms within a prescribed distance from British armed posts and lines of communication. If it was considered too severe, then he proposed the confiscation of arms and 50 lashes as punishment. The proposals again came from lieutenant-general commanding the forces in Punjab, in May 1897, when a 'fanatic' attacked some men of the 1st Punjab Infantry at Datta Khel, wounding a naik and two sepoys. To substantiate the argument, the cases of the Scottish Highlands, Punjab and Upper Burma were cited as instances where this policy or disarmament had been carried out successfully. The lieutenant-general pointed out that primary consideration should be the protection of Her Majesty's forces from 'treachery and fanaticism'.[66] However, in both the cases the commander-in-chief, George White dissented from the proposals and so they were not put into practice. Perhaps a more forceful plea for disarmament originated from a letter written in August 1897 by Brigadier-General E.R. Elles, who were commanding Peshawar district, by drawing attention towards bands of armed Pathans coming through the cantonments. After making inquiries, he was shocked to hear that legally all men in the district could possess and carry arms. In an interview with the commissioner of Peshawar Division, he urged that arms were not required by the people for defence except in the border villages, and that the carrying of arms should be prohibited and also the possession of arms in the interior of the district. A large number of possible dangers arising from the possession of arms by the 'natives' were listed by him; it rendered the cantonments the hunting grounds of armed burglars; paying a large sum for police protection and as Pathans were armed it rendered police work dangerous; to have double sentries at many posts; raids across the border against the British government; it made violent crime against Europeans easy, e.g. Lieutenant Steven's murder in 1896, Mr. Ross's murder in 1897; crime against sentries easy, e.g. a sentry, Devons, was shot in Nowshera in 1895. A sentry of the 20th Punjab Infantry was shot dead in Peshawar in 1897.[67] Elles' letter was put forward to the adjutant-general in India by the lieutenant-general commanding the forces in Punjab. He argued, 'the Afridi or Pathan who swaggers fully armed through the cantonment and city of Peshawar ascribes to our fears what is really due to our carelessness and leniency'.[68]

At the same time, hostilities started at the frontier and the whole question became more important than ever before. Orders were issued for the successful handling of the war effort. For instance, in October 1897, the general officer commanding Tirah Expeditionary Force recommended that the carrying of arms within a certain distance of the military posts and line of communication in the Kohat district should be prohibited under penalty of confiscation of arms. This proposal was carried into effect.[69] Even then, the crime committed by the 'natives' remained a serious concern for the army officials. For example, within one week in April 1899, the following crimes were reported in Peshawar: a burglary took place at Mr. Knotts house though the thief was shot dead by the *chowkidar*, another burglary took place at police *chowki*; a burglary at double-storeyed barracks took place in which 10 rifles were stolen; on another night a burglary was attempted at double-storeyed barracks; on the night of 6 April 1897 a European driver of 24th Field Battery was shot through shoulder. The scenario was so serious that the lieutenant-general commanding the forces in Punjab remarked that if the civil government at present is unable to keep any order then it is time for the military to act.[70] The crime against army personnel and the cantonments kept rising. That the situation was so serious can be judged from the information provided by the trans-border political diary of the Derajat Frontier for two months of 1898. On 18 February 1898, a rifle and a sword bayonet were snatched away from a sepoy of the 20th Madras Infantry, who was on duty at Miran Shah. The rifle was reported with Sardi Khan of Kaprai in Khost. Almost a month later some thieves seized a sepoy of the 8th Bengal Infantry and robbed him.[71] On the night of 18 April, a Martini Henry rifle was stolen from a man of the 1st Punjab Infantry while asleep in Saidgi Military Post. On the morning of the 30 April, about 2.30 a.m., a gang of 'thieves' visited Datta Khel Post. One of the thieves was killed. On the night of the 18 April, a carbine, 2 swords and 20 rounds of ammunition with a belt were stolen from the Cavalry Lines inside the Idak Post in *Daur* country. On the night of 22-3 April, a thief prowling near the Idak Levy Post was fired at by sentries and severely wounded.[72]

During this period, due to increasing insecurity of life and property as manifested in Peshawar district, the commander-in-chief brought to the notice of the Government of India the urgent necessity for protecting the officers and soldiers from the risk of assassination.[73] The commander-in-chief argued that all the ordinary methods of the civil government will prove powerless to restrain the 'fanatical spirit' as long as the population of these districts is allowed to keep arms. He further argued that most of the outrages which were committed at Peshawar were initiated or facilitated by disaffected villages in the neighbourhood of Peshawar city and the cantonment. So he strongly urged to bring these districts within the provisions of the Arms Act and exempting only the border villages.[74]

An important case arose after the murder of Colonel La Marchant. The general officer commanding Peshawar district argued in relation to La Marchant's murder that the murderer who had been hanged was an inhabitant of the village of Pooka within a mile of the Cavalry Lines and the second man who was arrested on the spot and was under trial was also an inhabitant of the same village. He also understood that the two men arrested by Mr. Plowden on suspicion of being associated with the murder case were also inhabitants of Pooka.[75] It was argued that village Pooka should be disarmed completely. However, as discussed below, the colonial state failed to go for this option due to expected dangerous consequences.

At the same time, the frontier risings provided a pretext for the disarmament of these districts. In case of frontier and trans-frontier areas, a careful surveillance was kept particularly in relation to arms. The district confidential diaries provided every detail to the foreign department. For example, one paragraph of the Peshawar confidential diary for the month November 1897 mentions: 'It is reported that Shakar of Hazarnao, the well-known rifle thief, went to Tirah about ten days ago with a party of 10 Utkhels of Lughman, 6 Khugianis and 15 Shinwaris, who, like Shakar, were all well known thieves and bad characters.'[76] In another case, it was reported: 'Information received from Dakka stated that Muhammad Husain Khan, Sartip of Dakka, purchased the other day a few rifles and one horse for Rs. 446 from certain Zakka Khel Afridis, who probably looted them from Tirah.'[77] Weapons collected from the independent tribes implicated in the risings at Malakand and Chakdarra during 1897-8 included 231 breech-loaders and 4,965 muskets and *jezails*.[78] At this time it was natural that the disarmament of the adjoining British territories was also considered. The political agent for Dir, Swat, Chitral and Chakdara argued that the present work of disarmament of the frontier will be thrown away if the present facilities for obtaining arms from our own districts was not removed; at present, there was practically nothing to prevent these people from purchasing arms from British subjects of Peshawar and Kohat, who should also be disarmed.[79]

Unfortunately too much emphasis on illicit arms trade does not take the above factors into account. However, it is not to deny that illicit arms trade was not an important issue. From time to time inquiries were held and legislations passed to stop this movement of arms.[80] For instance, the remarks of W.R.H. Merk, commissioner of Peshawar Division are very important in this regard. Merk argued that the breech-loaders and cartridges are not acquired by the trans-border tribes from the inhabitants of these districts because 'this population has not got them'.[81] In fact, the principal supply was from the cis-Indus districts via arms thefts. He argued: 'the average number of rifles stolen at present every month from the army all over India is 25 or about 300 a year. If this continues, the clans will be as well armed as ever, from this source alone, five years hence.'[82] In other words, the trans-Indus districts were not the major source of 'arms of precision' for the trans-border independent tribes.

Whatever may be the case, what is more important is that the disarmament of these districts was conceptualized at this point of time. But it was not an easy task. As Merk again argued, the disarmament of the Punjab and of Hindustan, after the annexation of this Province and after the Mutiny respectively, did not cause discontent; but these measures came after tremendous events which had thoroughly cowed the country; the people had not for 50 years, as they have here, enjoyed a highly valued privilege over other fellow-subjects; and what is the main point, the people were not Pathans.[83]

IV

The weapons of precision were important from the military point of view, while weapons of assassins, that is knives and pistols, were important from the police's point of view. But when the targets of the arms of assassin happened to be British military army and civil officials, then these two categories were no more separate and merged in one another. This was precisely the situation at the end of the nineteenth century in these districts.

The disarmament of trans-Indus districts was always argued by the local bureaucracy particularly police officials. They were quick to point out the alarming number of murders in these districts and raised the demand for disarmament of these districts. J. Wilson, Secretary to the Government of the Punjab argued, 'the recent extensive outbreak of hostility on the North-Western Frontier of the Province naturally attracted considerable attention to the state of the law in those regions as bearing on the carrying, possession and transport of arms by British subjects and by trans-border Frontier tribesmen within our border.'[84] However, at the same time he emphasized the marked recrudescence during the last three or four years [during commitment] of murders and serious crime of almost every description in Peshawar and Bannu. As a result, the objectives to be achieved were, firstly 'to prevent the supply of arms and ammunition . . . to trans-border tribes from British India'; secondly, 'to check violent crime in British territory'.[85] The increase in crime in frontier districts was linked to unlimited possession of arms, and especially of particular kinds of arms. The police statistics reflected that 'the knife and the pistol' were 'the favourite weapons of assassination', doubtless because they were most easily concealed. These objectives were to be achieved within the constraints. First, without destroying the efficiency of border defence or abandoning the system on which frontier revenue remissions are based, where the retention of this system was still necessary or desirable. Secondly, disarmament should be implemented without depriving peace-loving people in the frontier districts of their means of defence against criminals. Thirdly, without interfering with the issue of any instruments which were required otherwise than weapons. Fourthly, without arousing undue irritation in any community that had done nothing to deserve such special punishment. All the officers agreed that a 'war must be waged against these weapons'.[86]

In fact, in Derajat Division the 'carrying of weapons' was already prohibited although the 'possession of weapons' was legally allowed. While in Peshawar Division, there was no prohibition on both 'carrying' and 'possession' of arms, Mackworth Young, the lieutenant-governor of Punjab was of the opinion that 'for the prevention of crime it is [of] no use to prohibit carrying of pistols and daggers' rather 'the possession must be prohibited also'. But, there were also certain technical difficulties.[87]

Regarding the measures adopted for disarmament, it was suggested

> the confiscation of pistols and daggers should not be effected by wholesale searches, which might create much irritation and which could not be effected over a large area so simultaneously as to prevent evasion, but opportunities should gradually be taken to discover and confiscate hidden arms, especially in villages where crimes of violence came under investigation.[88]

Mackworth Young argued:

> the number of murders may be somewhat reduced by a systematic disarmament, but the Rawalpindi and Jhelum districts are standing proofs that such a disarmament as [we] can hope to accomplish will not stamp out murder. Then there is the discontent which might be aroused in a warlike race; the inequality between the members of the same race and tribe who happen to be in the interior and on the border of the district or just beyond it; the real need which exists in a society with a considerable sprinkling of assassins and braggadocios of allowing the peaceable members the means of defending themselves; the need for allowing villagers when turned out in pursuit of thieves or burglars to arm themselves; the difficulty in which we should be placed on application being made for a license to carry arms for defence on account a blood feud; the occasional requirement of arms of defence to enable villagers to till their fields, in case where there has been a boundary dispute, or water them [*sic.* then] others where there has been

a quarrel about irrigation rights, for it is no answer to these difficulties that others would be in same plight. They would take care not to be; they would beg, borrow or steal weapons from the border zone of from trans-frontier men; we cannot guarantee the absence of such opportunities.[89]

In fact, he was not in the favour of a general disarmament. He recommended disarmament in the case of daggers and pistols. It was argued that they have become 'disreputable on account of the frequency with which' they were used for murders and it was expected that the irritation resulting from their prohibition will be slight, if the disarmament was judicially managed. On the other hand, the cases of guns and swords were different. Such a prohibition, he feared, would turn all sections of the respectable population against the government.[90] He further recommended that the Peshawar Division should be brought at par with Derajat Division by prohibiting the carrying of weapons by trans-border Pathans and also prohibiting the carrying of weapons in cantonments and municipalities without license.[91]

On the whole it is clear from the very beginning that the need for disarmament of the trans-Indus districts arose from strategic reasons rather than due to the increasing number of murders. The demand for confiscation of knives and daggers had been raised earlier as well but it was ignored on one ground or the other. But, this time it was carried out. However, it needs to be recognized that the moral issues were brought into picture to substantiate the demand for the disarmament of the trans-Indus districts of British India. Whatever be the causes it was recognized by the colonial state that the task of disarmament was tedious as well as impractical for it directly affected the security of the 'natives' of the frontier districts. The presence of the border police as well as police in general was weak and it was not possible to provide security to these people without providing them weapons. The government kept on providing weapons to the border villages of these districts. In fact, weapons supplied to these border villages easily passed into the possession of trans-border tribes. As the Tucker Committee Report noted:

It must, however, be remembered that in both the Peshawar and Kohat districts, and probably also elsewhere on the Frontier some thousands of Enfields have been given to the villagers for self defence; and that, though at the present time a tally of these rifles is kept, and probably not more than 100 or so are known, to have been lost, in frontier days no such account was kept, and arms, which were given to frontier villagers for self-defence, were sold across the border, more or less as a matter of course. It would be impossible to estimate now how many arms have fallen into the hands of the tribesmen in this way.[92]

In fact, the arming of the border villages remained a part of the policy even after carrying out disarmament measures in these districts.[93] On 12 January 1898, L.W. King, deputy commissioner Kohat argued that disarming of the population will not help in checking the traffic in arms across the border as the trans-frontier population could obtain arms of precision from British subjects in the Kohat district. He argued that very few residents of Kohat possessed breech-loaders and all of them were entered in a register which was strictly checked by the officer of the border police. Further, security was taken from the owners of such weapons on the grounds that the weapons will not be sold across the border.[94]

The partial disarmament of the trans-Indus districts was notified in a Government of India notification dated 5 May 1899. Those persons not exempted from certain prohibitions of the Arms Act, were forbidden to go armed, except under a license in the Hazara district (excluding of feudal Tanwal and certain tracts and villages on Hazara Frontier), in the cis-Indus tahsils of the Bannu and Dera Ismail Khan districts and in the cantonments and municipalities throughout these scheduled districts. Outside in the areas specified, there were no restrictions upon going armed except that, in the trans-Indus tahsils of the Bannu and Dera Ismail Khan districts and throughout the Dera Ghazi Khan district and the Peshawar Division, a license was required for the trans-border Pathans for carrying arms of any kind and for other persons for carrying pistols and daggers.[95]

A few things are more important in this notification. First, there was an overwhelming emphasis on the restriction on carrying of weapons in the municipalities and cantonments. Second, there was an emphasis on the restriction on the carrying of 'pistols and daggers'. This reflects that the immediate context for this promulgation came from the increasing number of murders of the white officials and other forms of crime in these districts. But, in reality these measures failed to restrict these forms of crime. The army officials were particularly dissatisfied with the measures carried out and these had practically no impact on the number of murders reported in these districts nor it checked crime against the 'white' men. Particularly disturbing was the continuation of murders of European and 'native' soldiers. As the lieutenant-general commanding the forces in Punjab, argued in May 1900 that the application of the arms act in the Peshawar district had no effect whatsoever in diminishing crimes of violence against Europeans or 'native' soldiers. Within Peshawar district, between 1 January 1899 and 28 April 1900, 17 European officers were murdered. Most of them were shot while asleep in bed or on duty (see Table 10.3).[96] On the basis of this statistics, the general officer commanding of Peshawar district described the disarmament measures as a failure.

He strongly recommended in these cases collective responsibility. He argued that if a murderer from British side of the border was traced to any of the village then his family relations should be made to suffer either by banishment from the district or by the imposition of heavy fine. His whole property should be confiscated. He recommended house to house searches should be made in such villages and all arms should be confiscated and those who possessed arms should be prosecuted.[97] In case of a trans-border 'ruffian' being harboured by any village of the British side of the border, he suggested that the headmen of the village be held responsible and be severely punished. He argued that licenses for carrying knives and pistols should not be given even to the most respectable of the inhabitants, while he advocated that licenses for rifles for self defence should be freely granted. Further, if after the lapse of 6 more months, crimes with violence do not decrease, then 'arms act should be strictly enforced, and a house to house search made in every village, arms found to be confiscated, and those in possession prosecuted'.[98] The lieutenant-general commanding the forces in Punjab accepted his proposals and further advocated that the whole community along with the individual should be punished. He argued that crime may be prevented by the dread of the consequences which will entail if such criminal activities were committed.[99]

Power Palmer was of the opinion that for such criminals death sentence in the jail precincts was not any deterrent. He questioned that whether flogging followed by penal servitude for life in the

TABLE 10.3: AFFRAYS AND OUTRAGES COMMITTED BY THE INDIANS IN THE PESHAWAR AND DERAJAT DIVISIONS FROM 1 JANUARY 1899 TO 28 APRIL 1900

Rank	Name	Corps	Date	Particulars
Private	Whitmore	Dorsets	1.1.1899	Shot while asleep in bed
Private	Not known	Royal Scottish Fusiliers	21.9.1899	Assaulted by a Tum Tum Driver
Captain	Hall	12th Bengal Cavalry		Fracas with natives while returning from shooting
Lieutenant	Cummings	24th Punjab Infantry		
Lance Corporal	Wilson	Hants	16.1.1900	Assaulted in Billiard Room in *Sudder Bazaar*
Private	Brown	Royal Scottish Fusiliers	25.1.1900	Assaulted by a 'Native'
Private	Healy	Royal Scottish Fusiliers		Assaulted in *Sudder Bazaar*
Private	Boyd	Royal Scottish Fusiliers	5.3.1900	Shot while asleep in Bed
Corporal	Underhill	Dorsets	5.3.1900	Shot on guard while in bed
	Mr. Boyle	Assistant District Superintendent of Police	1.3.1900	Shot by Burglars in *Sudder Bazaar*
	Mr. Welsh	Head Clerk in the Commissioner's Office	28.3.1900	Assaulted on Circular Road
Lance Corporal	Edmunds	Hants	1.4.1900	Assaulted on Michni Road
Private	Worthington	Hants	2.4.1900	Assaulted in *Sudder Baazar*
Private	Hamilton	Royal Scottish Fusiliers	2.4.1900	Assaulted when on Police duty
Lieutenant	Green	No.1 Battalion Sappers and Miners	25.3.1900	Murdered by a 'fanatic'
	Mr. Gunter		25.3.1900	Murdered by a 'fanatic'
—	'Native' Sentry	King's Rifles	20.4.1900	Shot when on duty

Source: 'Disarmament of the Population of the Peshawar and Derajat Divisions', December 1900/613-15, Home (Public), p. 12, National Archives of India (NAI), New Delhi.

Andamans without any remission of sentence would not have much greater terror for the Pathans? He believed that public flogging in the man's village and in the marketplace at Peshawar administered by British soldiers would be infinitely more shameful to a Pathan than any other form of punishment.[100]

During this time two important cases came forward involving two villages. The army officials asked for disarmament of these villages as a punishment. The first important case came to forefront involving the village named Tahkal Payan. In this case, a man name Shabaz of Tahkal Payan (or Lower Tahkal) was arrested when in possession of a Lee-Metford rifle which had been stolen on 5 March 1899 from the Royal Inniskilling Fusiliers. On his house being searched, a government revolver, a muzzle-loading revolver and two pistols and a *gherra* full of Lee-Metford, Martini Henry and Snider cartridges were discovered. This was the only first case in which stolen government rifles have been recovered from Tahkal.[101] Another important case involved the village of Pooka of Peshawar district. The reason to disarm Pooka arose from the fact that Sultan, the man who has been hanged for the murder of Colonel La Marchant, came from Pooka. Though the village had a 'bad name' in criminal records, it was very difficult to prove that the murder was pre-mediated or that anybody in the village had idea that it was likely to take place.

Despite the pressure exercised by the army officials, in the end the civil government came to the decision that disarmament of these villages was not a wise decision. The deputy commissioner of Peshawar instead of disarming of the village advocated establishment of a police post in Pooka. He anticipated that this measure will have a good effect. Similarly, in case of village Tahkal Payan, he was of the view that disarmament of the whole village was not a prudent decision:

> I would deprecate as strongly as possible the punishment of a whole village on the sole ground that a single dealer of stolen arms has been found in the village. Such action . . . disposed people in the district against us and induce them to prevent (if possible) any similar discovery in any other village.[102]

Finally, A.H.L. Fraser, Officiating Secretary to the Government of India, agreed that the disarmament of these villages was not necessary.[103]

If the measures initiated on 5 May 1899 were a failure from the military point of view, same was the case from the police point of view. The civil authorities accepted that the number of arms submitted or those who applied for licenses were very small and at the same time the number of murders in the district did not decline. It was argued that the number of arms deposited was insignificant. Confiscated arms nearly constituted one-quarter of those given in voluntarily. An inference was drawn that a large number of persons escaped detection altogether. But, more important was the fact that no actual decline in the crime rate occurred; in fact, the incidence of crime increased. Rennie, deputy commissioner of Kohat, perhaps realized that it was possible only if pistols and daggers were also brought within the preview of the Arms Act.[104] Rennie argued that every murder committed since the publication of the notification was effected either with the pistol or the knife. With the exception of few wealthy people there was scarcely an application for permission to possess knives or pistols. The reason for this disinclination was understood: 'the people will not put themselves to extra expense of paying 8 *annas* for the privilege of possessing a weapon worth 12 *annas*'.[105]

In June 1899, in the Kohat district, measures were promulgated advising the inhabitants to hand in their arms within a month, commencing from 15 July 1899. This led to a total of 443 weapons being handed over to the civil authorities by 18 August 1899. On the other hand, the district superintendent of Kohat, noted on 6 April 1900 that compared with year 1898-9 with the last two years in which the lethal weapons were used there was an actual increase of 12 cases of murder, attempted murder and culpable homicide. In fact, there was an increase of 30 cases in which lethal weapons were used.[106] C.M. Rivaz in a note dated 29 July 1900 commented, that the number of pistols and daggers either voluntarily surrendered or seized by the police to the end of 1899 was 'ridiculously small' and compared this with 'acme of weakness and futility'.[107] According to Rivaz, in Peshawar licenses were given for nearly all the surrendered weapons, and even for some of those which were seized. He questioned the very idea of disarmament, because murder was possibie both with a licensed as well as unlicensed weapon.

In fact, on the official side there was a marked contradiction regarding the disarmament of these districts. They were vague and ambiguous regarding the utility of disarming these districts. The district superintendent of police at Kohat argued that the Gandiaur, Gurguri, Hangu, Karrak and Teri *ilaqas* (regions) have shown possibly a wise discretion in retaining their weapons. At the same time, he understood that throughout the district large numbers were sold or deposited with trans-border neighbours.[108] In fact, the commandant of the border military police suggested to the officiating deputy commissioner of Kohat, that total disarmament in this district is neither necessary nor advisable. For it was argued, 'exaggerated reports that their neighbours in British territory had been totally disarmed, trans-border offenders' will be 'encouraged to make raids'.[109]

It was soon realized by the colonial officials that the measures initiated on 5 July 1899 had completely failed and a new set of measures were required to deal with the situation. At this point of time an important role was played by the Tucker Committee Report which was submitted on 18 April 1899. While the findings and recommendation of the Tucker Committee Report were not taken into account during the promulgations of 5 May 1899, this time they were. It was understood that a more comprehensive set of measures were required. However, even at this point of time no complete disarmament of these districts was sanctioned.

The question of disarming the Baluchis was again considered at this time which was earlier not taken into account. C.P. Thompson, deputy commissioner of Dera Ghazi Khan cited the letter of Mr. Fryer dated 18 November 1898 and was of the opinion that no alterations were necessary or advisable with regard to the Baluchis.[110] M.W. Fenton was of the opinion that the disarmament of the Baluchis will arouse dissatisfaction, and might sow the seed of disloyalty among the members of a race who were thoroughly loyal to the British government. It was argued that the martial traditions of the race were preserved outwardly in the weapons which they were fond of displaying. However, he argued that non-Baluchi population of Dera Ghazi Khan should not be exempted from the provisions of the Arms Act.[111] Finally, the lieutenant-governor agreed that a Baluch living in any *tuman* should have the privilege of wearing arms, not only within his own *tuman*, but also within the limits of any other *tuman*. If he visits the town of Dera Ghazi Khan or any other place in the district outside *tuman* limits and desired to wear arms, it was expected that he should first obtain a license or permit from his *tumandar*.[112]

Earlier no comprehensive measures were initiated to deal with rifles and ammunition; rather the emphasis was on daggers and pistols. The objects of the GOI in extending the provision of the Arms Act to rifles and rifle ammunition in the Peshawar and Derajat divisions was perhaps to secure an effective control over long range fire-arms of precision on the frontier. But, in reality most of the arms in both of these divisions were muzzle-loading rifles of obsolete pattern. They were in no manners weapons of precision; they had a short range and even they were not worth stealing. On 5 November 1900, it was argued by C.E.F. Bunbury, deputy commissioner of Peshawar that these muzzle-loading rifles of which about 1,100 were in the hands of the 'natives' of border villages should be exempted from the new prohibition.[113] Mir Khan in a letter dated 30 June 1900 argued that there was no need to take any steps to limit the quantity of fire-arms, other than rifles, in the possession of persons in frontier districts. Regarding the source of illegal arms trade across the frontier he argued: 'the serious part of the trade in arms is located, not on the north-west frontier, but in India, cis-Indus. We see the results on the north-west frontier, which gives rise to the popular impression that most of the stolen or illicit rifles and ammunition are obtained [from] trans-Indus.'[114]

On the other hand, Captain C.B. Rawlinson, deputy commissioner of Kohat did not expect that many people will come forth for license. There was no idea about the exact number of rifles in the district. An estimate was made between 50 to 200 rifles. It was argued that most of the rifles were in the possession of persons who were already exempted from the operation of the Arms Act; 'while the remainder would appear to be in possession of persons who will probably prefer to risk detection and confiscation rather than surrender them on the chance of receiving a license or even on payment of any compensation which may eventually be fixed.'[115] Though, he did not expect much applications yet, he argued that in case of compensation being given, it should be so that 'to induce owners to surrender their weapons in preference to illegally retaining them or to disposing of them to trans-border purchasers, the sum must be in excess of the trans-border price . . . say ten per cent'.[116] The GOI accepted the proposal that compensation for rifles surrendered up to the 1 June 1901 was to be paid at the rates prevailing across the frontier and a reduction was to be made in the case of old and damaged weapons.[117]

J.M. Douie, the officiating chief secretary to the Government of Punjab, was of the opinion that the number of rifles in the Derajat Division was very small. On the other hand, even in Peshawar Division the number of rifles for which licenses were not given was very low. Even Mackworth Young, lieutenant-governor of Punjab was not in favour of 'any extensive confiscation of rifles immediately'.[118] Mackworth Young was cautious in his approach and argued that there was no ground for carrying out a general disarmament in districts of Peshawar Division at present.[119]

After the submission of the Tucker Committee, the government was in favour of prohibiting the possession of rifle ammunition. Accordingly, on 22 June 1899, the lieutenant-governor of Punjab proposed prohibition of the possession without licenses of all rifle ammunition in frontier districts; confiscation of all rifles and rifle ammunition not covered by a license after a fixed date. The governor-general accepted his views.[120] Finally, the lieutenant-governor proposed to prohibit: (i) the carrying and possession of rifles and rifle ammunition throughout the Peshawar and Derajat divisions, (ii) the possession of daggers and pistols throughout the Derajat Division, and (iii) the carrying of arms other than daggers or pistols throughout the Derajat Division except in the narrow strip lying between the frontier military road and the western boundary of British India. Mackworth Young proposed that the prohibition should be subject to two exceptions, viz.: (a) that armed grazing guards and armed pursuit parties shall be exempted from it, and (b) that it shall not apply to the Baluchis of the Dera Ghazi Khan district.[121] His proposals were accepted and a notification was passed on 16 November 1900 bringing these provisions into force.

What were the consequences of these measures? Were they really successful? By December 1901, the governor-general was of the opinion that the 'experimental measures' adopted in the Kohat and Peshawar districts had not proved successful. The Punjab government submitted its report on 1 May 1901 but the GOI was not of the opinion that all that was possible has been done to enforce disarmament in respect of pistols and daggers.

Mackworth Young considered that the introduction of a scheme for the total disarmament of these districts will imperil the viability of the frontier administration.[122] The views of H.A. Deane, chief commissioner of newly-formed North-West Frontier Province were perhaps most interesting. He questioned the very idea of relationship between the possession of arms and crimes committed. He argued the possession of any particular class of weapon amongst the Pathans is not necessarily any guide to the question of crime. In fact, he argued, amongst the Yusufzai clans (trans-border), where every man has his knife, murders were far less in number than amongst their fellow clansmen in British territory while burglaries and thefts were of rare occurrence. This was explained with the help of the fact that these clans across the frontier still had communal government, which used to prevail in British controlled Yusufzai region, but which was broken down by the British system of government:

> Disputes which, in one case, are settled by the community, in the other go to our courts, resulting mostly in embitterment of feeling. Across the border the petty official, the patwari, the petition writer and others who, with petty legal knowledge, ferment trouble, do not exist.[123]

In fact, he argued, that if there is any certainty of the death penalty being enforced by our courts for murder, the number of murders committed would be reduced immediately.[124] It was argued, once deprived of means of committing murder, the people from frontier districts would easily found other ways to commit such crimes with the help of poison such as arsenic.[125] Further, there was a large scale discontentment against the prohibition of knives and pistols, affecting poorer zamindars, who had to rely on themselves to cut their own wood and grass on the hills. It was this class who were mostly full of complaints. Their argument was that a man cannot carry his load of wood or grass and a rifle at the same time. That, if he could do so, he still could not be quick enough in bringing it into use when stopped by trans-border man, whom he is always liable to meet, whereas, in a knife or pistol, he has a handy weapon with which he can make some attempt to protect himself.[126] The difficulty was perhaps a real one for it was not possible to disarm these men without placing them at the mercy of gangs and thieves from both within and beyond the border. At the same time the trans-border neighbours were becoming better armed while the police and the border military police were inadequate to protect these districts. In fact, the weapons possessed by the latter were inferior to the weapons in the hands of the trans-border raiders.[127]

CONCLUSION

It will be perhaps incorrect to study this 'conjuncture' without understanding the position of arms in these districts in their pre-colonial past and how the colonial state moulded and controlled the material and discursive aspects of arms possession. Unfortunately, the available studies have focused more on the 'military' aspects of arms rather than writing a history of arms in civil society in general.[128] However, the large number of murders reported from these districts was always a concern of the colonial state. But, it was only during the end of the nineteenth century when disarmament, however, partial was finally carried out. This was for the reason that large number of murders committed came to include murder of European and Indian soldiers. The first set of measures for the disarmament of these districts originated on 5 May 1899 with an emphasis on restricting crime in cantonments and municipalities. The reports submitted on 1 June 1900 made it discernible that the measures have completely failed to achieve its objectives. Another set of more comprehensive measures were promulgated on 16 November 1900. The evaluation of these measures after almost one year dissatisfied the authorities. An important question at this moment may be that how disarmament of these districts was different from the disarmament of Hindustan after the Mutiny or the disarmament of the Punjab after the annexation in 1849. The one important difference was that these districts were at the frontier which was inhabited by the 'hostile' tribes. In this regard, the remarks of W.R.H. Merk were very important:

> London dominates the south-east of England. In a humble way Peshawar is the London of Yaghistan from the Indus to Tirah. If the Peshawar population, rural and urban, is violently against us, if there is a wave of discontent in the plains, it cannot fail to communicate itself to the hills. There is therefore every ground, if we wish the anti-English feeling in the Yaghistan to calm down, and if we wish the attitude of the clans to change, that we should cherish the contèntment of our own subjects

in the frontier districts. It may be asked if the feelings aroused by disarmament are likely to be lasting. I fear they will.[129]

As a result leaving aside the question of general disarmament of these districts at this point of time, the government was not even in a position to disarm a single small village like Tahkal or Pooka. At this time rather than going for a house to house search, the colonial state preferred waiting for the 'natives' to come forth and register their arms to get a license or to surrender them. But, this proved to be a futile effort. More important, the presence of hostile neighbours at the other side of the frontier also affected the state's response. It was not possible to provide security to the inhabitants of the villages on the frontier without providing them arms for self-defence. As a result official response for disarmament was fluctuating and they made half hearted efforts to disarm these districts.

But, one important question remains unanswered. How this disarmament should be assessed when we do not have any idea of actual number of arms in the hands of the 'natives'? The colonial state's position was that all disarmament efforts were a failure. Should we accept these disarmament efforts as a failure and agree with official response. Such a simplistic answer for this complex question will be a mistake. One should not answer this question in terms of failure or success. Rather one should understand the context in which these disarmament measures were carried out. Disarmament in this context was not sweeping arms from 'native' society through forceful measures. Rather it was a step by step approach in which these districts were brought within the purview of the Arms Act. It is acceptable that all the arms present in the hands of the 'natives' were definitely not removed by these measures but, these districts were legally now equivalent to their cis-Indus neighbours in relation to the Arms Act. The possession of arms and carrying of arms became a crime which was earlier was not the case.

NOTES

1. See Kaushik Roy, 'Military Synthesis in South Asia: Armies, Warfare, and Indian Society, *c.* 1740-1849', *Journal of Military History,* vol. 69 (July 2005), pp. 651-90; Daniel Headrick, *Tools of Empire* (Oxford: Oxford University Press, 1981).

2. *Political Diaries of the Agent to the Governor General, North-West Frontier and Resident at Lahore, Punjab Government Records*, vol. 3, *1st January 1847 to 4th March 1848* (Allahabad: The Pioneer Press, 1909), p. 149.
3. 'Exemptions of Soldiers, & C. in the Punjab from the Operation of the Arms Act', Home (Public), 1886/95-98, p. 6, Govt. of India (GOI), National Archives of India (NAI), New Delhi.
4. 'Supply of Arms and Ammunition to the Special and Ordinary Police Reserves in Madras, Bengal, Punjab and the Central Provinces', Home (Police), July 1890/68-80, p. 21, NAI.
5. 'Police Re-organisation and Reserves', Home (Police), March 1889/140-175, p. 187.
6. 'Armament of the Police', Home (Police), July 1900/64, p. 2.
7. For such cases see 'Armament of the Police', Home (Police), July 1900/64, p. 1.
8. T.R. Moreman, *The Army in India and the Development of Frontier Warfare, 1849-1947* (Basingstoke: Macmillan, 1998), p. 68.
9. T.R. Moreman, 'The Arms Trade and the North-West Frontier Pathan Tribes, 1890-1914', *Journal of Imperial and Commonwealth History* (*JICH*), vol. 22 (1994), pp. 188, 201.
10. William K. Storey, 'Guns, Race and Skill in Nineteenth Century South Africa', *Technology and Culture*, vol. 45 (October 2004), p. 688.
11. Foreign Department, 26 May 1849/65-7, p. 5, NAI.
12. Sukhdev Singh Sohal, *The Making of the Middle Classes in the Punjab, 1849-1947* (Jalandhar: ABS Publications, 2008), pp. 12-13.
13. *Political Diaries of the Agent to the Governor General, North-West Frontier and Resident at Lahore*, vol. 3, p. 153.
14. Foreign Department, 26 May 1849/65-7, p. 1.
15. *Report on the Administration of the Punjab and its Dependencies for the year 1867-68* (Lahore: Civil Secretariat Press, 1868), p. 38.
16. *Monograph on Murders for 1897.* Appendix C, p. xxvii, in *Report on the Police Administration in the Punjab for the year 1897* (Lahore: The Civil and Military Gazette Press, 1898) (*PARP*). For Western Circle it was written by C. Brown, deputy inspector general of police of Western Circle.
17. The 'Punjab' in throughout this paper means pre-1901 Punjab. Surya Kant, 'Administrative Space in the British Punjab', in Reeta Grewal and Sheena Pal (eds.), *Pre-colonial and Colonial Punjab: Society, Economy, Politics and Culture* (New Delhi: Manohar, 2005), p. 221.
18. R. Udny, Commissioner and Superintendent, Peshawar Division to Deputy Commissioner Peshawar, dated 7 October 1897 in 'Prohibition of the Possession of Arms in Peshawar Division', Home (Public), May 1899/225-40, p. 35.
19. The Committee consisted of Lepel Griffin, Secretary to Government, as

President, and five other members who were F.R. Pollock, commissioner, Peshawar Division, H.L. Miller, inspector general of police, Punjab, W.G. Waterfield, deputy commissioner, Hazara, P.L.N. Cavagnari, deputy commissioner Peshawar and J.M. Trotter, deputy assistant quarter master general. For the Report of the Committee on Police Administration, and Border Defence of the Peshawar Division, 1878 (henceforth Griffin Committee Report) see 'Rules under the Indian Arms Act, 1878, Home (Public), July 1879/192-277.

20. Griffin Committee Report, p. 72.
21. Ibid., p. 73.
22. Ibid., p. 74.
23. Both quotations from ibid., p. 115.
24. In a letter to the GOI, dated 23 March 1878, 'Prohibition of the Possession of Arms in Peshawar Division', Home (Public), May 1899/225-40, p. 35.
25. *Punjab Government Gazette*, dated 16 May 1878 (Lahore: The Civil and Military Gazette Press, 1879).
26. 'Prohibition of the Possession of Arms in Peshawar Division', Home (Public), May 1899/225-40, p. 35.
27. R. Udny, Commissioner and Superintendent, Peshawar Division to Deputy Commissioner Peshawar, dated 7 October 1897, 'Prohibition of the Possession of Arms in Peshawar Division', Home (Public), May 1899/225-40, p. 36.
28. Ibid., p. 37.
29. David Arnold, *Police Power and Colonial Rule, Madras: 1859-1947* (New Delhi: Oxford University Press, 1986), p. 39.
30. Denzil Ibbetson, *Punjab Castes* (1916, rpt., Patiala: Languages Department Punjab, 1995), p. 58.
31. Lieutenant-Colonel W.J. Keen, *The North West Frontier Province and the War* (Peshawar, 1928), p. 13. Library of NAI, Acc. No. 954.52,K.25N.
32. Ibbetson, *Panjab Castes*, p. 42.
33. Ibid., p. 43.
34. Ibid., p. 58.
35. Ibid., p. 59.
36. Griffin Committee Report, p. 78.
37. Ibid.
38. Ibid., p. 79.
39. Ibid.
40. *District Gazetteer of Rawalpindi, 1893-94* (Lahore: The Civil and Military Gazette Press, 1895), p. 150.
41. Ibid.
42. *General Report on the Administration of the Punjab Territories for the Year 1861-62* (Lahore: Government Press, 1861), p. 20.

43. *Report on the Administration of the Punjab and its Dependencies for the Year 1867-68,* p. 41.
44. *PARP*, 1898, p. 17.
45. See Tables 10.1 and 10.2.
46. 'Proceedings of the Lieutenant Governor of the Punjab, in the Home (Police) Department dated the 30 August 1900', *PARP*, 1899, p. 3.
47. *PARP*, 1900, p. 6.
48. *Report on the Police Administration of the North-West Frontier Province for the Year 1903* (Peshawar: Civil and Military Steam Press, 1904), p. 4.
49. Ibid., p. 8.
50. *PARP*, 1898, p. 9.
51. *PARP*, 1898, p. 17.
52. *PARP*, 1894, p. 14
53. *PARP*, 1898, pp. 21-2.
54. By H.P.P. Leigh, Commissioner and Superintendent of Peshawar Division in 'Disarmament of the Frontier Districts of the Punjab', Home (Public), July 1901/209-13, p. 39, NAI.
55. Gyanesh Kudaisya, '"In Aid of Civil Power": The Colonial Army in Northern India, *c.* 1919-42', *JICH,* vol. 32, no. 1 (2004), p. 45.
56. Rajit K. Mazumder, *The Indian Army and the Making of Punjab* (New Delhi: Permanent Black, 2003), p. 51.
57. Ibid., p. 61.
58. Robert Warburton, *Eighteen Years in the Khyber, 1879-98* (London: John Murray, 1900), p. 41.
59. G.D. Martineau, *Controller of Devils: A Life of John Paul Warburton, C.I.E. of the Punjab Police* (London: Dorset Press, 1910), p. 43.
60. 'Arming of the European staff employed on the frontier districts of the North-Western Railway with revolvers and swords', Foreign (External), June 1900/35-38, pp. 1-2, NAI.
61. In a letter dated 2 May 1900, 'Disarmament of the Population of the Peshawar and Derajat Divisions', Military, December 1900/613-15, Home (Public), p. 3.
62. *PARP*, 1894, p. 17.
63. In the year 1896 out of 52 cases of arms theft most of them were reported from trans-Indus districts, see *PARP*, 1896, pp. 23-5.
64. Tucker Committee Report, see 'Control of Certain Classes of Firearms and Ammunition in India', Home (Public), March 1900/231-52, p. 90.
65. For a detailed account and discussion of arms thefts and the concern of the colonial state see Tucker Committee Report, see 'Control of Certain Classes of Firearms and Ammunition in India', Home (Public), March 1900/231-52.
66. 'Disarmament of the Population of Peshawar District by Applying to it the Provisions of the Arms Act', Military Department (Arms), May 1899/2619-24, NAI.

67. E.R. Elles to Deputy Adjutant-General, Punjab Command, dated 6 August 1897, Peshawar, in 'Proposed Disarmament of the Peshawar and Kohat Districts', Home (Public), November 1897/78-96, p. 20.
68. Dated 11 September 1897, 'Proposed Disarmament of the Peshawar and Kohat Districts', Home (Public), November 1897/78-96, p. 19.
69. 'Disarmament of the Population of Peshawar District by Applying to it the Provisions of the Arms Act', Military (Arms), May 1899/2619-24, p. 2. 'Exemptions of Certain Border Villages in the Kohat and Hangu Tahsils of the Kohat District from the Operation of Certain Sections of the Arms Act', Home (Public), May 1898/384-90, p. 14.
70. 'Disarmament of the Population of Peshawar District by Applying to it the Provisions of the Arms Act', Military (Arms), May 1899/2619-24, p. 28.
71. Trans-border Political diaries of the Derajat Frontier for the month of March 1898, See Foreign Department (Frontier), September 1898/216-40, p. 12, NAI.
72. Ibid., pp. 17-18.
73. 'Disarmament of the Population of Peshawar District by Applying to it the Provisions of the Arms Act', Military (Arms), May 1899/2619-24, p. 1.
74. Ibid., p. 2.
75. In a letter dated 26 March 1899, Peshawar Division in 'Disarmament of the Population of Peshawar District by Applying to it the Provisions of the Arms Act', Military (Arms), May 1899/2619-24, p. 9.
76. Peshawar Confidential Diary No. 22, dated 23 November 1897, 'Peshawar Confidential Diaries, nos. 21 to 24 of 1897', Foreign (Secret F), January 1898/262-74, p. 3, NAI.
77. Peshawar Confidential Diary No. 23, dated the 13 December 1897, 'Peshawar Confidential Diaries, nos. 21 to 24 of 1897', Foreign (Secret F), January 1898/262-74, p. 2.
78. Captain H.F. Walters, 'The Operations of the Malakand Field Force and the Buner Field Force, 1897-98', Military Department (Miscellaneous), 1897-98/33, p. 77, NAI.
79. 'Proposed Disarmament of the Peshawar and Kohat Districts', Home (Public), 1897/78-96, p. 13.
80. See also Tucker Committee Report in 'Control of Certain Classes of Firearms and Ammunition in India', Home (Public), March 1900/231-52.
81. W.R.H. Merk, Commissioner, Peshawar Division to the Chief Secretary to the Government of Punjab, 2 February 1898, in 'Prohibition of the Possession of Arms in Peshawar Division', Home (Public), May 1899/225-40, p. 46.
82. Ibid., p. 47.
83. Merk to the Chief Secretary to the Government of Punjab, 2 February

1898, in 'Prohibition of the possession of Arms in Peshawar Division', Home (Public), May 1899/225-40, p. 48.

84. J. Wilson, Officiating Secretary to Government of Punjab to Secretary to the GOI, Home Department, 29 February 1899, Lahore in 'Prohibition of the Possession of Arms in Peshawar Division', Home (Public), May 1899/225-40, p. 29.
85. 'Prohibition of the Possession of Arms in Peshawar Division', Home (Public), May 1899/225-40, p. 30.
86. Ibid.
87. Ibid., p. 31.
88. Ibid.
89. Ibid., p. 32.
90. Ibid.
91. Ibid. p. 33.
92. Tucker Committee Report, see 'Control of Certain Classes of Firearms and Ammunition in India', Home (Public), March 1900/231-52, p. 71.
93. 'Arming of Certain Villages of the Bannu and Dera Ismail Khan Districts for purposes of Defence Against Trans-border Raiders', Foreign (Frontier), September 1900/ 55-7.
94. 'Prohibition of the Possession of Arms in Peshawar Division', Home (Public), May 1899/225-40, p. 44.
95. 'Proposed Disarmament of the Peshawar and Derajat Divisions', Home (Public), November 1900/322-29, p. 71.
96. 'Disarmament of the Population of the Peshawar and Derajat Divisions', Military (Arms), December 1900/613-15, NAI, GOI, p. 12.
97. In a letter dated 2 May 1900, 'Disarmament of the Population of the Peshawar and Derajat Divisions', Military, December 1900/613-15, p. 9.
98. Ibid.
99. In a letter dated 12 May 1900, 'Disarmament of the Population of the Peshawar and Derajat Divisions', Military, December 1900/613-15, p. 6.
100. In a letter dated 1 June 1900, 'Disarmament of the Population of the Peshawar and Derajat Divisions', Military, December 1900/613-15, p. 3.
101. 'Question of the Disarmament of the Villages of Tahkal Kalan and Pooka, Peshawar District, Punjab', Home (Public), September 1899/8-9, p. 2.
102. Ibid., p. 1.
103. In a letter dated 7 September 1899, 'Question of the Disarmament of the villages of Tahkal Kalan and Pooka, Peshawar District, Punjab', Home (Public), September 1899/8-9, p. 9.
104. F.P. Rennie, Deputy Commissioner, Kohat to the Commissioner and Superintendent of Peshawar Division, dated 27 April 1900, 'Proposed

Disarmament of the Peshawar and Derajat Divisions', Home (Public), November 1900/322-9, p. 22.

105. 'Proposed Disarmament of the Peshawar and Derajat Divisions', Home (Public), November 1900/322-9, p. 23.
106. District Superintendent of Police, Kohat to Deputy Commissioner Kohat, 'Proposed Disarmament of the Peshawar and Derajat Divisions', Home (Public), November 1900/322-9, p. 25.
107. 'Proposed Disarmament of the Peshawar and Derajat Divisions', Home (Public), November 1900/322-9, p. 2.
108. Ibid., p. 24.
109. In a letter dated 14 April 1900, District Superintendent of Police, Kohat to Deputy Commissioner Kohat, 'Proposed Disarmament of the Peshawar and Derajat Divisions', Home (Public), November 1900/322-9, p. 28.
110. In a letter dated 10 March 1898, 'Proposed Disarmament of the Peshawar and Derajat Divisions', Home (Public), November 1900/322-9, p. 45.
111. In a letter dated 6 January 1900, 'Proposed Disarmament of the Peshawar and Derajat Divisions', Home (Public), November 1900/322-9, p. 61.
112. In a letter dated 20 October 1900, 'Proposed Disarmament of the Peshawar and Derajat Divisions', Home (Public), November 1900/322-9, p. 75.
113. 'Disarmament of the frontier Districts of the Punjab' Home (Public), July 1901/209-13, pp. 13-14.
114. 'Proposed Disarmament of the Peshawar and Derajat Divisions', Home (Public), November 1900/322-9, p. 65.
115. In a letter dated 24 November 1900, 'Disarmament of the Frontier Districts of the Punjab', Home (Public), July 1901/209-13, p. 15.
116. Ibid., p. 15.
117. In a letter dated the 9 July 1901, 'Disarmament of the Frontier Districts of the Punjab', Home (Public), July 1901/209-13, p. 15.
118. In a letter to Secretary to the Government of India dated 15 March 1901, 'Disarmament of the Frontier Districts of the Punjab', Home (Public), July 1901/209-13, p. 9.
119. In a letter dated 16 July 1900, 'Proposed Disarmament of the Peshawar and Derajat Divisions', Home (Public), November 1900/322-29, p. 19.
120. 'Control of Certain Classes of Firearms and Ammunition in India', Home (Public), March 1900/231-52, p. 135.
121. 'Proposed Disarmament of the Peshawar and Derajat Divisions', Home (Public), November 1900/322-9, p. 71.
122. 'Disarmament of the Frontier Districts', Home (Public), May 1902/440, p. 7.

123. H.A. Deane, Chief Commissioner, N.W.F. Province to Secretary to GOI, Foreign Department, 5 February 1902, 'Disarmament of the Frontier Districts', Home (Public), May 1902/440, p. 8.
124 This is what I would like to call 'colonial criminology'. Certainty of death penalty was regarded as an indicator of efficiency and instrument of crime control.
125. Deane, Chief Commissioner, N.W.F. Province to Secretary to GOI, Foreign Department, 5 February 1902, 'Disarmament of the Frontier Districts', Home (Public), May 1902/440, p. 9.
126. Ibid., p. 10.
127. 'Disarmament of the Frontier Districts', Home (Public), May 1902/440, p. 11.
128. For instance work like Headrick, *Tools of Empire.*
129. W.R.H. Merk, Commissioner, Peshawar Division to the Chief Secretary to the Government of Punjab, 2 February 1898, in 'Prohibition of the Possession of Arms in Peshawar Division', Home (Public), May 1899/225-40, p. 48.

CHAPTER 11

What Paul Scott Might Have Read: *Our Indian Empire* and Anglo-Indian Ideology during Second World War[1]

CHANDAR S. SUNDARAM

> . . . [T]he young men, for instance, who are out here as soldiers, young Englishmen . . . as we all know, have absolutely no idea about India except that it is a long way from home and full of strange, dark-skinned people.

The above passage comes not from a history book but from a work of historical fiction. It was uttered by Mohammed Ali Kasim, a fictional Muslim politician from a fictional province of British India, who appears in Paul Scott's novel *The Day of the Scorpion.*[2] The British have jailed Kasim in their attempt to nip the 'Quit India' agitations of mid-1942 in the bud. Yet, though Kasim is a fictional creation, we know from the work of Christopher Thorne that his opinion of the British 'Tommy' sent out to India during the Second World War (ostensibly to defend the Raj from the Japanese, but also to ensure the Indian Empire's continued co-operation in the Allied war effort) dovetail rather well with the views of actual leaders of the Indian Nationalist Congress, such as Mahatma Gandhi and Jawaharlal Nehru.[3] We also know from the work of Robin Moore and Hilary Spurling[4] that Paul Scott, who was later hailed as the 'prose-poet of the Raj in decline',[5] was himself one of the 'young English soldiers' to whom Kasim referred.

Scott was not quite twenty-three when he sailed for India in February 1943.[6] It was to be his first experience of a country that was to be the focus for almost all of his fiction. There is no evidence, however, that he had any prior knowledge of India beyond the usual schoolboy fare. He himself admitted as much many years later: 'I knew that there had been a Mutiny,[7] a Black Hole [of Calcutta],[8] and that sometimes the chaps had to have a scrap with the Pathans and settle the Khyber.'[9] It is therefore quite possible that before leaving England or just after arriving in India, he might have picked up a copy of *Our Indian Empire.* This work, 'a short review and some hints for the use of soldiers proceeding to India', was expressly designed for people like him. During the British Raj in India, it was not expected, nor encouraged, that officer cadets like Scott should take any serious interest in India, her history, or culture.[10] General Sir Philip Chetwode, one of the last commanders-in-chief of the British Army in India, criticized this attitude, which encouraged officers with 'narrow interests . . . bounded by the morning parade [and] the game they happen[ed] to play . . . [and] quite unaware of the larger aspects of what . . . [was] . . . going on in India around them.'[11]

Our Indian Empire went through four editions in a span of twenty-eight years. The first edition was published in London, in 1912, and ran to 119 pages. Twenty-three years later, in 1935, the second edition was published. This edition was an expanded one, 137 pages long. Either due to its popularity among British troops sent to India, or perhaps because it was mandated to be part of their standard 'kit', the next edition of *Our Indian Empire* was issued in 1938, and was 149 pages in length. The onset of the Second World War a year later was the most probable impetus for issuing the fourth edition, in 1940. In this edition, the book underwent its most considerable expansion, rising to 170 pages. Another noteworthy feature is the fact that, while the 1912 edition lists the author as A.G. Stuart, the second and third editions were issued under the authority of the Manager of Publications, Delhi, while the fourth edition was authored by the General Staff of the Army in India.[12] The authorship of the fourth edition lends weight to my assumption that the book was indeed made part of the standard kit of every British soldier posted to India. The present paper is an examination of the fourth edition of *Our Indian Empire.* It will analyse its content, evaluate how this content reflected Anglo-Indian ideology in the waning years of the Raj, and finally, evaluate the book's utility to a young Englishman like Scott.

Imperial powers, by their very nature, have to develop ideologies to justify their rule over their colonial territories, both to themselves and to their subject peoples. The British colonial state in South Asia was no exception. Anglo-Indian ideology, however, demands a more nuanced scrutiny than postmodern and post-colonial cultural theorists, coming at the topic from an essentially a-historical literary background, are willing to admit.[13] This is because the Anglo-Indian ideology was never a simple, coherent monolith of ideas. Rather, as Thomas Metcalf convincingly argues, the ideas buttressing the British imperial enterprise in South Asia were complex, and, at times, contradictory and inconsistent. This was because Anglo-Indian ideology was animated by two fundamental and opposing principles: those of 'similarity' and 'difference'. Metcalf postulates that: '[a]t . . . times, and for some purposes, the British conceived of the Indians as people like themselves, or as people who could be transformed into . . . a facsimile of themselves; while at other times, they emphasised what they believed to be the enduring quality of Indian difference.'[14]

The situation was further complicated by the fact that, at times Anglo-Indian ideology simultaneously accommodated both similarity and difference, which renders any attempt at straightforward demarcation difficult. Despite this, however, a general periodization is possible: during the Company Raj, varieties of similarity, such as the Orientalism of William Jones and Hastings, and the Utilitarianism of Lord William Bentinck and the elder James Mill, predominated Anglo-Indian Ideology. Thomas Macaulay's pronouncement that the ultimate aim of British rule in the subcontinent was the creation of an independent India firmly based on English moral, political and legal precepts bespoke a belief in a Briton and an Indian having innately similar aptitudes and abilities.[15]

In sharp contrast, during the Crown rule, the differences between Briton and Indian were highlighted, almost to the extent of becoming institutionalized, by Anglo-Indian ideologues such as James Fitzjames Stephen, Henry Maine, and Alfred Comyn Lyall. Indians, far from being rational and intelligent people, capable of assimilating English values and practices, were now characterized as a people '. . . steeped in idolatrous superstition and ignorant to the last degree. . . .'[16] Such a people, therefore, were deemed incapable of improvement. Significantly, this view essentially stemmed from the Anglo-Indian inability to comprehend why, during the 'mutiny', Indians in large numbers had rejected that obvious embodiment of 'modernity and human progress': British rule.

Though Orientalism, as conceptualized by Edward Said, is rather one-sided in its approach, seeing Western imperialism as a grand predetermined project or conspiracy, to control and conceptualize the non-Western 'Other',[17] the anthropologist Bernard Cohn fine-tunes the original Saidian concept in ways that are especially applicable to the analysis of *Our Indian Empire.* In his book *Colonialism and its Forms of Knowledge: The British in India,* Cohn proposes a framework of six interpretive modalities to analyse Anglo-Indian knowledge of India. Only four of these are pertinent to this essay. The historiographic modality constructed the nature of Indic civilizations in conformity with Anglo-Indian ideological lines.[18] The enumerative modality refers to the imperialist's passion for gathering and taxonomically ordering information concerning imperial possessions on a large scale for basically revenue-gathering purposes.[19] Closely allied to this was the survey modality, which covered '. . . a wide range of measuring practices . . . and the process by which these vast amounts of knowledge were transformed into textual forms . . . that were deployed by the state in fixing, bounding, and settling India.'[20] The museological modality was one by which India was represented as a vast museum, its people typifying past ages.[21] These modalities provide us with the tools by which we can analyse *Our Indian Empire.*

Our Indian Empire consists of ten chapters. It begins with an introductory chapter, which states the aim of the book as being to introduce the British newcomer to India. 'Everyone going abroad', it intones, 'will find it a great advantage to learn beforehand something about the people likely to be met in the new land'.[22] Almost immediately, it stresses how different India is: '. . . the customs, character, dress, language, colour, and religions of the inhabitants have but little resemblance to that of the Englishman.'[23] By stating this, the book not only justifies itself, but also sets up India as a place which is ineluctably and insurmountably different. An interesting feature of this chapter is that, although only three pages long, it devotes a full one-and-one-third pages to the Indian princely states ruled indirectly by the Crown. It speaks of them in the most laudable terms. The 1940 edition of the book says that the ancestors of the ruling princes remained loyal to the British cause during the 'mutiny', and that: 'Big or small, [the] internal affairs [of these princely states] are not interfered with except insofar as it is necessary to maintain some slight control over such legislation and action as might involve imperial interests.'[24]

This echoes an instruction manual for officers of the Foreign and Political Department of the Government of British India—whose job it was to liaise between the princely states and the Crown—which told political officers to '. . . leave well alone; the best work of a political officer is very often what has been left undone.'[25] *Our Indian Empire* also tells us that most of the more prosperous princely states have established and maintained, at their own expense, armies or 'State Forces' to '. . . aid in the defence of the country and the upholding of the national honour, thus showing in practical form their desire for the continuance of British authority.'[26] This last statement is noteworthy because of the contradiction it reveals. It speaks of a 'national honour', but during the early part of the twentieth century, most official and unofficial Anglo-Indians were emphatic in their rejection of the concept of India as a nation. The question therefore must be raised as to what kind of 'national honour' *Our Indian Empire* was talking about. Was it the national honour of the states themselves? If it was, then how could this be reconciled to their desire for the continuance of Anglo-Indian overlordship, which, undoubtedly, would tend to diminish the autonomy of the princely states, and therefore be inimical to their 'national honour'? Then again, the book's concept of the national honour of India could be wholly Anglo-centric, arguing that, as the British were the first of India's myriad conquerors to weld the whole subcontinent into one political entity, they were the ones who created the country's nationality and therefore, its national honour. Thus, it would be perfectly logical to expect British rule to continue.

Yet, the reality was somewhat different. 'Native' princes knew fully well that they were subordinate partners in an unequal relationship that left them suffering from a sense of inferiority that was integral to the whole concept of paramountcy. They knew they needed the support of Anglo-India if they wanted to remain as they were. And yet, this basic fact rankled because every nawab or raja, proud as he was of his lineage, traditions and the fact that he ruled a kingdom, nevertheless resented the prying eyes of Anglo-India in the form of the Foreign Department.[27]

The emphasis on princely states in *Our Indian Empire* is quite remarkable. Eleven pages of the book are devoted to a table detailing the statistics of 120 princely states—there were over 500. For each of these states, we are given its area in square miles, its population, and its permanent salute—that is, the number of artillery pieces that would be fired in honour of the state's raja or nawab on state occasions.[28] Although it is conceivable that this information would be of use to British officers of higher grade, why it should be part of the ordinary British soldier's introduction to India escapes this author.

Geographical information in *Our Indian Empire* is contained, appropriately enough, in a chapter titled 'The Frontiers of India'. Frontiers and their defence was an abiding preoccupation of Anglo-India. By the time the first edition of the book came out in 1912, the Raj had reached the 'natural' frontiers of the subcontinent. One did not have to worry much about the Himalayan 'roof of the world', but the north-west was another matter. For Anglo-Indians, and for that matter, for most Britons, the word 'frontier' in connection with India conjured up images of a mountainous land, harsh and barren, fissured by the Khyber and Bolan passes, sparsely peopled by the wily, treacherous, and fiercely independent Pathans, Afrīdis and Afghans.[29] It also meant 'The Great Game', a Cold War waged from the late-1820s onwards between Anglo-India and Imperial Russia, and its successor, the Soviet Union.[30] Whatever the reality of this view, it was imprinted on the Anglo-Indian psyche by writers such as G.A. Henty and Rudyard Kipling. It is no accident that the short bibliography appended to chapter four of the 1940 edition of the book includes Kipling's tale of frontier derring-do, *Kim*.[31] That this overly romanticized view is still with us is evidenced by the work of contemporary authors such as Peter Hopkirk.[32] The 1940 edition of *Our Indian Empire* still made great hay of the north-west frontier, an area that, apart from some paltry operations against the Wali of Swat and the *Faqir* of Ipi,[33] was mostly quiet during the Second World War. In contrast, India's north-east frontier is dismissed with these few words: 'In the north-eastern frontier tracts dwell various semi-civilized tribes. They possess firearms, and, in general, do not give much trouble.'[34]

Thus, the 1940 edition is completely oblivious of the fact that, by that time, a very tangible Japanese threat to Anglo-India was developing. Within two years, this threat on the eastern frontier of the subcontinent would become very real indeed.[35] This oversight, in the face of evidence that was all too apparent, leads this author to suspect that the compilers of this edition were blatant racists of the old school, men like Air Chief Marshall Sir Robert Brooke-Popham, the commander-in-chief of the British and Commonwealth forces in the Far East in 1940-1. While on an inspection tour of Hong Kong in October 1941, Brooke-Popham ventured up to the border between the British colony and Japanese-occupied China. There, he '. . . had a good, close up look . . . [at] . . . various sub-human specimens, dressed in dirty-grey uniforms, which . . . [he] . . . was informed were Japanese soldiers.' He later wrote to a friend that '[i]f these represent[ed] the average of the Japanese Army . . . [he] . . . could not believe that they would form an intelligent fighting force.'[36] The fact that even the Punjab Art Press edition of *Our Indian Empire*, printed in 1942, did not even mention the Japanese threat and the consequent serious defence concern in eastern India seems downright scandalous. In this respect, Scott would have found the book to be quite useless as he spent the war as an air supply officer on the north-eastern frontier—and wrote about it brilliantly in his first novel *Johnnie Sahib*.[37]

However, the geographical chapter contains a fairly straightforward way of describing the geography of South Asia that modern teachers of things South Asian might appreciate. It states that:

> India may be described as two triangles having a common base running from Karachi in the west to Calcutta in the east. The greater triangle points to the south with its apex in the Indian Ocean and its two sides washed by the sea. The smaller one points to the north, with its apex stretching to the Pamirs. Its north east side is formed by the almost impenetrable barrier of the mighty Himalayas; while, along its northwest side runs the barren slopes of the Suleiman mountains.[38]

This two-triangle concept might come in handy as an explanatory tool, though only for colonial India as Karachi, the Pamirs, and the Suleiman range, are not part of independent India.

As expected in a book designed for military men, *Our Indian Empire* contains a great deal of information about the army in India. Most of this information is straightforward and technical, detailing such things as the organization of the army and including a brief run-down of the functions of each of the army's departments.[39] However, in introducing the subject, a few remarks revelatory of Anglo-Indian ideology are made. Much is made of the Indian Army's role in aiding the civil power. The aid to civil power function of modern armies has two components: the humanitarian, where armed forces personnel and equipment are deployed to aid people affected by natural disasters (floods, earthquakes and the like); and the coercive, '. . . which relates to the internal use of the . . . army for the maintenance of the . . . state and for dealing with challenges to its authority. . . .'[40] The book states that this is a most important function in India. This has been corroborated by a recent article which states that

> [i]n colonial times, the 'internal security' functions of the army were considered absolutely vital by colonial officials, as upon them ultimately depended the ability of the Raj to uphold its authority. As much as one-third of the resources and manpower of the Army in India were committed almost exclusively to these functions.[41]

In keeping with its dominant ideological tone, *Our Indian Empire* employs difference to explain why. 'The population of India is made up of widely different races . . . [between whom] . . . religious feeling and party feeling are apt to create ill-blood.'[42] It is then maintained that Indians of one class are prone to exhibit uncalled for jealousy if the Anglo-Indian government grants a concession to another class, and that Anglo-Indian efforts 'to do right'[43] were sometimes misunderstood. That the obvious solution to this problem was to grant concessions to all communities on an all-India basis was not entertained. The jealousy argument was a common one in Anglo-Indian circles. It was used to great effect by all opponents of liberal measures to forestall any forward movement on those issues. This passage concludes by pinpointing another major area of fundamental difference between Indian and Briton by stating that '. . . the great majority of the [Indian] people are uneducated, excitable, and are easily led astray by those who are able to play upon their feelings with a ready tongue. . . .'[44] The picture presented, therefore, is of a people who are, for the most part, inert and docile, incapable of collective agency until provoked by rabble-rousers.[45] These passages dismiss Indian nationalism out-of-hand. They were indicative of an Anglo-Indian ideology that never really came to grips with the nature of Indian nationalism.[46] From the military perspective, however, such depictions of Indians were essential if the army was to be an instrument of civil power, because they dehumanized the Indian, or, at the very least rendered them child-like, not fully responsible for their own actions. What is noteworthy here is that the swiftness with which the British put down the Quit India disturbances of August 1942 was due, in no small part, to the deployment of 57 British battalions, notably in Bihar and the United Provinces.[47]

Though this chapter deals with the Indian Army, no mention is made of the Indianization of the Indian Army's officer corps. This is indeed surprising, for, by the time the fourth edition of *Our Indian Empire* was published, Indianization—the admission of Indians into the Indian Army's officer corps—had been underway for a little over twenty years.[48] Moreover, it was accelerating quickly, as the Indian Army expanded from a peacetime to a wartime strength. In fact, in 1940, the Government of India ruled that, henceforth, Indian Commissioned Officers were to be '. . . available for posting throughout the Indian Army, where[ever] their services can best be used', instead of being segregated into certain specified units, as had previously been the case.[49] That *Our Indian Empire* ignores Indianization is particularly egregious when taken together with the fact that the wartime expansion of the Indian Army entailed the recruitment not only of increased numbers of Indian officers, but also of British officers unfamiliar with the Indian Army.[50] The inclusion of a section on Indianization and Indian officers would have undoubtedly gone a long way in defusing some rather fraught situations in Indian Army units in the Far East prior to the Japanese attack.[51] It also might have helped maintain, or even bolster, unit morale and cohesion once the battle was joined in Malaya in December 1941.

Paternalism is one of the elements that arises out of difference, and, in this regard, *Our Indian Empire* comes across as a highly paternalistic document. This is most apparent in the chapter on relations between British soldiers and Indians. It is assumed that the type of Indians with whom the British soldiers will have the most contact were the servants. The book outlines the various types of servants that the British soldier is likely to encounter, and gives the Hindustani/Urdu name for each. There is the cook or *bawarchi*, the sweeper or *mehtar*, the water-carriers or *bhistis*, the horse-tenders, or *syces*, and the washermen, or *dhobis*. It also dilates on the basic characteristics of the servant class, with special emphasis on their deviousness. Indian servants, when questioned by their masters, will generally tell them what they think the master wants to hear, irrespective of whether it is the truth or not. Indian servants were also in the habit of taking bribes. No 'innocent-until-proven-guilty' here! Rather, it was the other way around: Indian servants were guilty until proven innocent. Finally, the British soldier, when he has caught a servant red-handed, is warned not to be manipulated into striking or assaulting that servant. In reading this, one gets the impression that whoever wrote it was wistfully reminiscing about earlier times, for by 1940, for a Briton to assault an Indian was definitely not 'politically-correct'.[52]

This chapter also deals with shooting, a great pastime of Anglo-Indians, civil or military, official or unofficial; though doubt must be cast on whether a British soldier going out to India in the desperate days of Second World War would have had much time for *shikar* (hunting). Clearly then, this section is a survival from an earlier edition of the book. Contained in the book is the 12-page section: 'Some Notes on Indian Game', a clear example of Cohn's enumerative and museological modalities. This includes sub-sections on big game, horned game, hill birds, and swamp birds. The soldier is warned that there is a formidable list of birds and animals that must not be shot at, for fear of offending the religious sensibilities of 'natives'. A series of permits and licenses had to be obtained by British servicemen before going on *shikar*. He is also warned of a series of *don'ts* in dealing with Indian villagers and womenfolk while out hunting.[53]

The book's remarks on the subject of railway accommodation are quite illuminating, from the point-of-view of difference. The book acknowledges that Britons are sometimes to blame for thinking it below their dignity to share railway carriages with Indians.[54] Yet, deliberately emphasizing difference, it maintains that, if an Indian had his choice, he would not want to share his carriage with Englishmen either. After rather sternly warning that '. . . so long as an Indian has paid for his ticket, he has as much right as you have in a carriage . . .'[55] the tone becomes more conciliatory, urging the Briton to make the best of a bad situation. 'To take matters into your own hands never pays: you immediately put yourself legally in the wrong and are lucky if you do not have to suffer pretty severely in consequence.'[56]

Designed as it was for British soldiers destined to command, and also serve with, Indian soldiers, it is not at all surprising that a significant part of *Our Indian Empire* should focus on the sepoy. Difference is highlighted—almost fetishized—in this chapter. The British soldier is told that difference exists between not only him and Indians in general, but between Indians themselves. The Indian soldier was a breed apart, '. . . resembl[ing] neither in birth, character, or mode of living, the barrack servants or residents in bazaars with whom he [the British soldier] . . . comes into most frequent contact.' Caste was characterized as an occupational hierarchy, and the four Hindu *varna*s were somewhat likened to the 'orders' of *Ancien Regime* Europe, and were conflated with contemporary European ideas of class. Therefore, *Our Indian Empire* tells us that '. . . the soldier ranks next to the priest at the top of the social scale'. The emphasis on difference, however, blinded the Anglo-Indian to any possibility of fluidity and social mobility in Indian society, which modern research has demonstrated actually was more close to the truth.[57]

The Martial Race ideology was one of the most unequivocal manifestations of difference in Anglo-India. It held that in Indian society, as opposed to European society, only certain 'castes', 'classes', and 'races'—what we today would term ethnicities—possessed the physical courage to bear arms. Four factors made an Indian 'race' martial. The first was its physical bearing. Tall, physically imposing men were preferred over shorter, slighter-built men. The second was the climate of its region. Men from temperate and bracing climes were preferred, because those climatic conditions best approximated the European climate, which was deemed essential for advanced civilization. Third were its credentials as comprising fiercely independent and proud 'yeoman' farmers, unaffected by, and uninterested in book-learning. Finally, there was the increasing shift of recruiting towards the north-west of India, where the climatic factor generally operated, and which was nearer the scene of the 'Great Game'.[58]

Our Indian Empire enumerates a total of sixteen martial races, and describes their characteristics.[59] Five different Muslim classes are described, from the following regions of India: Hindustan, the Deccan, Rajputana and Central India, Madras, and the Punjab. The 'Punjabi Muslims' are identified, somewhat incorrectly, as '. . . more largely employed in the Army than any other class.'[60] However, the class the book most focuses on is the Sikhs. We are presented with what amounts to a brief history of Sikhism, right down to the 'five Ks', or outward manifestations of faith, that the baptized 'Singh' male is supposed to wear. The British soldier is also enjoined never to forget '. . . that the profession of arms is a kingly one in the East.'[61]

In conclusion, *Our Indian Empire* is largely a document that promotes difference between Briton and Indian, and even amongst Indians themselves. It also exemplifies some of the interpretive modalities postulated by Cohn. The list of princely states is an example of the enumerative modality. The notes on Indian game combine the enumerative and the museological modalities. Similarly, the table of the martial races merges the survey and the museological modalities. Indeed, this table seems to be an example of a modality Cohn does not include: the ethnographical. An average British soldier like Paul Scott would have found the practical information contained in Chapter 8, on climate and health, and the latter part of Chapter 10, where a pronunciation guide to the Hindustani language, a glossary of Hindustani words and phrases, and weights and measures tables are provided, the most useful.[62] Other sections of *Our Indian Empire* are redolent of an earlier age, and do not take into account the changes that were occurring in India in the 1940s under the stresses of Second World War.

NOTES

1. The term 'Anglo-Indian' is used in its historical sense, to denote the British community in India during the Raj. This essay is based on a paper presented at the 2000 New York Conference on Asian Studies, as part of a panel titled 'Empire and India: Reading in an Imperial Context'. Thanks

are due to DeWitt Ellinwood for inviting me to participate in the panel. This article is dedicated to the memory of my mother.

2. Paul Scott, *The Day of the Scorpion* (London: Panther Books, 1973), p. 43.
3. Christopher Thorne, 'The British Cause and Indian Nationalism in 1940: An Officer's Rejection of Empire', *Journal of Imperial and Commonwealth History* (*JICH*), vol. 10, no. 3 (1982), pp. 344-57.
4. H. Spurling, *Paul Scott: A Life of the Author of the Raj Quartet* (New York: W.W. Norton, 1991).
5. See Robin Moore, *Paul Scott's Raj* (London: Heinemann, 1990), p. 116.
6. Ibid., pp. 10, 16.
7. This is a reference to the 1857 Uprising when most of northern India rebelled against British rule. It was put down with great difficulty. It passed into British imperial mythology as 'The Mutiny' because it was sparked by an insurrection of the sepoys and sowars (infantry and cavalry soldiers) of the Bengal Army.
8. In 1756, Siraj-ud-Daulah, the nawab (Muslim ruler) of Bengal imprisoned Europeans in a small room, without proper ventilation. Most of them died in this 'Black Hole of Calcutta'. Though in Anglo-Indian memory this event was vilified as an intentional atrocity, and proof-positive of Indian despotism, recent research has shown that it was due to indifference rather than premeditation on part of the nawab.
9. Paul Scott to Dame Freya Stark, 14 February 1976, cited in Moore, op. cit., p. 20.
10. Ibid.
11. Thorne, 'The British Cause', p. 355.
12. Email communication to author from Tim Thomas, Oriental and India Office Collections, British Library, 3 March 2000, in response to a query regarding the publication history and authorship of *Our Indian Empire*, 26 February 2000.
13. A scathing critique—with which the present author generally agrees—of historians subscribing to the postmodern 'literary turn', most of whom generally produce arcane, unreadable, theory-laden, and empirically poor work, is: Richard J. Evans, *In Defence of History* (London: Verso, 1997). See also Dane Kennedy, 'Imperial History and Post-Colonial Theory', *JICH*, vol. 24, no. 3 (1996), pp. 344-63.
14. Thomas R. Metcalf, *The New Cambridge History of India*, III.4: *Ideologies of the Raj* (Cambridge: Cambridge: Cambridge University Press, 1993), pp. x-xi.
15. See S.N. Mukerjee, *Sir William Jones: A Study in Eighteenth Century British Attitudes to India* (Cambridge: Cambridge University Press, 1968); and Javed Majeed, *Ungoverned Imaginings: James Mill's History of British India and Orientalism* (Oxford: Clarendon Press, 1992), pp. 11-46.

16. J.F. Stephen to *The Times,* 4 January 1878.
17. See Edward Said, *Orientalism: Western Conceptions of the Orient* (London: Routledge & Kegan Paul, 1978). An even-handed critique of Said—even though its author professes to be 'polemical'—is J.M. MacKenzie, *Orientalism: History, Theory and the Arts* (Manchester: Manchester University Press, 1995).
18. Bernard Cohn, *Colonialism and its Forms of Knowledge* (Princeton: Princeton University Press, 1996), pp. 5-6.
19. Ibid., p. 8.
20. Ibid., pp. 7-8.
21. Ibid., pp. 9-10.
22. General Staff, India, *Our Indian Empire: A Short Review and Some Hints for the Use of Soldiers Proceeding to India* [*OIE*], 4th edn. (Delhi: Manager of Publications, 1942), p. 1.
23. Ibid.
24. Ibid., p. 3.
25. Lawrence James, *Raj: The Making and Unmaking of British India* (London: Abacus, 1997), p. 328.
26. *OIE,* p. 3
27. On the 'vigilance' of British residents in the princely states, who were members of the Indian Political Service, see: James, *Raj,* p. 330; and T. Creagh Coen, *The Indian Political Service: A Study in Indirect Rule* (London: Chatto & Windus, 1971). For the Indian princes' perspective, see: S.H. Rudolph, L.I. Rudolph and M.S. Kanota (eds.), *Reversing the Gaze: Amar Singh's Diary, A Colonial Subject's Narrative of Imperial India* (Boulder: Westview Press, 2002), pp. 241, 411-85; and S.R. Ashton, *British Policy Towards the Indian States, 1905-1939* (London: Curzon Press, 1982), pp. 1-32.
28. *OIE,* pp. 30-40.
29. This frontier area, now lying between Afghanistan and Pakistan, is unstable even today. If we are to believe what is being said in the news media these days, it is the main sanctuary for both the Al-Qaeda and the Taliban.
30. There is no scholarly treatment covering the 'Great Game' in its entirety. The early period is covered in M.E. Yapp, *Strategies of British India: Britain, Iran and Afghanistan, 1798-1850* (New York: Oxford University Press, 1980), while, for the later period, see: S. Mahajan, *British Foreign Policy, 1874-1914: The Role of India* (London and New York: Routledge, 2002).
31. See Paul Scott's 1976 comment to Frey Stark, cited in n. 6 above; *OIE,* p. 42.
32. Peter Hopkirk, *The Great Game* (London: Hamish Hamilton, 1989).
33. See Alan Warren, *Waziristan, the Faqir of Ipi, and the Indian Army: The North West Frontier Revolt of 1936-37* (New York: Oxford University Press, 2000).

34. *OIE*, p. 11.
35. See R. Callahan, *Burma, 1942-1945* (London: Davis-Poynter, 1978), p. 105; L. Allen, *Burma: The Longest War, 1941-1945* (London: J.M. Dent, 1984); and C. Thorne, *The Issue of War: States, Societies, and the Far Eastern Conflict, of 1941-1945* (London: Hamish Hamilton, 1985).
36. J.W. Dower, *War Without Mercy: Race and Power in the Pacific War* (Toronto: Random House, 1986), p. 99.
37. Published: London, 1952. The action of this novel revolves around Captain Johnnie Brown, a bit of a free-spirit who chafes at the bureaucratizing interference of his superiors into the way he runs his unit. The backdrop for this is the closing year of the war in Burma, where air supply played a major role in Allied strategy there. See Daniel P. Marston, *Phoenix from the Ashes: The Indian Army in the Burma Campaign* (Westport and London: Praeger, 2003).
38. *OIE*, p. 4. This description may be compared with the 1:15 million scale map of 'The Indian Empire, with Burma and Ceylon', in *The New Illustrated Atlas, with a Section Dealing with World Problems* (London: Oldhams Press, n.d., but *c.* late 1944 or early 1945), pp. 42-3.
39. *OIE*, ch. 5.
40. G. Kudaisya, '"In Aid of Civil Power": The Colonial Army in Northern India, 1919-1942', *JICH*, vol. 32, no. 1 (2004), p. 41.
41. Ibid.; *OIE*, Ch. 5.
42. *OIE*, p. 48.
43. Ibid.
44. Ibid.
45. Such perceptions persist to this day. For example, the front cover of G.P. Chapman, *The Geopolitics of South Asia*, 2nd edn. (Aldershot: Ashgate, 2002) depicts a South Asian bearded 'tough', in a T-shirt and jeans, caught in mid-yell, and wielding a metal bar like a sword. In the background, we see orange flames and black smoke.
46. Kudaisya brings this point out well by citing the Simon Report of 1929, as well as a military report on the Meerut District, written in 1928. See Kudaisya 'Civil Power', pp. 42, 64.
47. A. Bhuyan, *The Quit India Movement: The Second World War and Indian Nationalism* (New Delhi: Manohar, 1975), p. 95.
48. See Chandar S. Sundaram, 'Grudging Concessions: The Officer Corps and its Indianization, 1817-1940', in Daniel P. Marston and Chandar S. Sundaram (eds.), *A Military History of India and South Asia: From the East India Company to the Nuclear Era* (Westport and London: Praeger, 2007), pp. 88-101; P.S. Gupta, 'The Army, Politics, and Constitutional Change in India, 1919-1939', in Gupta, *Power, Politics, and People: Studies in British Imperialism and Indian Nationalism* (New Delhi: Permanent Black, 2002), pp. 219-39.
49. Government of India, Defence Department, Communiqué of 17 June 1940, L/MIL/7/19112, India Office Records, British Library, London.

50. On this wartime expansion, see G. Sharma, *Nationalisation of the Indian Army, 1885-1947* (Delhi: Allied Publishers, 1996).
51. See K.K. Ghosh, *The Indian National Army: Second Front of the Indian Independence* (Meerut: Meenakshi Prakashan, 1969), pp. 67-8; and Chandar S. Sundaram, 'Seditious Letters and Steel Helmets: Disaffection Among Indian Troops in Singapore and Hong Kong, 1940-1941, and the Creation of the Indian National Army', in Kaushik Roy (ed.), *War and Society in Colonial India, 1807-1945* (New Delhi: Oxford University Press, 2006), pp. 126-60.
52. There is a bit of a double standard here, because, in India today, for an Indian employer to strike his servant seems to be all right. I even witnessed one such incident on my last research trip to India. Probably, the 'race-card' is the decisive factor.
53. *OIE*, Ch. 7.
54. Remarks by Anglo-Indian officers, such as 'There's room here, isn't there . . . there are some niggers (Indians) in the only other first [-class cabin].' were quite common, even as late as the 1920s. See Edward Thompson, *An Indian Day* (1927, rpt., Harmondsworth: Penguin, 1938), p. 11; Kenneth Ballhatchet, *Race, Sex, and Class under the Raj: Imperial Attitudes and Policies and their Critics, 1793-1905* (London: Weidenfeld and Nicholson, 1980), *passim*. Thompson was the father of the renowned social historian and activist E.P. Thompson, and Ballhatchet, a historian at the University of London's School of Oriental and African Studies, had, like Paul Scott, been posted to India during World War II, as was Eric Stokes.
55. *OIE*, p. 89.
56. Ibid.
57. See Susan Bayly, *Caste, Society, and Politics in India from the Eighteenth Century to the Modern Age* (Cambridge: Cambridge University Press, 1999).
58. For a brief summary, see Chandar S. Sundaram, 'Reviving a Dead Letter: Military Indianization and the Ideology of Anglo-India, 1885-1891', in P.S. Gupta and A. Deshpande (eds.), *The British Raj and its Indian Armed Forces, 1857-1939* (New Delhi: Oxford University Press, 2002), pp. 48-52; A more in-depth consideration, which develops similar themes, is: Douglas M. Peers, 'The Martial Races and the Indian Army in the Victorian Era', in Marston and Sundaram (eds.), *A Military History of India and South Asia*, pp. 34-52. A very good comparative work is: Heather Streets, *Martial Races: The Military, Race, and Masculinity in British Imperial Culture, 1857-1914* (Manchester: Manchester University Press, 2004).
59. *OIE*, pp. 66-72.
60. Ibid., p. 69.
61. Ibid., p. 72
62. Ibid., pp. 153-70.

CHAPTER 12

The Indo-Japanese Collaboration during Second World War: The INA's Spy Schools in Malaya

AZHARUDIN MOHAMED DALI

The story of the Indian National Army (INA) during Second World War (WW II) is well known and has secured a firm place in the historiography of modern India. The bravery and undying commitment of those involved in the INA movement is remembered as one of the most heroic episodes in the struggle for India's Independence. One of the many interesting stories in this battle against imperialism is that of the INA-Japanese collaboration. It was an instance whereby two parties worked together in partnership against a common enemy: the Allies. The willingness of the Japanese to participate in secret operations alongside INA secret agents was a drama that calls for a more in-depth exploration.

Indeed, the history of the Indo-Japanese collaboration during WW II has been dominated by studies on the role of Subhas Chandra Bose at the expense of other insightful perspectives. This is unsurprising given the strong connections between Subhas Chandra Bose and the Japanese. Hugh Toye's work entitled *Springing Tiger* depicts the INA as an extension of the life of Subhas Chandra Bose whilst Leonard A. Gordon's *Brothers Against the Raj* is a comprehensive account of the life of Sarat and Subhas Chandra Bose. The most recent work by T.R. Sareen devotes a large part of its attention to Subhas Chandra Bose.[1] One is left to wonder whether it is at all possible to do a study on the INA without giving much attention to Subhas Chandra Bose.

This article focuses its attention on a lesser known story of the INA: the role and activities of the INA secret agents. Surprisingly, previous studies of the INA and the Indo-Japanese collaboration ignore the importance of the Indo-Japanese spy schools. One respected study of the INA, K.K. Ghosh's *The Indian National Army: Second Front of the Indian Independence Movement* does not discuss INA spying activities but concentrates on establishing that the INA was a combatant army.[2]

So far, very little has been done on the subject of spy schools, one of the most recent studies being the book titled *Indian National Army Secret Service* written by Motilal Bhargava and Americk Singh Gill. Bhargava has compiled various intelligence documents related to INA's secret activities and provided some of the movement's operations. This book is very valuable because Bhargava was able to collaborate with Americk Singh Gill and include his diary as a separate chapter. It is a useful reference to any study on Japanese Indian intelligence activities during WW II, opening a new dimension on the history of the INA such as the organization of the spy schools.[3] Joyce C. Lebra's *The Jungle Alliance* allocates only four pages to the subject, and like Bhargava, she mentions only the spy schools in Malaya.[4] Let us shift our attention to the beginning of spying activities by the INA and the Japanese.

THE EARLY COLLABORATION

It were the Japanese who trained Indians as secret agents before sending them to India. With their massive experience, the Japanese initiated various spy-schools in Malaya and Burma to fulfil their imperialist objectives. From the history of the INA's secret war, it can be ascertained that agents sent to India had received formal training at the special schools. Before any secret agent could be sent to India to carry out secret operations, it was necessary to train them in special techniques of espionage, sabotage and other fifth column activities. The creation of skilled spies was often the outcome of outstanding training, not just extensive experience. With this in mind, the training of spies becomes necessarily important, especially during war. To be sure, the training of spies was an important part of military operations of all countries directly involved in WW II.[5] The INA also regarded the training of spies as necessary and vital to its operations.

Along with the direct offensive campaigns in the frontline, the INA-Japanese collaboration entailed the establishment of intelligence schools to train suitable and willing Indians in espionage work. After graduation, these spies were sent to India for infiltration work as a prelude to an attack on India by the INA in collaboration with the Japanese armed forces. The Japanese allocated substantial amount of funds and elaborate sets of training courses, while the INA provided agents selected from Prisoners of War (POW) and civilians. Various locations in Malaya (now known as Malaysia) were identified for spy school centres.[6] The schools in Malaya were crucial to the Indo-Japanese collaboration plot, particularly in terms of strengthening the *espirit de corps* among the Indians in East Asia. The selected secret agents were implanted with the psychological belief that theirs was a holy war for the independence of India from British rule. The establishment of spy schools throughout Malaya was the solution to the INA's lack of experience in intelligence matters. As most of the INA recruits were uneducated, specific and intensive training was necessary to prepare them before any missions could take place.[7]

Before the outbreak of the Pacific War in 1941, Japan had actively promoted the idea of the Asia co-prosperity sphere. It not only propagated this idea throughout the Asian countries but also participated in providing military training and training of spies from among the subject populace of the West European maritime empires in Asia. Prior to WW II, the priority in Tokyo was to encourage disaffection among various communities in the Asian countries, particularly among local indigenous nationalists and expatriate Indians in Malaya and Burma. This was supported by intelligence operations as part of a propaganda and espionage war directed primarily against Britain and the United States. Japanese intelligence officers were assigned to Burma and Malaya, where they contacted nationalist leaders and fostered anti-British sentiment. Ultimately, the Japanese maintained that one of the best ways of achieving their objective was to organize and train military units and espionage parties among the anti-British elements in these places. These men would later engage in joint military operations against the Allies.[8] The training of 30

Burmese nationalist comrades, who later became the nucleus of the Burmese National Army, at a training centre near the Naval Training Centre on Sanya Hainan Island and the establishment of the INA are examples of the Japanese attention to this scheme.[9]

The secret war against the British in India before WW II was a part of the Indian revolutionary activity abroad, particularly among those who resided in East Asia. As war became inevitable, the Indian revolutionaries based in East Asia collectively regarded Japan as the only practical power that could help them in their long awaited war against the British in India. More important, the decision to collaborate with the Japanese had been influenced by several Indian revolutionaries in Japan, particularly Rash Behari Bose, Anand Mohan Sahay and A.M. Nair. There is no doubt that they were used by the Japanese to influence the Indian population in East Asia into accepting the Japanese as their savior in the war against the British. Japan also looked at the Indian revolutionaries as allies who could help them to take control of the Asian regions from the Western imperial power. At the same time, to the non-revolutionary Indians, Japanese success in China, Hong Kong, French Indo-China, Malaya and Burma left them with little alternative but to collaborate with the Japanese in order to secure a better condition. According to a Weekly Intelligence Summary of July 1942, the Japanese were involved in training Indians even before WW II began in 1941. Prior to Japan's entrance into the war, a fifth column school financed by the German Embassy in Tokyo and staffed by German military officers, had been established at Tsingtao in China. Recruits were drawn from the Asian 'races', with special attention paid to Indians.[10]

After succeeding in the Malaya campaign and capturing Singapore and Rangoon, the Japanese and Indians entered into a new collaborative phase. Their focus shifted to India. To guarantee success, the INA's fifth column planned to send its secret agents to India to engage in activities such as espionage, sabotage and propaganda. The INA learnt from the Japanese that to achieve victory in war, they needed not only excellent troops and equipment but also information relating to British strengths and weaknesses. This information could be used to great advantage in their preparations for war against the enemy. The

Sino-Japanese and Russo-Japanese wars had taught the Japanese that information was vital to winning a war. To collect vital information particularly that related to the military and defence of the enemy, the Japanese used various spying methods. According to Joyce C. Lebra, 'intelligence operations and training had existed in the modern Japanese army since at least the time of the Russo-Japanese war'.[11] Training was given under the auspices of the Army's Training Affairs Office, which sent officers abroad as military attaches. Their contribution was very important to Japan's involvement in WW II. For example, before the Malaya campaign, the Japanese already had large amounts of information relating to the positions of the British Army in Malaya. Japanese secret agents had also worked diligently to collect information related to the dispositions of the British Army units as regards the defence of Malaya. This information was used effectively during military operations from 8 December 1941, and helped the Japanese to successfully occupy Malaya within 70 days.[12]

During WW II, the Japanese military establishment emphasized on field operations and secret operations and both involved intelligence and clandestine activities. Japan's military operations depended heavily on intelligence inputs and before any offensive operation was initiated intelligence gathering was given importance. The military collaboration between the Japanese and the Indians took place after the INA and the Indian Independence League (IIL), with Japanese help, had successfully propagated the idea of Indian liberation within the British Indian Army. As thousands of Indian troops of the British Indian Army eagerly changed their attitudes, it was not difficult to absorb them into war operations in the name of the Indian independence movement. In the words of Sumit Sarkar: 'Indian prisoners of war in Japanese camps provided a ready recruiting ground for the INA, which was able to rally about 20,000 out of the 60,000 prisoners of war and financial aid and volunteers came from Indian trading communities settled in South-East Asia.'[13]

From the very beginning, the INA's objective was to get involved not just in regular offensive operations, but also in a secret war against the British in India. The infantry regiment (Hind Field Force Group) and guerrilla regiments (Gandhi, Azad and Nehru) were to act together with the INA's secret service in dealing with sabotage and espionage activities. The Special

Service Group (SSG), which afterwards became known as the Bahadur Group and the Intelligence Group, played an important part in INA secret intelligence services.[14] Even though spying activities within the Indo-Japanese collaboration began before the war, the more serious activities began soon after Japan occupied Malaya. This essay looks at two fundamental areas: first, the relative importance of spy schools in Malaya in the INA's secret war, and second, the extent of the Japanese Kikan involvement and the role of Bose in the establishment of these schools.

INA-JAPANESE SPY SCHOOLS IN MALAYA

During the second half of 1942, there were several Japanese-Indian spy schools in Penang, namely the Hind Swaraj Institute, the Penang Propaganda School, the Chopra School controlled by Hari Singh, and a small school for Sinhalese.[15] Some of these schools were closed later, but others were set up, such as the one named Sandicroft.[16] The Iwakuro Kikan sponsored and supported the establishment of the Japanese Indian spy schools both in Malaya and Burma.[17] Iwakuro Hideo, head of the Iwakuro Kikan, was an officer whose principal experience was in espionage, and he regarded organizing fifth column activity in India as his primary task. With this end in mind, he set-up Indian spy schools in Japanese occupied-territory, responsible for training Indian youths in sabotage, propaganda and espionage activities.[18] In Malaya, Penang and Port Dickson became important spy school centres along with a number of other centres in Burma. Graduates of the Burmese schools infiltrated into India by land or were dropped by air, while those from Malaya were sent by submarine to India's coastal areas, as well as overland. Recruits for these centres came from various Indian communities in East Asia and were contacted through the IIL and INA branches.[19] To quote Lebra: 'Under Iwakuro the training schools for intelligence activities expanded and turned out graduates, some were sent into India by Iwakuro. Penang was a special center for training in propaganda and espionage. . . .'[20]

The earliest Japanese-Indian spy school in Malaya was the Hind Swaraj Insititute (HSI), situated on Penang Island in northern Malaya.[21] Penang was chosen for its strategic position.

Geographically, it is close to the Burmese and Indian frontiers, and it had an excellent naval and submarine base, vital for submarine training and for transporting INA agents to India.[22] Japanese intelligence community was familiar with the Penang Naval Base even before WW II, as Japanese ships were granted port facilities by the British from 1914 to 1918, and had patrolled the Indian Ocean against German submarine and surface raiders from there during First World War.[23] The HSI became an INA spy school after two small parties of agents were formed and sent to India for secret operations. Ram Sarup headed the first party and N.S. Gill the second. The parties were sent overland to India with the main objective of setting up a base as a spy-web for intelligence activities, and to report information on British Indian troop movement to INA headquarters in Burma and Singapore. Later, Rash Behari Bose felt it was necessary to send more capable and better-qualified Indian agents on future operations. With this in mind, the HSI was established and the move was initiated by Colonel Bhonsle and R. Rangavan.[24] During the war, Rangavan became leader of the Malayan branch of the IIL, and was appointed a member of the Council of Action (COA) at the Bangkok Conference.

The HSI began at the Penang Free School on Green Line Road in Penang on 3 August 1942 with R. Rangavan as director,[25] and several other Indian members of staff, including Colonel Allangapan,[26] D.P. Gupta,[27] Dallal and Colonel Sangha. There were about 32 students in the first intake, and they became a 'nucleus of material for training purposes, through which the future Japanese espionage plans would radiate'.[28] All were civilians, largely Tamils and Malayalis (south Indians), who were recruited in various parts of Malaya by Rangavan himself.[29] The main objective of the school was to train Indians in espionage and propaganda activities. At the opening ceremony, Rangavan addressed the trainees and explained the objectives of the school. The trainees took on oath and signed the pledge. The oath read:

> I dedicate my life for the cause of Indian independence and will carry out the duties allotted to me from this institute to the best of my abilities and even at the risk of my life. When serving my motherland and India, I would not seek any personal advantage. I will treat everybody as brothers and sisters without any [regard to] caste, creed or religion.[30]

The HSI might be regarded as the first test of Japanese sincerity and trust in assisting the Indians in the scheme against the British in India. As the first batch of students were civilians, it was clear that the school would rely on Japanese instructors and intelligence to train them. There was a real difference in opinion between the Japanese and Rangavan concerning the emergence of the HIS. The Japanese planned to send graduates of the school to India as secret agents, while Rangavan regarded the school as a training ground for 'a cadre of national workers'.[31] In Rangavan's mind, training national workers certainly did not involve spying or espionage. This difference of opinion later became a major factor in the strained relationship between Rangavan and Iwakuro Kikan. Japanese intelligence expertise was a vital ingredient in the functioning of the HSI. More importantly, the HSI also involved graduates of Japanese spy schools in the Indian spy-training scheme. Lieutenant Kaneko Noburo, a graduate of Nakano Gakko and an expert in intelligence and sabotage techniques, led the Japanese group at the HSI. Noburo was assigned to head the Penang Island Branch of the Japanese Hikari Kikan, with several staff members including graduates of the Nakano Gakko, Lieutenant Ishimura and a civilian interpreter, Watanabe.[32] Their main task was to establish contact with the INA and to supervise the institute.[33] The Japanese gave military and financial assistance to the institute. Kaneko referred to the Swaraj Institute as 'actually a Nakano Gakko to train Indians for espionage, propaganda and counter-intelligence'.[34]

The students at the schools received various types of physical and mental training. Courses included instruction in preparing and deciphering secret letters, telephone tapping, wireless monitoring, developing secret ink, forging documents, making explosives and fighting guerrilla war and were supervised by graduates of Nakkano Gakko. Training in propaganda via broadcasting, leaflets and films were given. Field training included techniques of infiltration, secret communications, disguise, shadowing, ambush and escape, use of firearms, swimming, wireless and other methods of communications, scouting, photography, mapping and surveying, the use of rubber boats and physical drill. The students developed considerable ingenuity for their missions into India. Besides the physical training, they also attended lectures on world affairs and political science, propaganda and defence intelligence. Rangavan prescribed political and cultural training.[35]

Besides the director's office, headed by Rangavan, there were four departments—the General Department, the Department of Propaganda and Intelligence, the Department of Special Services and the Department of Defence Intelligence. The General Department was responsible for teaching, and the general curriculum included world affairs, principles of war in Greater East Asia, the geography of India and its frontiers, first aid and medical treatment.[36] The Department of Propaganda and Intelligence was responsible for teaching propaganda and increasing the students' affection for and spiritual devotion to India.

Colonel Alaggapan headed the Department of Special Service, which was responsible for teaching the science of topography and communication, and the science of propaganda. The most important department was the Department of Defence Intelligence, where students were exposed to the arts of handling arms, using vehicles, operating communications apparatus, disguise, and scouting, and the manufacture of explosives. Ishimura of the Iwakuro Kikan gave instructions on military and physical training, including the art of self-defence. Five officers from the INA were appointed for the military training. One of them was Colonel Sangha.[37] G.P. Ramachandran argues that the students had two handicaps: first, they had other occupations and worked for the INA's fifth column on a voluntary basis; and second, they had no training in political activities such as propaganda.[38] These disadvantages impaired the students' efficiency.[39]

The students were enrolled for two different courses, one short and one long. Both started in August 1942, but only the students trained in the short course were despatched by land and submarines to India. The long course dragged on in a desultory fashion and with no fixed purpose until May 1943, when it was disbanded. Some of its students took up duties at various IIL institutions in Malaya, while many more joined the INA officers' training centre in Singapore in September.[40] The duration of the short course was about six weeks, while the long course lasted for about three months.[41] Students of the short course were required to attend lectures for ten hours a day in politics, espionage, methods of infiltration, methods of secret correspondence, propaganda, Indian history and Indian geography. They also learnt Morse code, the use of small arms and elementary cipher work, and were trained to generate pro-Japanese propaganda and organize strikes. There was also an additional fortnight's training in swimming and the use of canvas and rubber boats. The aim of the short course was to train parties of secret agents to be sent especially by submarine to India to link up with other parties from the institute, who would be sent overland. Two submarine parties of five agents from this course landed on the west coast of India in September 1942. The long course was originally intended to last for three months and appears to have had been organized systematically on a larger scale. As with the short course, the majority of the students were from the south of India. Students were divided into three divisions, specializing in political work, propaganda, and special operational espionage. Besides physical training such as drilling and the use of rubber boats, students were given lectures in economics, hygiene and first aid. According to British intelligence reports, students on the long course were instructed to supervise the work of those on the short course and to prepare for submarine landing on the west coast of India early in 1943.[42]

The activities of this spy school first came to notice when two separate parties, each of five Indian civilians, were arrested on landing in India from Japanese submarines at the end of September 1942. One party landed at Dwarka on the extreme west coast of Kathiawar and the other at Tanur, also on the west coast in Malabar district. These parties had been despatched from the Penang School after a month's training.[43] After completing their training, the objective was to send the agents secretly to India.[44] They were charged with collecting information about the strength, composition, disposition and morale of the fighting services, communications and transport facilities, the labour situation, workshops and factories and the state of public morale in India relating to the war situation.[45] The school was closed on 29 November 1942 when Rangavan fell under suspicion and was put under house arrest by the Japanese.[46]

Rangavan had discovered that the Iwakuro Kikan had sent the first batch of students to carry out secret work in India without his permission. After realizing that the students were missing, Rangavan demanded an explanation from the Japanese. However, as the students were being used as spies, the case was considered top secret by the Japanese, who denied any knowledge of the missing students. The Iwakuro Kikan also ignored Rangavan's questions about the students' whereabouts. However, after Rangavan protested, the Iwakuro Kikan finally revealed that they were already in India. This news upset Rangavan so much that he closed the school, angering the Japanese, who kept him under house arrest.[47] In Malaya, the Japanese started meddling with Ranghavan's Penang Swaraj School, which was being run on socialist lines, which the Japanese could not tolerate. The Japanese then were running their own spy schools at different places where the men, who were receiving wireless and sabotage training, were directly under their own supervision and many unwilling men too had to work under them.[48] The INA received further information from various Japanese occupied territories that the Japanese were preparing and training young Indians at various locations for infiltration into India without informing their Indian colleagues. There was also an accusation that some Indian revolutionaries worked directly under the guidance of the Japanese such as Hari Singh Rathore alias Usman Khan, Babu Banerji at Rangoon and Master Chopra at Penang.[49] According to K.L. Palta:

> The Japanese attitude became clear when the Japs started keeping their own intelligent [*sic*] service to spy on the League's activities. In Burma in spite of the IIL's protest the 'Swaraj Young Men's Training School' as well as the Indian Cadets School of Indo-Japanese interpreters at Mymyo was being virtually run by Japanese officers for their own benefit. When Mr B. Prashad, the president of the Burma Territorial Committee, raised his voice in protest, the Japs maliciously dubbed him as a British spy.[50]

A personality closely associated with INA spy schools in Penang was Hari Singh Rathore, alias Usman Khan, alias Roda Singh. He was involved with at least three of the Penang schools and one in Rangoon. The first in Penang was the Penang Propaganda School (PPS), located at Dr. Jagat Singh's seaside residence at 34 Northam Road, and also known as Jagat Singh School (JSS). The school opened in mid-September 1942. Sukhchain Nath Chopra of Gujranwala and Mohammad Zahir Khan, a graduate of Banaras University, also known as Moti, were responsible for administering the school. Recruiting appears to have been carried out mainly through IIL branches all over East Asia, particularly in Shanghai, Bangkok and Malaya. Most of the recruits were POWs, including Gurkhas. The PPS had a greater combination of Indian students than the Hind Swaraj Institute, which focused only on Indians in Malaya, mostly Tamils and Malayalis. In comparison, the students of the Hari Singh schools were outstanding in terms of courage and dedication than HIS as Hari Singh not only recruited specially selected individuals or small courageous groups but most of the students, particularly those from Shanghai, had revolutionary experience and had undergone extensive training before leaving for India.[51] Hari Singh was assisted by R.L. Avasthi, alias Niranjan Singh, and S.N. Chopra. In early 1943, the PPS students were known in British intelligence reports as Avasthi's party after their leader, R.L Avasthi, a former wireless operator in the Andaman Islands.[52]

More students from Singapore now came to the school together with Hari Singh's followers from Shanghai, who had been arriving in Penang since December 1942. This development was closely related to the fact that Hari Singh had worked in Shanghai before the war. Accordingly, the Hari Singh schools, including the PPS, became a shelter for and part of the long training course for Indian students from spy school in Shanghai. A document states: 'A Japanese Indian training centre established in Shanghai at East Paoshing Road. Headed by Major B. Narain, a former officer in British Army . . . and the students who came to Hari Singh School were part of the long courses held in that centre.'[53] In December 1942, about 70 students attended the training, and, until April 1943, there were about 132 students at the school.[54] Avasthi's party, however, faded from the Penang picture in May 1943 when 30 of them, together with 10 of Sangha's group, left for Burma to form the Swaraj Young Men's Institute at Thingangyun, the other Hari Singh school.[55] A Japanese document refers to this group as 'O' Butai. The objective was to work for the elimination of British influence in India. The Japanese held this group in high regard.[56]

The second school under Hari Singh's umbrella was the Chopra School, administered by Sukchain Nath Chopra, alias Choudhary, a former headmaster of the Batu Pahat High School in Johore, southern Malaya.[57] The school was opened near the

PPS at 82 Northam Road, Penang. The permanent staff was few, consisting only of Chopra and a Japanese named Sakuma. This school opened in January 1943 with 24 students. Not long afterwards, Hari Singh brought several students back to Penang from the Thingangyun Institute for special training. In February 1943, 19 other joined this group, bringing the total to 24. Of them, six were from Shanghai and 18 from the Jagat Singh School where they had attended submarine and boating courses. Alongside physical training under the instruction of Sakuma, the students also attended political lectures given by Chopra. The duration of the course was three months.[58]

In March 1943, all the students at this school began intensive training until September 1943 when Chopra and some of his students left for Burma.[59] Lecture topics included espionage and politics, and practice was given in the handling of small arms. Avasthi gave training in wireless telegraphy in the early stages and, from July 1943 onwards, this part of the course was taken over by N.G. Swamy.[60] Swamy had been trained in intelligence techniques in Germany and was charged with inspecting the Japanese-Indian spy schools in Penang and Rangoon and improving the students' technical training.[61] It is still a mystery how Swamy reached Singapore. According to one reference, he was one of 30 specially trained men who reached Singapore from Berlin at about the same time that Bose arrived.[62] However, not much is known about these 30 comrades except that one of them, Swamy, was attached to the Penang spy school.

The Sangha or Palace School was another spy school in Penang.[63] It was named after Major Balwant Singh Sangha of the 5th Battalion of the 2nd Punjab Regiment, a protégé of Hari Singh and a graduate of the long course at the HSI. This school was opened at the Raja of Alor Star Palace, a seaside palace belonging to the King of Kedah in Penang in January 1943. The earliest recruits to the school were the remaining members of the Avasthi party when Avasthi moved to Burma. There were four Japanese and six Indian part-time instructors. Among the Indian instructors were S.N. Chopra, R.L. Avasthi alias Niranjan Singh and Gurbuk Singh.[64] During its early days, the students were made up of half a dozen volunteers sent from Singapore, three or four students from the HSI, and some fifteen POWs. Most of the personnel were Punjabis.[65] There were ten Gurkha POWs among the students who for a short period had to undergo preliminary training. The students joined the school at the beginning of March 1943, and for 20 days the training consisted mainly of swimming, learning Hindi, English and Urdu and elementary instruction in wireless transmission (W/T). Sangha was assigned to lecture on the independence of India. In the middle of April 1943, Sangha selected a party of ten to establish a wide network of secret agents in what he called the 'southern Punjab scheme'. The actual area of operations was three static headquarters at Jalandhar, Shimla and Ferozepore or Ambala and four mobile headquarters throughout Punjab. These agents were responsible for the murder of high military officers and for sabotaging the British war effort in India, particularly in Punjab. Arms were procured for their use through the North-West Frontier Province probably from Afghanistan. The agents were instructed to contact the Indian National Congress leaders to ask them to prepare sketches of important military targets within the spheres of the agents' operations and to carry out general intelligence and propaganda activities. Each agent was to receive Rs. 20,000 and Sangha himself was to receive Rs. 50,000. However, the departure of this party was delayed until May 1943 when three of the original members of the 'southern Punjab' mission and seven others were moved to Rangoon. Though Sangha's 'southern Punjab' plan was ambitious, no detailed preparations seemed to have been made.[66]

SUBHAS CHANDRA BOSE AND THE SPY SCHOOLS

More systematic INA spy schools were established after Subhas Chandra Bose became the leader of the INA. Spy schools such as Usman Camp, the Amar Singh Sabotage Training School at Rangoon and the wireless training schools in Burma and Malaya were placed under the direct supervision of the *Azad Hind* government. An officers' training school was established in Rangoon under the command of Lieutenant-Colonel Thimayya and another school for training Bengali youths in revolutionary and sabotage work was set up under the able and experienced leadership of Lieutenant-Colonel Jiwan Singh.[67] Bose drastically re-structured the INA spy schools in Penang. After discovering that there were several independent INA spy schools in Penang, he decided to consolidate them all under one roof and the end result was the Sandycroft School. A Japanese record states: 'This school was formed out of an amalgamation of several small groups that were undergoing training of one kind or another and of varying efficiency.'[68] Originally, Sandycroft was a rest camp for British convalescent soldiers on the beach and was situated at Tanjung Bungah, $4\frac{1}{2}$ miles from Georgetown, Penang.

Several parties attended this school. For example, in January 1944, 20 Muslims belonging to Colonel Gilani's party were brought to Sandycroft. Gilani's party was originally from the Gilani's School, established to train Muslim POWs as secret agents in September 1942. Its early students numbered about 70 but, due to dissension amongst them, Subhas Chandra Bose recalled the whole party to Singapore in November 1943. However, not long after that, some 20 personnel were returned and given further training at Sandycroft School.[69] About five members of Chopra's school were also transferred to this school. This indicates that the Sandycroft School offered extensive training for the INA spies. All the parties at the school were trained in modern methods of espionage and supplied with modern equipment. For example, the Pasni Party, consisting of 12 Gilani students, was supplied and trained with wireless equipment, revolvers and hand grenades.[70] In May 1943, Sandycroft received some Bengali students transferred from a Bengali spy-training school in Port Dickson. Secret agents trained at Penang were regarded, by far, the most dangerous and most capable of carrying out their operations effectively compared to all other spy school graduates.[71]

At Sandycroft, special attention was paid to wireless and propaganda methods. Interpreters were responsible for translating instructions from Japanese into Indian languages. The staff was made up of six Japanese, one of whom was a naval officer, and five Indian instructors. The administrative head was Dr. Sudhirananda Roy, and Captain B.S. Sangha was responsible for supervising the courses, particularly those devoted to wireless training. Another member of staff was N.K. Bannerji, who was at one time the second most senior instructor at the spy school in Thinggangyun in Rangoon. Associated with Sandycroft were several houses on the Batu Feringgi and Ayer Rajah Roads, which had been selected as training bases and shelters for students before they received their final instruction.[72]

There is also evidence indicating that the Indians in Malaya played a major part in the establishment of the INA spy schools. They were involved as graduates as well as teaching instructors, as in the case of Harbak Singh who was senior technical assistant at a wireless station in Kuala Lumpur before the war. After the Japanese occupied Kuala Lumpur, he received instructions from the Hikari Kikan to teach Indian secret agents to operate and maintain wireless transmitters and receivers. For this purpose, he was transferred to the spy schools in Penang. According to his statement, the Japanese supplied the schools with portable wireless transmitters and receivers, which he was responsible for teaching the students to operate and maintain. This incident illustrates the Hikari Kikan's direct involvement in the spy schools in Malaya. It sought out persons in Malaya with experience and knowledge in wireless transmission and asked them to join the schools. According to Harbak Singh, he was able to recruit many of his friends, such as Makhan Singh, Niranjan Singh, R.J. Somasundaram and Gajinder Singh to work at the spy schools.[73]

A British intelligence record stated:

> At first the training given to these agents was superficial and the men chosen for the work were not capable of carrying out the tasks allotted to them. By the end of 1942, however, the instruction was becoming more efficient and detailed, and the last party of agents reported to have arrived in India in December 1943 was considered to be better trained and superior in every way to those which had come before. . . . It has become customary to refer to these agents as the Japanese Inspired Fifth Column which has been abbreviated to JIFS.[74]

CONCLUSION

The Indians and Japanese concentrated their efforts on developing a capable Indian Army that could be sent on offensive operations at the Indian frontier and establishing an efficient spy network inside India. This was vital for the supply of secret information to the Japanese units in Burma and as a method of linking with the local revolutionary parties in different regions within India.

The network was to be used as a tool to spread pro-Japanese propaganda as well. However, it is obvious that the task of establishing the spy web inside India was not at all simple.

The establishment of spy schools in Malaya could be regarded as the INA's most ambitious programme. It was intended to produce capable and skilful secret agents within a short time, and demonstrates that the collaboration between the Indians and the Japanese had culminated to a common understanding between the parties. The INA and IIL, as well as the Japanese Kikans, agreed that only special agents, trained and highly nationalistic, were to be sent to India. Agents of this quality needed to be produced through strenuous and specific training by experts at special schools. This idea became a principal point in the existence of the INA spy schools. The Japanese contributed their expertise in the spy-training scheme, while the Indians, through the INA and IIL, supplied the schools with students recruited from either POWs or civilians. After their graduation, the agents were sent to India on their secret missions. The Japanese Kikans did more than mere intelligence work. They looked after the POWs and Indians and built up a close relationship with the IIL, of which the INA was an intergral part. This combination was responsible for harnessing and coordinating the subversive effort of Indians in South-East Asia, including the establishment of spy schools. The Japanese officers assisted the IIL and INA officials in the spy schools. Their involvement was evidence of their commitment to support the INA's secret war against the British in India.

Establishing competent agents within a limited time frame was a formidable task for both the INA and the Japanese. Though the Japanese had a long history and experience of training spies, limited time and inadequate resources made this task very difficult. It was more complicated because most of the Japanese intelligence officers responsible for training the INA spies spoke neither the Indian languages nor English, and communicated with students through interpreters. However, the existence of Indian spy schools in Malaya showed the commitment of both parties and could be considered as one of the most important episodes of the Indo-Japanese collaboration.

The INA spy schools in Malaya played a significant part in Indo-Japanese affairs during WW II. In the beginning, the existence of INA's spy school experienced many fluctuations. Rangavan, who was in charge of the HSI, the earliest spy school in Malaya, complained of uncalled for Japanese interference in the institute. On account of this state of affairs, which he found intolerable, Rangavan threatened to resign from the movement. During Subhas Chandra Bose's era, all the spy schools were placed under the jurisdiction of the provisional government of *Azad Hind* and their members were enrolled as soldiers in the INA or as members of the *Azad Hind Sangh*, an organization with strong pro-Japanese leanings.

One fundamental question remains unanswered is how effective were these spy schools in producing excellent and capable Indian secret agents? The answer to this question is mixed as available evidence comes only from British records.[75] On the one hand, the infiltration of Indian secret agents into India has produced enough evidence for historians to surmise that the graduates of the INA spy schools did cause enormous trouble for the British authorities in India, who became cautious and alert to the existence of INA spies. All possible entry routes into Indian territory, either by land or sea, needed to be guarded carefully. The British authorities had to ensure that their counter-intelligence worked effectively to counter the threats posed by INA agents. However, as we have seen, the INA secret agents' potential for producing excellent results was unfulfilled, as most of them were apprehended by the British authorities or surrendered.

NOTES

1. T.R. Sareen, 'Subhas Chandra Bose, Japan and British Imperialism', *European Journal of East Asian Studies*, vol. 3, no. 1 (December 2004), pp. 78-80. See also Leonard A. Gordon, *Brothers Against the Raj: A Biography of Indian Nationalists: Sarat and Subhas Chandra Bose* (New Delhi: Rupa & Co., 2000), pp. 441-547 and Hugh Toye, *The Springing Tiger: A Study of a Revolutionary* (London: Cassel, 1959), chs. 1-5.
2. K.K. Ghosh, *The Indian National Army: Second Front of the Indian Independence Movement* (Meerut: Meenakshi Prakashan, 1969).
3. Motilal Bhargava and Americk Singh Gill, *Indian National Army Secret Service* (New Delhi: Reliance Publishing House, 1988), chs. 1 and 2.
4. Joyce C. Lebra, *Jungle Alliance* (Singapore: Donald Moore for Asia Pacific Press, 1971), pp. 71-4.

5. The British intelligence agency considered the training of spies as a principal component of its operations. See *SOE Syllabus: Lessons in Ungentlemanly Warfare in World War II,* 2001, Public Record Office (PRO), Kew, London. See also Charles Cruishank, *The Fourth Arm: Physiological Warfare, 1938-45* (London: Davis-Poynter, 1977); Amletto Vespa, *Secret Agent of Japan: A Handbook to Japanese Imperialism* (London: V. Gollancz, 1938); and Cyril Cunningham, *Beaulieu: The Finishing School for Secret Agents, 1941-1945* (London: Leo Cooper, 1998) and 'Indian Army: Schools and Training', in *Indian Information Service Magazine*, January 1942 and HS 7/115, Document on Special Operation Executive: Training of the Force 136 (PRO).
6. Bhargava and Gill, *Indian National Army Secret Service*, p. 10.
7. K.S. Giani, *Indian Independence Movement in East Asia* (Lahore: Singh Bros, 1947), p. 38.
8. Joyce C. Lebra, *Japanese-Trained Armies in Southeast Asia; Independence and Volunteer Forces in World War II* (New York: Columbia University Press, 1977), pp. 3-4.
9. Minami Kikan under the command of Colonel Keiji Suzuki was responsible for smuggling 30 comrades out of Burma. In Hainan, they were organized into three major groups, namely high command and administration works, guerrilla tactics, and training for troop command. See Won Z. Yoon, 'Japan's Scheme for the Liberation of Burma: The Role of the Minami Kikan and the Thirty Comrades', Center for International Studies, Ohio University, no. 27, 1973, p. 31. See also Lebra, *Japanese-trained Armies in Southeast Asia,* p. 176. In the Burma scheme, Japanese army instructors worked under the direction of Captain Kawashima Takenobu, a graduate from the Nakano School. The Burmese were given military training and tactics, war games were held and lectures were given to strengthen the spirit of self-sacrifice, as well as regular training in command of soldiers and special training in guerrilla tactics and espionage were carried out. See Joyce C. Lebra, 'Japan and the Genesis of the Burma Independence Army', *Papers on Far Eastern History,* vol. 5 (March 1972), p. 20; Louis Allen, 'The Nakano School', *Proceedings of the British Association for Japanese Studies,* vol. 10 (1985), p. 14. Colonel Suzuki entered Burma from Bangkok in May 1940, and opened a secret office at Rangoon. This became the base of the Minami Kikan, the Japanese secret subversion organisation. See also Louis M. Allen, 'Fujiwara and Suzuki: Patterns of Asian Liberation', in William H. Newell (ed.), *Japan in Asia* (Singapore: Singapore University Press, 1981), p. 86 and Christopher Bayly and Tim Harper, *Forgotten Armies: The Fall of British Asia, 1941-1945* (London: Penguin Books, 2004), p. 8.
10. See L/WS/1/1433, Weekly Intelligence Summaries (WIS), no. 39 (pt. III), 31 July 1942, India Office Collection (IOC), British Library (BL), London.

See also 'The Nazi Fifth Column in Japan', *Oriental Affairs* (September 1940), pp. 114-17 and 'Nazi Firms in the Far East', *Oriental Affairs* (November 1941), pp. 244-6.

11. Lebra, 'Japan and the Genesis of the Burma Independence Army', p. 9.
12. See CO 537/1353, Report by Major-General A.E. Percival during 8 December 1941–15 February 1942, PRO.
13. Sumit Sarkar, *Modern India, 1885-1947* (Basingstoke: Macmillan, 1989), p. 411.
14. See K.R. Palta, *My Adventures with the I.N.A.* (Lahore: Lion Press, 1946), pp. 34-5.
15. F. no. 601/19869, General Staff to GOC Indian Army, 18 December 1943 Appendix I, American Intelligence Report on the activities of INA and the working of Hikari Kikan, in T.R. Sareen, *Indian National Army: A Documentary Study*, vol. 2 (New Delhi: Gyan Press, 2004), p. 288.
16. According to Ghosh, there were only three spy-training schools in Penang, namely the Swaraj Institute, Penang High School and the Sandycroft School. This study was unsuccessful in locating any references to Penang High School. It is important to note that Ghosh only refers to this subject with reference to the memoir by Durrani's *The Sixth Column* and Shah Nawaz Khan, *My Memories of the INA and its Netaji*, without any support from other primary documents. See Ghosh, *The Indian National Army*, fn. 35, p. 104.
17. According to a Japanese document captured in Burma, under Japanese Military Intelligence regulations strict rules were applied concerning recruiting local inhabitants as agents, including:

    a. Agents must be physically strong, intelligent and pro-Japanese.
    b. Family background and social contacts must be investigated thoroughly.
    c. Agents will be personally selected by the unit commander and agents must not know one another's identity.
    d. Selection areas will be determined according to each unit's local zone of responsibility.
    e. Those selected will be paid 3 *yen* per day while employed. From time to time they may be given special bonuses and priority on civilian goods.
    f. The agents will be given sufficient training. They will be despatched within their local zone of responsibility as investigators for the unit commanders.
    g. The agents will be paid 1 *yen* (20 cents) per day during their training period.

    Documents on the Indian secret agents' training syllabus and activities show clearly that these regulations were accordingly followed. This evidence identifies the fact that Japanese military intelligence relations were used to train Indian secret agents and set general guidelines in the secret agents' operations to India. See WO 208/4817, SEATIC Bulletin no. 119, item 1380. This document was captured at PS 1351, South of Pauktaw on 10 February 1945. Cyclostyled sheet in Public Record Office.

18. See Yoji Akashi, 'Chandra Bose to Nihon', *Journal of Asian Studies*, vol. 29, no. 1 (November 1969), p. 147.
19. G.P. Ramachandra, 'Indian Independence Movement in Malaya', M.A thesis, University of Malaya, 1970, pp. 130-1.
20. Joyce Lebra, 'Japanese Policy and the Indian National Army', *Asian Studies*, vol. 7, pt. 1 (April 1969), p. 42.
21. Ramachandran stated that this school also been recognized as Hind Swaraj Vidyalaya. Ramachandran, 'Indian Independence Movement in Malaya', p. 104.
22. Penang Naval Base became important in the Japanese naval planning during the war. See Michael Wilson, *A Submarine War: The Indian Ocean 1939-45* (Gloucestershire: Tempus Publishing Ltd., 2000), ch. 10.
23. Peter Elphick, *Singapore: The Pregnable Fortress* (London: Coronet Books, 1995), p. 16.
24. Bhargava and Gill, *Indian National Army Secret Service*, pp. 100-13.
25. Rangavan hailed from Guruvayar (South Malabar) and studied at the Christian College, Madras. He studied law in England and went to Penang as a lawyer in 1920, becoming a distinguished lawyer and leader of the Indian community there during the pre-war period. See Bhargava and Gill, *Indian National Army Secret Service*, p. 2; Ghosh, *The Indian National Army*, p. 54.
26. Colonel Alagappan, Indian Medical Service Officer of Madras and second in command of the school. See Partha Sarathi Gupta (ed.), *Towards Freedom: Documents on the Movement for Independence in India 1943-1944*, pt. 3 (New Delhi: Oxford University Press, 1997), p. 2697.
27. D.P. Gupta was an interpreter in the HSI; see L/WS/1/1433, WIS, no. 53, 6 November 1942, IOR. He also acted as supervisor of the school. See 'The Swaraj Institute', in *Netaji Subhas Chandra Bose: A Malaysian Perspective* (Kuala Lumpur: Netaji Centre, 1992), p. 95.
28. Bhargava and Gill, *Indian National Army Secret Service*, p. 2. Other sources mention only 30 students. The second batch started with 150 students. See L/WS/1/1433, WIS, no. 53, 6 November 1942, IOR.
29. L/PJ/12/513, Survey No. 5 of 1944, 14 February 1944. See also WO 208/803, JIF Training Centre in Burma and Malaya. Rangavan travelled from place to place throughout Malaya to recruit students. His job was easy as during that time many Indian youths were unemployed. See

Gupta (ed.), *Towards Freedom: Documents on the Movement for Independence in India 1943-1944*, pt. 3, p. 2697.

30. Bhargava and Gill, *Indian National Army Secret Service*, p. 3.
31. 'The cadre of national workers' could be anything but secret agents. Rangavan believed that the students would be sent to India purely as political workers whose responsibility would be to close the gap between Indian National Congress and the Indian movements in East Asia. See Ramachandran, 'Indian Independence Movement in Malaya', p. 128. However, Rangavan's understanding on this subject obviously differed from the Japanese and this difference became a major factor for the closure of the school.
32. L/WS/1/1433, WIS, no. 53, 6 November 1942. In other documents, Watanabe was known as Watanabhi. See Gupta (ed.), *Towards Freedom: Documents on the Movement for Independence in India 1943-1944*, pt. 3, p. 2697.
33. Joyce C. Lebra, *Jungle Alliance: Japan and the Indian National Army*, (Singapore: Asia Pacific Press, 1971), p. 71.
34. Ibid., p. 71.
35. L/PJ/12/510, Survey no. 36 of 1942, 3 October 1942, IOR.
36. Bhargava and Gill, *Indian National Army Secret Service*, p. 18.
37. Ibid., pp. 3-21.
38. Ramachandra, 'Indian Independence Movement in Malaya', p. 105.
39. Altogether 30 of the HIS graduates were arrested and faced trial in India in 1942-3 and 20 of them were executed. A summary of cases involving the Hind Swaraj Institute is below:

Place	Total Tried	Executed
Madras—Case 1	19	11
Madras—Cases 2 & 3	6	6
Delhi	5	3
Total	30	20

Source: 'The Swaraj Institute', in *Netaji Subhas Chandra Bose: A Malaysian Perspective* (Kuala Lumpur: Netaji Centre, 1992, p. 98).

40. L/PJ/12/513, Survey no. 29 of 1944, 4 September 1944.
41. L/PJ/12/51, Survey no. 18 of 1943, 8 May 1943.
42. L/PJ/12/513, Survey no. 5 of 1944, 14 February 1944.
43. WO 208/804A, GHQ, India to HQ NW Army, The General Survey of the IIL and INA, 30 March 1943.
44. The first party of students was sent to India by submarine and a party of ten landed on the Malabar Coast, Tanur, on 27 December 1942. The next party landed on the Gujarat coast of Okhawadi on 29 September 1942. The third party moved into Akyab.
45. L/PJ/12/513, Survey no. 5 of 1944, 14 February 1944 and L/PJ/12/510, Survey no. 36 of 1942, 3 October 1942.

46. See Bhargava and Gill, *Indian National Army Secret Service*, p. 17. See also Japanese Secret Intelligence Service (JSIS), Prepared by General Staff (Intelligence), Australian Military Forces, February 1947, pt. II, p. 73, Imperial War Museum, London. The decision to close HSI was coincidental with the crisis in INA and IIL at the end of 1942. A letter from A. Yellapa to Rangavan concerning the HSI after its closure reads: 'I discussed the future of the Institute with Bose [Rash Behari]. He intends to make this an Army Officers' Training School, but that cannot start until the reorganization of the INA is completed. In the meantime, he does not want the buildings to be vacated lest the Navy or Osman [Osman Khan alias Hari Singh] should take possession of them. As to the students who are still in the Institute, Bose requests you to reconsider the question of sending them to infiltrate India. He suggests that you should take this matter over with Lieutenant Kaneko and find out if infiltration is now possible. He thinks it is better that persons who are reliable should infiltrate rather than people of the type Osman or others who were considered 'disloyal'. The latter may do a lot of harm to our cause.' See letter from A. Yellapa to Rangavan about developments in Singapore regarding the reorganisation of the INA and IIL, 12 February 1943 in T.R. Sareen, *Indian National Army: A Documentary Study*, vol. 2, p. 27.
47. See Ramachandran, 'The Indian Independence Movement in Malaya', p. 129.
48. Palta, *My Adventures with the I.N.A.*, p. 39.
49. Bhargava and Gill, *Indian National Army Secret Service*, p. 14. See also Chapter 4.
50. Palta, *My Adventures with the I.N.A.*, p. 39.
51. L/PJ/12/513, Survey No. 5 of 1944, 14 February 1944; Bhargava and Gill, *Indian National Army Secret Service*, p. 32.
52. R.L. Avasthi or R.L. Awasthi was in the Andaman branch of the IIL who played a significant role in the INA. He was wireless operator in Andaman, but when the Japanese came, he was arrested, then released and employed as a schoolteacher. He became one of the most enthusiastic members of the INA and was responsible for recruiting and training new members of the INA. He was taken to Penang and involved in the INA spy school. See Jayant Dasgupta, *Japanese in Andaman & Nicobar Islands: Red Sun Over Black Water* (New Delhi: Manas Publications, 2003), pp. 60-2.
53. Bernard Wasserstein, *Secret War in Shanghai* (Boston: Houghton Mifflin Company, 1999), p. 183. See also Palta, *My Adventures with the I.N.A.*, p. 75.
54. L/PJ/12/513, Survey no. 5 of 1944, 14 February 1944.
55. L/PJ/12/513, Survey no. 29 of 1944, 4 September 1944.
56. JSIS, pt. II, p. 74.

57. In December 1943, S.N. Chopra led eight secret agents on the secret missions to India and landed by submarine on the Dwarka coast. See Government of Baroda, Huzur Political Office, File no. 206, 'Landing of Enemy Agents on the West Coast', in Gupta (ed.), *Towards Freedom*, pt. 3, p. 2745.
58. L/PJ/12/513, Survey no. 5 of 1944, 14 February 1944.
59. L/PJ/12/51, Survey no. 29 of 1944, 4 September 1944.
60. JSIS, pt. 2, p. 72.
61. See Tapan Banerjee, *Mystery of Death of Subhas Chandra Bose* (New Delhi: Rajat Publication, 2002), p. 215. N.G. Swamy was recognized as one of the earliest Indians to join Subhas Chandra Bose to establish the 'Free India Centre' in Berlin to carry out propaganda for Bose's activities. He arrived from Germany with wireless equipment, which was used in secret operations in India. See also Rudolf Hartog, *The Sign of the Tiger: Subhas Chandra Bose and His Indian Legion in Germany, 1941-45* (New Delhi: Rupa & Co., 2003) (2nd edn.), p. 82. Swamy was also suspected as being amongst the earliest Indians to receive specific military training in radio transmission, sabotage techniques, parachute landing and collecting of military intelligence in the Indo-German Commando Group. This group was established in 1940 and commanded by Walter Harbich. See Alexander Werth, 'Planning for Revolution', in Sisir K. Bose et al., *A Beacon Across Asia: A Biography of Subhas Chandra Bose* (London: Sangam Books, 1996), p. 113.
62. See Stephen P. Cohen, 'Subhas Chandra Bose and the Indian National Army', *Pacific Affairs*, vol. 36, no. 4 (winter 1963-4), p. 412. Swamy arrived from Germany with wireless equipment which was used in secret operations to India. See Tapan Banerjee, *Mystery of Death of Subhas Chandra Bose*, p. 215.
63. In one report, this school is also known as Hari Singh Palace School. See WO 208/803, The JIFS Training School in Malaya and Burma.
64. L/PJ/12/513, Survey no. 5 of 1944, 14 February 1944.
65. Ibid.
66. Ibid.
67. Ibid., p. 114.
68. JSIS, pt. 2, p. 74.
69. Ibid., p. 73.
70. L/PJ/12/513, Survey no. 29 of 1944, 4 September 1944. See also WO 208/4817, SEATIC Bulletin no. 119, item 380.
71. WO 203/516, HQ Allied Land Forces, SEAC, Supplementary Guide to JIFC (Indian) Activities (Malaya), 16 August 1945.
72. JSIS, pt. 2, p. 74.
73. See Statement of Harbak Singh to Security Branch on 29 September 1945. This document was held in the University of Malaya Library,

University of Malaya, Kuala Lumpur and never extensively used in any INA history before.

74. WO 208/804B, IPI to MI2, Japanese espionage directed against India, 12 June 1944.
75. Until the availability of Japanese documents or the INA/IIL documents (if available, which we assume did not), the conclusive answer to this question will not be found. We do hope in the near future, evidences could be obtained to show the effectiveness of these spy-training schools in the Japanese Indian fifth column's secret war against the British in India.

CHAPTER 13

How Colonial is the Post-Colonial Indian Army?

SABYASACHI DASGUPTA

This paper attempts to look at the disjuncture, breaks and continuities of the post-colonial Indian Army with respect to the colonial Indian Army. Here one seeks to investigate these issues on the basis of indices such as recruitment, civil–military relationship, and so on. The Indian Army is unique among developing nations in the sense that it has remained under firm civilian control despite periodic stress and strains. To what factors should we ascribe this phenomenon? If it is a consequence of the inherited legacy from the colonial army, then how do we account for the fact that the Pakistani and later the Bangladeshi armies adopted divergent paths despite sharing a common past with the Indian Army?

This paper therefore initially looks into the evolution of civil-military relations in the initial years of independence. Was the legacy of the colonial army as regards civil–military relations actually a hindrance in the efforts of the political establishment to subordinate the military to the civilian establishment contrary to Stephen Cohen's argument? What then were the indices around which civil–military relations evolved? Was then the establishment of a culture of civilian supremacy over the military a clean break with the past? Was the process of civilian control over the military a trend contested by the military?

This paper deals with the recruitment policies of the pre- and post-Independence Indian Army. The Indian Army on paper has broadened its recruiting base and all citizens of India irrespective of caste, creed, religion and ethnicity are liable to serve in the military. There exist recruiting zones that apparently have a zonal base. Why is it so? Is it simply the weight of colonial legacy or are there very deliberate political reasons behind it? Will the rapidly changing political dynamics of the Indian political scenario exemplified for instance by the rise of regional parties leave its impact on the recruitment policies of the Indian Army by making it more inclusive as regard recruitment in the infantry regiments?

POLITICS AND MILITARY PROFESSIONALISM

This paper addresses the important question of the Indian Army being avowedly apolitical and loyal to the civilian establishment despite being surrounded by armies having contrary traditions. The Pakistan Army is the prime example. If political neutrality of the Indian Army is a colonial legacy being carried over then one is at a loss trying to explain the Pakistan and Bangladesh armies' frequent political interventions. Does the credit for sustaining an apolitical Indian Army therefore go to the post-Independence Indian political establishment? One could for instance argue that the history of the colonial army was replete with instances of the army identifying itself with the aims, aspirations and grievances of civilian society, and the 1857 'Mutiny' being a prime example. This chapter examines the formative years of the modern Indian Army, the period from 1947 to 1950, and the kind of discourses prevalent at that time. How was the transition from being a soldier serving a colonial master to that of a soldier of an independent army effected? What difference if any did this transition make for the Indian soldier? How was the role of the army in independent India visualized? These are some important issues one seeks to address.

THE TRANSITION FROM COLONIALISM TO INDEPENDENCE

On 15 August 1947, India became an independent country. Officially, Indians were in charge of their own affairs. The organs of government such as the bureaucracy, police and the object of our attention here, the army, came under the control of an independent Indian government. This meant inheriting a rich and chequered legacy of traditions, ethos, values, etc. Of course, this is not to say that the legacy inherited was a static one which had stood the test of time. The legacy which myriad commentators on the army talk about had constantly evolved over a period of 200 years, which was roughly the period during which the colonial army had existed.

One allegedly important aspect of this legacy was that the British bequeathed a tradition of military professionalism and political neutrality on the part of the army. The values which the colonial army apparently imparted to the average Indian recruit were the culture of strict disassociation from politics. The personnel of the armed forces were expected not to be partial to any particular political ideology. The army was supposed to serve with loyalty the political masters of the day though the colonial government could not possibly say that the army should serve the legitimate political masters of the day. For even the simple *jawan* knew that the colonial government was a usurper whose ultimate source of legitimacy lay in the near monopoly of force it had at its disposal. The colonial armed forces were of course the most prominent manifestation of this monopoly of force which the colonial state enjoyed.

However, one is inclined to question the assumption that the colonial armed forces passed on a tradition of political neutrality to the modern Indian armed forces. The history of the colonial army was replete with instances of rebellion against their civilian masters, the 1857 Uprising being the most glaring example. One is, however, alive to the possibility that 1857 might have represented a definite break in the history of the colonial army and could have led to the gradual forging of a new ethos in the colonial army. One could then advance the counter-argument that in terms of legacy, it was the ethos prevailing in the colonial army in the few decades preceding Independence which was really relevant.

If one is to buy this argument than it becomes imperative to examine critically the culture prevailing in the colonial army in the few decades prior to India's Independence. Let us for instance take the post-First World War years as a benchmark, the assumption being that notions like cultural ethos change slowly unless confronted with grand events like for example the Second World War. While one would be tempted to equate Second World War with 1857 in terms of its disruptive potential on the dynamics of the colonial army, calmer reflection would reveal the obvious fact that 1857 presented a more direct internal challenge to the colonial army while the sheer breadth of events like First World War and Second World War presented the colonial army with formidable challenges of different kinds.

Coming back to our original focus on the prevailing culture of the colonial army in the few decades pre-dating Independence, the colonial armed forces operated in a scenario where nationalist politics had entered a stage of extensive mass mobilization. While the nationalist leaders did not make any covert attempts to tamper with the loyalty of Indian recruits in the colonial army, certain nationalist leaders were very much concerned about the career prospects of Indians serving in the colonial army and the broader question of Indianization of the armed forces. Prominent among them were Bal Gangadhar Tilak, Mohammed Ali Jinnah in his early years, and Motilal Nehru to speak of a few.

Nationalist opinion had been in general emboldened by wartime promises made by British government in times of stress. The Great War (First World War) had generated an unprecedented demand for manpower and resources, straining even the immense demographic potential and assets of the British Empire. Manpower from the colonies became an urgent necessity even if it required making promises which they had no intention of keeping. Consequently, the colonial government gave the assurance that Indian Viceroy's Commissioned Officers (VCO) would be granted the King's Commission. In reality the promise meant nothing because Indian VCOs would usually be of an advanced age and could realistically never hope to become a King's Commissioned Officer (KCO). Moreover, they were not trained to assume the responsibilities of even a junior commissioned officer.[1]

Certain limited rewards for the Indian war effort followed and in 1919 the British government announced that there would be 10 vacancies for Indian cadets, hailing mostly from the landed and aristocratic classes loyal to Britain at Sandhurst each year. The Indian cadets passing out of Sandhurst would be directly eligible for the King's Commission.[2] In addition, such commissions would be given to specially selected candidates from the Cadet College at Indore. One very prominent recruit from the Cadet College at Indore was the first commander-in-chief of the Indian Army.[3] Thus, the first tentative steps were taken in the formation of an Indian officer corps. Their numbers would steadily increase with the setting up of the Indian Military Academy (IMA) in 1932.[4]

What was the culture these officers imbibed? This is an important question since these officers formed the nucleus of the post-Independence Indian Army. Their professional ethos to a large extent decided the professional ethos of the Indian Army. It to a large extent decided the important issue of attitude of the army towards the civilian establishment. Would they adopt a neutral stand with a studied loyalty to any democratically elected government? Or would the army abrogate to itself a political role where they felt constrained to interfere or takeover the running of the nation under certain circumstances?

A perusal of the career graph of a few of these officers reveals that at least some of them were susceptible to the propaganda of the national movement. B.M. Kaul for instance was open in his sympathies for the national movement. He maintained active contact with a host of nationalists. He even gave a speech in an anti-British rally organized by students during the 1942 Quit India movement. Kaul also threw in his lot with the accused officers of the Indian National Army (INA) by procuring a vital document in possession with the British intelligence authorities for the perusal of defence counsel under Bhulabhai Desai. Kaul also claims that he helped Colonel M.S. Himmatshinghji, a nominated military representative in the legislative assembly to draft a speech in defence of the accused.[5] By all accounts, he was an ardent nationalist doubling up as an army officer. Significantly, he would carry his attitudes of dabbling with political leaders even during his career as an officer in the post-Independence Indian Army. It would be perfectly reasonable to argue that he remains the only officer in the history of the Indian Army who openly flirted with the political establishment during his service career.

The case of Field-Marshal K.M. Cariappa is intriguing. He by all accounts maintained a carefully cultivated distance from the national movement. He was known as a 'brown sahib' and there are anecdotes about his inability to address the *jawans* in Hindi. Yet, indifference to the nationalist movement and its leaders did not automatically translate into indifference for matters political. Lord Ismay claims in his memoirs that Cariappa, alarmed at the possible chaos arising out of the imminent Partition of the subcontinent, urged Lord Ismay to dwell on the possibility of a military form of government. The British, however, were not prepared to seriously ponder this option.[6]

Yet, the same Cariappa became a passionate advocate for an apolitical role for the Indian Army. His speech for instance at the passing out parade of the graduates of the IMA in 1948 strongly emphasized that involvement in politics was a cancer for the army. On his part, Cariappa maintained a scrupulous apolitical stance refusing to be embroiled in any affairs of a political nature. He was successful in forging an identity as the man who established the apolitical traditions of the Indian Army on a sound basis.[7]

Such accounts seem to ignore Cariappa's utterances after his retirement. In an article in the newspaper *The Mail*, he advocated military rule under the overall control of the president to tide over certain difficulties the nation was facing. Cariappa was of the opinion that the nation was facing a dangerous situation and military rule under the overall command of the president was needed for a certain period of time in certain states where the situation had spiralled out of control. He was, however, at pains to emphasize that he was not advocating a military coup. He claimed that he had never harboured such notions though he could have effected a coup in 1949 if he had desired. Cariappa, however, does not enlighten us about the particular circumstances arising in 1949, which afforded him the opportunity to execute a coup. One suspects that it was more a case of his vanity imagining a scenario where he could have masterminded a coup.[8] Cariappa then proceeds to contradict himself by arguing that India was too big and diverse a country for a coup to be staged successfully. There was also the problem of three separate service chiefs with little coordination among themselves. Therefore, there was little chance for a homogeneous outlook to develop among the service chiefs, the obvious implication being that a successful military coup was not possible without the concurrence of the three service chiefs.[9]

Cariappa concludes by stating that military rule was a possibility under two conditions. The first was his fond hope that politicians would cede power voluntarily as in the case of a neighbouring country, the obvious allusion being to Pakistan. One calls this a fond hope because there was simply no reason why Indian politicians rooted in a reasonably mass-based political system possessing enough experience, ability and lastly but not the least important stakes in polity would voluntarily cede power to the military who they held in a certain degree of contempt. This is an issue which we will discuss later. Secondly, Cariappa again dreams of a scenario where the people of the country demand military rule as a temporary measure to tide over a critical situation. He would in such a case support the demand of the people for military rule if they felt it would provide them security and better administration.[10] Tragically for Cariappa, his fond dream was never realized. While middle-class drawing room discussions occasionally veers around the country requiring a strong dictatorial form of government, there has never been any serious degree of support for military rule in India. People though they are apt to critique the system and complain of corruption, lawlessness and nepotism, take the parliamentary form of government for granted. A strong movement or demand for military rule emanating from the mass, despite Cariappa's optimism, was not a feasible proposition.

A nuanced analysis of Cariappa's propositions makes it amply clear that he was apprehensive of the lack of popular support for a military coup. He implicitly was acknowledging the existence of a vibrant public opinion and a strong political culture in India. Cariappa possibly sensed that a military coup would not only fail to draw support from the general mass but would face strong resistance. Hence, military rule and not a coup, Cariappa makes a pointed distinction, could only ensue out of popular sanction. For a successful and enduring military regime, the people had to endorse it and make it their popular choice. Cariappa was also alive to the strength of the political class in India and their ability to mobilize public support in the event of any attempted takeover by the military. As stated above, the politicians despite their abuse of the system have also a strong stake in the system and would defend any fundamental assault on the system with all resources at their disposal. And the resources they possess are not exactly meagre. The only way the military could take over is the very slim possibility of a consensus emerging in the political class of the need for the military to take over. One says consensus because no political class in independent India had ever approached the military to settle scores with their political rivals.

We turn our focus on K.S Thimayya, the man who would be at the forefront of one of the biggest controversies surrounding the post-Independence Indian Army. Like some of his other colleagues, Thimayya was a Sandhurst graduate. By all accounts he was extremely popular in Sandhurst and did not face socialization problems unlike many of the early recruits. He was good looking, articulate and good at sports. All these qualities were handy if you desired to be a hit among the British 'gentry'. Thus, Thimayya had all the trappings of the archetypical Westernized Indian Army officer who hailed from a privileged background. He was also from what the British would dub as a loyalist background. But, as the social profiles of many of the Indian national movement leaders would reveal, this was no guarantee against possessing strong anti-colonial statements. While Thimayya lacked deep-rooted contacts with nationalist politicians, he had his sympathies for the cause. Thimayya is reported to have asked Motilal Nehru whether it was the right thing for him to join the colonial army. Motilal replied that a time would come when the independent Indian state would need experienced army officers. That is when their experience as officers in the colonial army would come in handy.[11] Thimayya and many other officers of his ilk were perturbed by the jibes they had to endure from racist British officers. Thimayya and some other officers then sought out nationalist leaders to solicit their opinion about the possible course of action they had.[12] Thus, it was not a case of Thimayya or many others like him being isolated from the nationalist cause. Service in the colonial army did not inure him from the pulls and strains of his parent society. He had his political sympathies firmly in place.

We now come to the final officer under our scanner. Unlike the other officers discussed above, Harbakhsh Singh was one of the earliest graduates of the IMA. He came from a prosperous farming family. Though he did not exactly come from a aristocratic or princely background, his forefathers were established gentry for generations and had served in the Sikh army as gentlemen troopers. His own father was the first doctor in his village.[13] Harbakhsh Singh's memoirs discuss extensively about his experiences in the academy and then as a young officer with the colonial army. It is, however, initially silent on the Indian national movement. There is also no rancour towards the British officers in the academy. In fact, he had nothing but effusive praise for Brigadier Collins, the Commanding Officer of the IMA. There is a British instructor who is portrayed as an ill-tempered man but no mention is made of the fact that racism had anything to do with his foul temper. His nationality seemed to be a non-issue for Harbakhsh Singh. He in fact displays a complete lack of antagonism for the British. One could be pardoned for assuming that Singh was lukewarm about the anti-British independence struggle which was raging on in the country at that time.[14]

However, this would be an erroneous perception as his account of his travails as a prisoner of war (POW) with the Japanese in Singapore shows. Singh like many others was exhorted by the Japanese captors to join the INA. Singh and his brother, who was also an Indian Army officer and a POW like him, were taken to meet Mohan Singh, the supreme commander of the INA. Mohan Singh, however, failed to persuade Singh to join the INA. Harbaksh Singh and his brother told Mohan Singh that they had no faith in the credibility of the INA leadership. They would join only if M.K. Gandhi and Jawaharlal Nehru asked them to do so as they were the legitimate leaders of the national movement.[15] The Japanese tried one last time. This time Harbakhsh Singh and his brother were taken to meet Netaji alias Subhas Chandra Bose who had just arrived in Singapore to take charge of the INA. Even he failed to persuade them. Harbaksh Singh's elder brother acted as a spokesman for both of them and he replied with candor that they were not prepared to join the INA as they simply did not possess the patriotism to do so. If they had possessed it they would not have joined the colonial army in the first place. He also said that they were skeptical of the future of the INA as the Japanese were suffering reverses everywhere.[16]

At this moment, we could pause and also reflect on some of the writings in the *Journal of the United Service Institution* on the eve of Independence. For instance, Lieutenant-Colonel Rajendra Singh in a prize winning article argued that it was only during a revolution when the authority of the government had been compromised that undisciplined armies showed political bias.

He was of the view that those who considered the army as an instrument to achieve freedom did not understand the function of the army.[17] Another officer Major Gurbachan Singh argued that it would be disastrous to have a politically oriented army. Politics should be taboo if the army is to carry out its functions efficiently. He therefore firmly advocated the view that officers from politically active classes are unwelcome in the army. He argued that soldiers had the right to vote and they were entitled to hold their political opinions as long as they did not make a public exhibition of it. Their loyalty and duty were to their country and not the political party in government. In the same vein, Lieutenant-Colonel B.L Raina writing in 1949 argued that the military had no business meddling in civilian matters. It did not matter whether a government was republican or socialist as long it was representative in nature. The military, he argued, owed its allegiance to any elected government and by extension to all the people of the country.[18]

While we have chosen a small sample of officers, these were some of the most influential officers of the day who rose to pre-eminence in the post-Independence years. These were some of the officers who could have hypothetically plotted a coup. Therefore, their take on civil military relations was vital. This is not to say that the opinions of countless nameless officers and *jawans* were not vital. Nevertheless, these were officers who at various stages wielded massive influence over the army.

All these officers had sympathy with the national movement in varying degrees. They in spite of their service in the colonial army were committed to the idea of a free India. What the accounts of Harbakhsh Singh's meetings with the INA leadership show was that even ostensibly apolitical Indian officers were conscious of the national movement and had definite sympathies for it. Apart from Singh's admiration and respect for Gandhi and Nehru, the brothers were effusive in their praise of Netaji who they described as a great leader. The case of Harbakhsh Singh shows that most Indian officers, even those seemingly indifferent to the nationalist cause, were nationalists at heart. However, the likes of Kaul were an exception in the sense that most of these officers possibly possessed a sense of duty or professionalism which would have barred them from actively hobnobbing with the nationalist leaders beyond a point. But, that did not stop them from being sympathetic to India's national struggle. In that sense they were all politically conscious officers, the partial antithesis of the ideal the British desired. I say partial because most of the Indian officers of the INA in spite the desertion of the colonial army and possessing soft corner for the nationalist cause did not violate their oath of loyalty. This is despite the fact that the officers had been aware of the illegitimacy of the government they were serving.

The British were aware of the moral weakness in their position. Therefore, they made strenuous though partly unsuccessful efforts to insulate Indian recruits from political turmoil. Incidents like the refusal of the Garhwal Rifles to fire at Indians alarmed them. This is not to say that feelings of fraternity with the civilian populace were the only reason for their refusal to open fire. There were other reasons too. But, the British were suitably alarmed.[19]

Such incidents probably strengthened their resolve to insulate the army from politics. While the discourses emanating in the training schools are not available to us, the address by Roy Bucher, the last British commander-in-chief of the Indian Army, to the cadets of the IMA, Dehradun, a few months after Independence is educative. Some of the excerpts from his speech read thus:

> The interests of the Army come first in your thoughts and in your actions all the time. Remember that the army, and indeed all the services, are the servants of the Government in power and the political complexion of particular Government makes not the slightest difference to this fact. As soldiers you are not concerned with politics. . . . No army which concerns itself with politics is ever of any value. . . . It follows therefore that the army therefore has never the slightest right to question the policy of the government. Implicit obedience to the orders issued by Government is essential, and only in this manner will the interests of the country be fully served.[20]

Influential commentators on the Indian Army such as Stephen Cohen have cited this speech as epitomizing British military culture and their commitment to keep the military aloof from politics. They have argued that the Indian military inherited this legacy of political aloofness and that went a long way in making the Indian Army one of the few apolitical armies in the developing world.

One would beg to disagree. First, the insistence on political neutrality was a pragmatic necessity. The British were all too aware of the moral weakness of their position. Secondly, the legacy they bestowed on the Indian Army officer was not exactly that of the politically sanitized officer. On the contrary some of the most prominent Indian officers were politically aware and had a definite bias for the national movement and its leaders. This was definitely at odds with the ostensible British military culture of political aloofness of the soldier. Some of these politically aware officers had ironically been trained at the holy grail of the British military establishment, Sandhurst. These included B.M. Kaul, who would in his later career, be at the heart of one of the most bitter political controversies afflicting the post-Independence Indian Army.[21]

Of course, this is not to say that the British effort to insulate the Indian Army from nationalist politics was an abject failure. There was the archetypical Westernized Indian officer who scrupulously kept aloof from the Indian national movement. These officers would ascribe everything good in the Indian Army to the positive legacy inherited from the colonial army. Some officers who were commissioned before Independence or in the initial years after Independence exhibited this trend. They concurred that the apolitical nature of the Indian Army was the consequence of the good traditions inherited from the British. Of course, they took into account other factors too but the positive legacy of the British they felt was a determining factor.[22]

One high profile example of the Anglicized officer having no truck with the nationalist movement was Cariappa. Yet, Cariappa paradoxically also reflected the failure of the colonial state to ensure the political neutrality of its Indian officers. As his later career as a public figure would testify, Cariappa was always on the lockout for situation in which the military could step in. The imminence of Partition and the impending chaos it promised emboldened Cariappa as mentioned before to make a bid for power though he was averse to hold the reigns in his own hands. He as stated before preferred Nehru or M.A. Jinnah as the head while he no doubt visualized himself as a kingmaker. There seemed to be an apprehension about the people not accepting a military man at the helm even in times of extreme crisis, hence he possibly preferred the role of kingmaker.

THE POST-INDEPENDENCE DECADE

The same man would, however, become a passionate advocate of the detachment of military from politics. Why was it so? That brings us to the political conditions which prevailed during the initial years of Independence and its impact on civil–military relations. What were the advantages which the Indian state possessed? The post-colonial polity for example had experienced and stalwart politicians at the helm. While Nehru was the tallest of them all there were a host of experienced and capable politicians with a basic commitment to democracy. While there were dissenters of the parliamentary form of government, these dissidents were not interested in infiltrating the army.

A strong democratic political culture as stated before developed based on mass participation at the ground level. The successful conclusion of the first general elections in 1952 further cemented the foundations of a mass-based participatory democracy. India was rapidly evolving what S.E Finer would term a mature political culture.[23] Such cultures were usually inimical to military involvement in politics. Even if certain members among the officer corps harboured political ambitions, they were apprehensive of encountering widespread popular resistance to their attempts to capture political power. Also often the political systems in mature political cultures enjoy a legitimacy which extends to the military too.

However, even a popular leader with a mass base can provoke a coup if he ceases to observe certain rules of the game. Z.A. Bhutto of Pakistan for instance despite his demagogic popularity was not above using dubious and undemocratic means to eliminate his political opponents. The repression had reached such an impasse that his political opponents would have welcomed intervention by the military. And that is precisely what happened. India too has seen the system manipulated by cynical politicians of all hues. However, I argue that there is a threshold level till which the system can withstand its abuse. Anything beyond that could invite a fundamental challenge to the legitimacy of the system. This challenge could come in the form of an attempted military takeover of the country. This has been the case in Pakistan where civilian governments have often failed to establish firm democratic practices and traditions when they have had the opportunity to do so. The 1971 military debacle of the Pakistan Army gave Bhutto a unique opportunity to subordinate the brass hats. He was popular while the army was discredited. Yet, Bhutto missed this golden opportunity. In the first place his rise to power came at a heavy price to the Pakistani nation. Bhutto in flagrant violation of democratic principles blatantly disregarded the democratic mandate; the Awami League had secured votes to form a government. In fact, much of the blame for the 1971 fiasco should be laid at the doorstep of Bhutto though he was shrewd enough to cast the blame on the shoulders of the army. Thus, the foundations of his ascension to power were blatantly undemocratic and so. Bhutto could not be the bedrock on which democracy might have acquired deep roots in Pakistan.

In contrast, violation of democratic principles never crossed a threshold level, especially in the early years of Indian democracy. While there were instances of rigging, intimidation of opponents, and so on such malpractices never got institutionalized. Rigging and election violence were hardly a coordinated nationwide phenomenon. The results of general elections and even local state-level elections were on the whole fair even though there might be pockets marked by rigging, violence and intimidation. The political process witnessed mass participation even though sections of the mass would often exhibit undemocratic tendencies.

There are also areas like Nagaland which have been marked by insurgency from the 1950s. While the situation has been grave enough for the army to be deployed occasionally, there was never any serious threat of the province seceding. Consequently, the army though witnessed the failure of the civilian administration to tackle the problem politically, possibly retained an overall confidence in the capability of the political establishment. The tall reputation enjoyed by the politicians at the helm bolstered no doubt by their glorious record in the independence movement went a long way in ensuring that this confidence was unimpaired. Of course, this confidence in the civilian establishment would have been severely tested if Nagaland had seceded from the Indian Union.

However, there were certain irritants present which could have imperilled civil–military relations. The years after Independence saw the steady whittling away of the privileges of all ranks of the military. The rank of the chief of armed forces was pegged several notches down. In addition, the chief of the army was no longer the commander-in-chief. The president of India became the commander-in-chief. Various monetary privileges and perks of the officer corps were also done away with.[24]

Adding grist to the mill was the fact that most of the bureaucrats in the defence ministry and the politicians holding the defence portfolio had no experience of military matters. Consequently, they often ended up rubbing the armed forces the wrong way. The armed forces might have perceived that they were given short shrift. Who could they turn to? Nehru was an option but then he was reputed to be disdainful and distrustful of the military. In Nehru's words: 'The soldier is bred in a different atmosphere where authority reigns and criticism is not tolerated. So he resents the advice of others and when he errs, he errs thoroughly and persists in error. For him the chin is more important than the mind or the brain.'[25] However, Nehru was not blind to the necessity of possessing a strong and reliable military. He wrote in *Young India* to the effect that the armed forces would have to be altered to suit the needs of independent India. Therefore, an independent India's army would have to be completely reorganized to 'create out of fresh material, a truly national army with a national outlook'.[26] Thus, Nehru was visualizing the complete overhaul of the army after Independence. That would have meant large-scale tampering with the existing structure, a job easier said than done.

A Tornado by the name of Krishna Menon

To be fair to Nehru, despite his contempt for all things military and the existing structure of the army, he did not interfere in the internal organizational matters of the armed forces. The real challenge to the military's hierarchy and corporate interests came with the appointment of Krishna Menon to the post of *raksha mantri* (defense minister). Menon gradually played havoc with the hierarchy and internal organization of the army. He had his own coterie of favourite officers and he used them to get things done informally bypassing the proper procedure. B.M. Kaul was the most prominent among his clique. This soon led to a crisis when General Thimayya resigned citing Menon's interference in the internal functioning of the army. Menon was accused of ensuring the out-of-turn promotions of his favourite officers. Nehru initially lent a sympathetic ear to Thimayya and persuaded him to withdraw his resignation. In a sudden abrupt turnaround he castigated Thimayya in Parliament, calling his grievances childish. Thimayya though no doubt stunned by this abrupt alteration in Nehru's position on this matter choose to remain silent. On his retirement, the government ignored his anointed successor, General Thorat and instead chose General Thapar a more pliant general who was firmly in the Menon–Kaul camp.

Things took a particularly vicious turn when the Menon–Kaul camp sought to victimize Sam Manekshaw, the outspoken major-general of the Indian Army and a star in his own right. He was accused of disloyalty to the country and conspiring against it. Fortuitously for Manekshaw, he was saved on account of two factors, one the investigating officer Daljit Singh, was a conscientious man and refused to kowtow to any dictates. Secondly, the debacle in the Chinese War resulted in the ouster of the Menon–Kaul nexus from the corridors of power.[27]

With the fall of Kaul and Menon, a sordid saga in the history of the Indian Army came to an end. These were the years when the corporate structure of the army and its internal autonomy were severely threatened. No army in the world takes kindly to interference in solely military matters. The army considers its internal hierarchy sacred. Any attempt to bypass and meddle with it is viewed with extreme revulsion. Such attempts could provoke an army to act against the government with the aim of simply protecting its own turf.

Yet, the Indian Army did not rebel. What could be the reasons for it? One, the army might have viewed certain individuals such as Menon and Kaul as responsible for the murky state of affairs rather than laying the blame on certain endemic faults within the system. Secondly, the bureaucracy acted as a big safety valve. The grouse of the army officers was mostly directed against the bureaucracy. The bureaucracy became a punching bag, which shielded the political class from the ire of the army.[28]

There were also the oft-repeated other important factors such as the widespread legitimacy of the political system and the politicians at the helm. People like Thimayya were hardly apolitical but they had a lifelong admiration for Nehru and what he stood for. For Thimayya, parliamentary democracy was the logical culmination of the national movement, a movement he despite his lack of active participation had been fervently committed to. In a speech to the gentleman cadets of the IMA, he made it clear that they would under no circumstances be involved in taking part in the political life of the country.[29]

The other general in this black episode, Kaul had been a lifelong nationalist and had actively hobnobbed with the national leadership even during his days as a colonial army officer. He and Thimayya had vital differences. Their sense of duty and their professional ethos were of a completely different order. Service in the colonial army did not deter Kaul from giving anti-colonial speeches or handing over confidential files to the nationalists. Thimayya's sense of duty stopped him from going for an intensive interface with the nationalist leaders, even though he had admiration for them. He would still make a fine distinction between the call of duty and his political leanings. Kaul did not possess the ability to make this fine distinction. Hobnobbing with his political masters was a matter of habit for him. While he too believed in the legitimacy of the political system, he possibly wanted a stake in it too. He was not content to acknowledge the legitimacy of the political culture and the politicians' right to run the show. He wanted to be a part of the show. He was in effect a politician in a soldier's garb. One suspects he was among politicians a soldier in civilian clothing. He traversed both worlds and in effect perhaps belonged to no world completely.

Kaul's attitude was anathema for many of his contemporary officers. The writings of Major-General D.K. Palit and Air Marshal P.C. Lal are indicative of this fact. Lieutenant-General S.N. Sharma who was about a decade junior to Kaul says that save for Kaul there was never any instance of an army man trying to gain power. Kaul according to him was a protégé of Nehru and was fancying his chances of succeeding the latter.[30] Palit had a slightly different take on this issue. He believes that Thimayya was equally culpable of unofficial and informal channels between him and Nehru. Timmy as he was known did not hesitate to exploit his easy access to the prime minister. Palit does not spare Nehru either and says that he was equally guilty of breaches of procedure. Menon therefore had a precedent before him which he exploited for his own ends.[31]

In conclusion, India's apolitical army was definitely not a British legacy. On the contrary the peculiar experience of serving in an alien army in a time of intense anti-colonial struggle had made many of them political creatures despite their lack of participation in the national movement. The surcharged political climate made them aware that they were essentially serving an illegitimate regime. And one is not talking of the officers who joined the INA. One is referring here to the ubiquitous Indian officer who apparently performed his duties with devotion. The apolitical nature of the Indian Army, I argue, is a complete break with the colonial tradition. Most commentators on the army like M.L. Chibber, S.K. Sinha, Apurba Kundu, and so on, all agree on certain parameters which went a long way in establishing the supremacy of civilian rule in India on a firm basis. Some of the oft-repeated reasons are the presence of able politicians, the evolution of a mass-based democratic political culture, experienced bureaucracy, the existence of a vibrant middle class, and so on. Kundu lists a host of factors which have contributed to the absence of a coup in India basing himself on responses to questionnaires by him. Apparently factors such as the professionalism of the armed forces, diversity of people and cultures, initial political stability and quality of democratic rule are listed as some of the pre-eminent factors that explain the absence of a coup in India.[32]

While I do not discount any of these factors, one could argue that the very choice of the army to stay out of politics was a conscious political choice. It was made because for many officers the birth of democracy was their lifelong political dream, and they had in that sense a stake in the political order. And the men who were at the helm of affairs, the likes of Nehru and Sardar Vallabhbhai Patel, the officers respected and were in awe of. In fact, the towering personality of Nehru is very important for the history of civil–military relations in India because it enabled the man to get away with his often shabby treatment of the military. While granting legitimacy to reasons like the popularity of Nehru and the Congress party, the creation of a mass-based political culture and the sheer force of Nehru's personality cannot be underestimated. One needed to overcome a psychological barrier to contemplate overthrowing Nehru, and that was not easy. For most, one could think of succeeding him, but the idea of replacing him was not easy. I also argue against exaggerating the strength of India's democratic ethos. This country experienced multiple abuses of the democratic system. It is just that the threshold level was never breached. While India may have been luckier than Pakistan in the initial decades, the ills attributed to the Pakistani political culture are not totally absent in India. It is just that some basic ground rules had not been breached and ills like rigging and muscle power were regional affairs rather than coordinated nationwide affairs. The legitimacy of the results of no general election had been questioned in spite of aberrations.

Also, there are certain barriers to the military acting as a unified body. That is where the emphasis on regimental ethos, the legacy of colonial recruitment policies comes in play. The colonial army had designed regimental recruitment policies with the objective of tiding over the possibility of national feelings in an essentially mercenary army. As a result regimental ethos and recruitment were interrelated. Their partial continuance in the post-Independence period ensured that broader coordination would remain an issue in the Indian Army. Let us first investigate the recruitment policy of our armed forces.

RECRUITMENT IN THE POST-INDEPENDENT INDIAN ARMY: CONTINUITY OR BREAK?

The armed forces epitomize the ideals of service, sacrifice, patriotism and our country's composite culture. The recruitment to the armed forces is voluntary and every citizen of India irrespective of his caste, religion and community is eligible for recruitment into the armed forces provided he meets the required physical, medical and educational criteria. Recruitment into the army is carried out according to the Recruitable Male Population (RMP) of each state.[33]

Intent and Practice

The section on recruitment in the annual report of the Ministry of Defence begins with the above paragraph. This paragraph in some ways illuminates the basic tenets of the recruitment policy of the Indian armed forces. However, the above paragraph hides as much as it reveals. The Ministry of Defence does not release data regarding the ethnic composition of its rank and file in the army in particular. It is anybody's guess as to the ethnic and religious composition of India's troops. There are no direct ways of knowing save for oral interviews and autobiographical accounts by retired army officers.

The report though mentions the system employed to recruit *jawans*. The army revised its recruitment procedure in 1998 Aspirants now are recruited through open recruitment rallies staged all over the country. The Ministry of Defence report informs us that at least one rally is held each year seeking to reconcile the different districts, areas and regions and the anticipated response. The rallies are tailored in such a way that every citizen of India gets at least a single opportunity to enroll him/herself in the army.[34] The new system was adopted to tide over the difficulty of aspirants having to traverse long distances to present themselves for enrolment. Apparently, the new system had the advantage of presenting potential recruits an opportunity to enroll themselves in the vicinity of their area of domicile. Aspiring youths of the locality in question would be informed through the medium of hoardings and advertisements in local newspapers.

Again the layman is left in the dark as regards the initiation rites of the successful recruits. How are they posted? Are they posted in a regiment to the vicinity of their area of domicile? Or can they be posted in any unit? Or are there particular regiments which will be the home of certain specific ethnic groups? Sadly, the army will not deign to provide us the answers. The question naturally arises as to why is there a conspiracy of silence on these matters? Is the silence linked to the security concerns of the Indian state?

While the official answer to such a query may be on the lines of that, the truth may lie somewhere else. One would dare venture the proposition that there are considerable anomalies between the ground realities and the officially stated policy on recruitment of *jawans* in the Indian Army. Despite the protestations about the armed forces being open to Indian citizens of any caste, creed and religion, the recruitment base of the Indian Army might be much narrower than is thought to be. While admittedly all infantry units formed after Independence have been mixed units the Indian Army is heavily weighed down by the weight of regiments with long and hoary histories.

It is the actual recruiting practices of such regiments which may run foul of the professedly noble intentions of the army to be as broad based as possible. For many of these regiments originated as single class regiments, i.e. to put it in a nutshell these were regiments which were dominated by recruits of a particular ethnic type. The inevitable outcome would be that regimental traditions would often evolve around the religious and social customs of the community constituting a particular regiment. Policy makers would often be loath to tamper with the composition of these regiments as they feel these traditions have served them well and contributed immensely to the cohesion and the battle worthiness of a regiment. Therefore, the reality is that many mixed regiments might be mixed only in name. There would be certain companies or at times battalions which would possess recruits from a different community.

Recruitment in the post-1857 Colonial Army

To understand this phenomenon one needs to go back to the recruitment policy of the colonial army in the post-1857 phase. The great revolt of 1857 had involved nearly three-fourths of the predominantly high-caste Bengal Army infantry regiments. This naturally necessitated some drastic measures and extensive plans were drawn up for reorganization after 1857. The British now sought to follow a balanced policy of recruitment. The idea was to balance different communities from various regions against each other. Kaushik Roy terms this the Balanced Recruitment Policy.[35] The balancing was affected at two levels, among the three presidency armies and the presence of diverse groups within an unit.[36]

Certain officers from the Punjab Frontier Force favoured balancing the *Purbiyas* (high castes of Bihar and Awadh) with the Sikhs and the Pathans. Some officers wanted to balance the *Purbiyas* with communities such as the Marathas. There were still others who argued that the army should recruit all possible groups from a particular region so that there would be no predominance of groups from any particular region.[37] This state of affairs continued till the 1880s, when the British perceived a threat along the North-Western Frontier of India from Russia. The Great Russian 'bear' cast its shadow over the formulation of recruitment policy for the British-Indian Army. It was then felt that the army should be composed of the best possible recruits as they possibly might have to contend with a formidable foe like the Russian Army. The army therefore had to be geared to meet external threat overriding its preoccupation of warding off internal threats on the lines of the great conflagration of 1857, the memories of which were still fresh in the minds of the British.

In the 1880s, influential officers like Field-Marshal Roberts argued that there were certain communities in India who possessed innate martial qualities. He argued that unlike in the West, there were only some communities in the Orient who had the requisite qualities. He dubbed certain communities such as the Sikhs, Pathans, Dogras, Gurkhas as martial. Roberts attributed their warlike demeanour to the interplay of geographical factors. In his opinion people living in cooler regions were apt to be martial in contrast to people living in hot regions such as southern India.[38] Roberts also drew a connection between the existence of a frontier and warlike qualities. Ironically, peaceful and settled frontiers were dubbed as inimical to the martial spirit of communities residing in such areas. Communities such as the Tamils and Telugus had become soft due to prolonged absence of conflict whereas a 'live' North-West Frontier meant that the Sikhs and the Pathans maintained their war like spirit.[39]

Roberts' followers further fine-tuned and modified his theory. Major-General George Macmunn seemed to think that the descendants of the so-called Aryan invaders were martial in spirit whereas the descendants of the defeated Dravidians lacked any martial spirit. Accordingly, the Muslims and the Rajputs of Punjab were termed as martial.[40] There were other followers of Roberts who established further connections between agriculture and soldiering. Consequently, the Jats were included as a martial group.[41]

Thus, was born the Martial Race Theory which increasingly influenced recruitment. A table prepared by Kaushik Roy bears ample testimony to this. Around 1885, no particular group was dominating the military. By 1912, groups such as the Gurkhas, Dogras, Garhwalis, Pathans, Punjabi-Muslims and Sikhs together constituted 44 per cent of recruits. The share of these groups was 25 per cent in 1885. In contrast, the intake of south Indians declined drastically from 25 per cent to 8 per cent in 1912, thus setting the tune for the future. This was the result of the Martial Race Theory operating in all its venom. The ethnic composition of the colonial Indian Army would roughly hover around the above-mentioned percentages barring emergencies such as First World War and Second World War. These two wars saw the induction of new 'classes' (actually communities) but these classes were again retrenched following the conclusion of these two wars.[42]

Recruitment in the Post-Colonial Indian Army

So, this was the legacy the post-Independence Indian Army inherited. The moot question before the Indian Army was, should it go for a wholesale alteration in the recruiting by opting for a truly broad base or should it choose a compromise? The army opted for the later option. On one hand it was officially stated that henceforth all new regiments would be mixed. At the same time the order stated that regiments composed of particular classes, through several decades, had developed a certain kind of cohesion and while steps would be taken to broad base their composition, it was essential to ensure that the sentimental attachment arising out of such composition must not be suddenly disturbed.[43]

There lay the crux of the matter. The loophole integral to the order ensured that a large number of single class units made only an ostensible change to a mixed unit even though the Gorkha (the British designated them as Gurkha) regiments remained for the record the only single class regiment. In practice, there were certain regiments which remained single class units. Major-General Ian Cardozo dwelling on the role of religion in the army argues that it was a very important motivating factor in single class units such as the Sikhs, Gorkhas, Garhwalis, Kumaonis and other one class units.[44] This was circumstantial evidence that there still existed single class units in the Indian Army. Sinha writing in 1986 asserted that about 30 per cent of India's regiments were still single class or fixed class units.[45] Statistical data is unfortunately lacking but we do get some state-wise data which might tell us something (Table 13.1).

While these are not official statistics, these have never been refuted by the army. We therefore operate on the thumb rule that they are genuine unless proved otherwise. Operating on the premises set in Table 13.1, one can assume that while certain

TABLE 13.1: REPRESENTATION OF VARIOUS STATES IN THE INDIAN ARMY

State	Population percentage	Percentage of recruits in the army
Punjab	2.6	15.30
Haryana	–	7.82
Jammu & Kashmir	00.6	4.68
Rajasthan	00.9	2.92
Kerala	03.7	7.04
Assam, Nagaland, Meghalaya, Tripura, Manipur	3.4	5.38
Uttar Pradesh	16.5	4.05
Maharashtra	9.2	15.58
Tamil Nadu	7.3	7.64
Bihar	10.2	5.09
Madhya Pradesh	7.6	5.13
Andhra Pradesh	7.8	4.08
Karnataka	5.3	2.81
West Bengal	8.2	3.63
Orissa	3.9	1.27

Source: Chitra Sudarshan, 'Continuity and Change: The Story of Integration in the Indian Army', *Strategic Studies*, vol. 12, no. 12 (March 1989), p. 1390.

states still sent a lot of recruits, their share in the overall percentage of recruits supplied to the army had declined. Punjab in spite of the decline still supplied a disproportionately high number of recruits when compared to their population percentage. One of the most celebrated fighting groups in India is still making its presence felt. Most other states, however, sent troops to the Indian Army either in proportion to their population percentage or supplied troops below their population percentage. For example, Uttar Pradesh while supplying a large percentage of troops still supplied below its population percentage. To put things in perspective, Punjab was by far the largest supplier of troops to the Indian Army even in 1989.

The scenario gets complicated when we look at Table 13.2. What is marked is the massive decline of troops supplied by Punjab between 1965-6 and 1996-7. The decline if we are to compare the statistics of both the tables is particularly marked between 1989 and 1996. In 1989, Punjab despite a considerable decline in percentage of troops supplied to the army, in real terms was still supplying far higher than its population percentage. In 1996, it was steadily inching towards parity with its population percentage. Whether this was the result of the insurgency

TABLE 13.2: STATE-WISE COMPOSITION OF THE INDIAN ARMY: 1965-6–1996-7

Percentage of Total	1996-7	1965-6	Resulting+ or-
Uttar Pradesh	20.6	18.1	+
Rajasthan	7.9	7.2	+
Punjab	7.6	31.6	–
Maharashtra	7.3	7.0	+
Bihar	7.0	5.4	+
West Bengal	5.7	2.8	+
Haryana	5.1	—	+
Andhra Pradesh	4.9	4.0	+
Tamil Nadu	4.9	5.4	–
Himachal Pradesh	4.4	1.4	+
Jammu & Kashmir	3.7	2.0	+
Karnataka	3.1	3.5	–
Kerala	3.1	4.6	–

Source: Manoj Joshi, 'Ethnic Group Recruitment in the Indian Army: The Contrasting Cases', *India Today*, 12 April 1998, p. 29.

movement that had raged there during the 1980s is anybody's guess.

Again combining the statistics of Tables 13.1 and 13.2, we find that men from Uttar Pradesh or the Hindustani soldier registered an increase in the percentage of recruits sent to the Indian Army though it was not substantially higher than its population percentage. The 1990s, however, saw a marked increase in the percentage of recruits which resulted in Uttar Pradesh contributing a higher percentage of recruits compared to its population percentage. The Hindustani soldier if one were to also include the percentage of soldiers from Bihar was thus making a comeback post-Independence but this was to be expected. The question of reaching the dizzy heights of pre-1857 Mutiny era did not arise. This comeback was also reflected in the number of officers hailing from Uttar Pradesh passing out from the IMA. M.L. Chibber's book provides us two extremely revealing tables. (Tables 13.3 and 13.4). These have a greater authenticity since Chibber is a reputed army officer and he presumably got this statistics from army sources.

However, these statistics do not enlighten us about certain finer points. It does not for example tell us what percentages of recruits were made up of Sikhs. It is possible that Punjabi Hindus were being recruited in larger numbers after Independence. We have no way of telling the ethnic composition of recruits from

TABLE 13.3: SUMMARY OF STATE-WISE BREAK-UP OF OFFICERS IN FIRST TEN IMA REGULAR COURSES 1932 TO 1936

State/Union Territory	Number of Officers	Percentage
Punjab	165	41.77
Princely States	88	22.22
North-West Frontier Province (now in Pakistan)	50	12.62
United Province (after 1947 Uttar Pradesh)	41	10.35
Delhi	13	3.28
Bombay	9	2.27
Central Province	6	1.50
Madras	6	1.00
Bengal	4	1.00
Gujarat	4	0.75
Kerala	3	0.49
Bihar	2	0.25
Orissa	1	0.25
Bassein	1	0.75

Source: Lieutenant-General M.L. Chibber, *Military Leadership to Prevent Military Coup* (New Delhi, 1986), p. 40.

TABLE 13.4: SUMMARY OF STATE-WISE BREAK-UP OF OFFICERS IN LAST TEN IMA REGULAR COURSES (64 TO 73RD), (1978-82)

State/Union Territory	Number of Officers	Percentage
Uttar Pradesh	721	19.18
Delhi	416	11.03
Haryana	375	9.94
Maharashtra	321	8.51
Kerala	237	6.28
Bihar	207	5.49
Andhra Pradesh	196	3.20
Rajasthan	138	3.66
Himachal Pradesh	137	3.63
Madhya Pradesh	127	3.37
West Bengal	115	3.05
Tamil Nadu	105	2.78
Karnataka	103	2.73
Jammu & Kashmir	93	2.47

Note: The rest about 20 per cent roughly is from Punjab.
Source: Lieutenant-General M.L. Chibber, *Military Leadership to Prevent Military Coup* (New Delhi, 1986), p. 41.

Punjab. It is also not possible to concur from these statistics that Sikhs were being posted in homogeneous units. The Sikhs could well be dispersed in various regiments.

Certain groups such as the communities from south India made a moderate comeback after Independence though they do not acquire the halo that certain ethnic groupings continue to have. In fact, it could be argued that their numbers have declined slightly in recent times. Clearly they have some way to go before they can get over the baggage imposed on them by the Martial Race Theory.

This was clearly at variance with the fortunes of one of the most celebrated 'non-martial races', i.e. the Bengalis. There was a clear increase in the number of recruits from Bengal. While it was still lower than the population percentage, the 1990s had seen a further increase in the percentage of recruits in the army. There was also a slight increase in the number of Bengali officers. It seems Bengali recruits had been able to persuade the army to an extent that they are capable of exhibiting a certain amount of martial prowess.

These tables do tell us that recruitment patterns have changed to an extent, especially with regard to Punjab. The government seems to have made a conscious decision to lessen the reliance on Punjabi or Sikh troops though the equalization of both the categories is open to question. Political instability may have complicated the situation and further strengthened the resolve to lessen the number of Sikh troops. Till this date, there was only one Sikh chief of army staff in post-independent India. Whether it was a coincidence or it constituted a deliberate policy option by the political establishment is an open question.

Apart from the above-mentioned deviation, there seems to have been no drastic change in the recruitment pattern. Old values and ethos surrounding recruitment had its takers and many Indian Army officers continue to be nostalgic about these notions. Sam Manekshaw in his foreword to Joginder Singh's book says about the proposed move to organize the Indian Army into mixed units on the basis of state populations:

> Are the past glories and achievements of regiments like the Marathas, the Sikhs, the Dogras, the Rajputs, the Madras, the Gorkhas, etc. to be forgotten? Are their famous rallying slogans to be done away with? In the name of god desist from such folly. Neither god nor future generations will forgive the perpetrators of such misplaced nationalism, which will lead to the destabilization of the famous Indian Army. . . . If this imprudent proposed political decision is accepted by sycophantic generals, I forecast doom and calamity.[46]

Manekshaw is probably summing up a widely held view in the Indian Army. The enmeshing of ethnic, religious and regimental traditions had served the Indian Army well. Single class regiments had won many famous victories. These traditions were in fact the bedrock on which the ethos of the Indian Army evolved. Such time tested traditions, many felt should not be trifled with. Otherwise disaster would befall the army.

Interviews of certain other officers have revealed the same sentiments. Mathew Thomas was of the view that the present recruitment policy has stood the test of time and should not be meddled with. Echoing Manekshaw's views, he reckons that religion and ethnicity are very important for building morale in the combat arms. Other corps such as the signals, electrical and mechanical corps should be on an all-India basis.[47] Mathew Thomas is espousing strong support for what he considers to be a time-tested mode of recruitment in spite of serving in a mixed newly-formed post-Independent unit of the Indian Army. In contrast, Lieutenant-General S.N. Sharma of the Armoured Corps seems to think that mixed units are the future of the army. He feels the army has to evolve into a totally indiscriminate mixture of ethnic, regional and religious communities as the rest of Indian society.[48] Sharma's reply makes it clear that many old regiments were mixed only in name. While they had a small mixed element based on geographical source of recruitment, the core ethnic composition remained. Sharma though is not clear about what he means by geographical source of recruitment. He is probably referring to the *jawans* recruited through the 11 zonal recruitment centres in India excepting the Gorkha recruiting centres in Nepal and Bhutan. Recruitment rallies are organized in all these zones. The rallies are planned in such a way that each district in the country is covered by at least one recruitment rally in a year.[49] The emphasis here is ostensibly on geography because all males from a particular geographical area are eligible to enlist.

However, tradition has ensured that particular communities with a genealogy of service in the army should continue to supply majority of the recruits. Many of these zonal recruiting areas are based in the traditional supply grounds of the Indian Army. In fact, the only fresh recruiting ground in the post-Independent period is West Bengal. That is the sole clean break the recruitment practices have achieved in the post-Independence Indian Army. Of course, the option exists of indiscriminately mixing regiments with men recruited from different geographical zones but clearly this is one option which has not been taken despite some noises to this effect.

Retired army officers during interviews also spoke in detail of the changing profile of the officer corps. Almost all of those interviewed say that the officer corps has increasingly acquired a middle class character. Some like S.N. Sharma and Mathew Thomas are very appreciative of this widening of the social base of the officers. They feel that this gives the officer corps a truly Indian character. Sharma takes pride in the remarkably free and impartial selection system which he terms unique for a society marked by prejudice, influence and favouritism. Thomas implies that the officer corps has over the years evolved a truly pan-Indian ethic.[50]

CONCLUSION

To conclude, one would say that the recruitment policy of the army shows considerable continuities from the past. Sure, there have been some changes like the considerable reduction in the number of Punjabi/Sikh recruits; these decisions have been essentially dictated by political considerations rather than military compulsions. The political factors involved are a two-way affair, i.e. the reduction in the number of Sikh recruits could have been equally the product of certain choices the Sikh youth may have made in the heydays of militancy. Equally, the perpetuation of certain recruitment practices has been determined by a combination of political exigencies and the sentimental associations of a considerable number of army officers. The weight of traditions is never easy to shed. Modes of recruitment and regimental traditions are intertwined in the Indian Army.

While the government has raised only mixed units after Independence, it is hesitant to tamper with the traditional composition of old single class and mixed class regiments. The unimpeachable logic given is why tamper with a system which has produced outstanding results over the decades. One never knows what might come in place of it. The fear of the unknown is a formidable one and leads to a tendency not to rock the boat.

The regimental system also helps to absorb ethnic communities whose emotional bonding with the Indian state remains loose.

The colonial government chose to construct a regimental ethic to supplement the lack of national ties. Then, what purpose does it serve for the Indian state? Does the regimental system enable groups like the Nagas with loose emotional bonding to the nation to fit into the army? The answer is not an easy one. One of my respondents S.N. Sharma seemed to think that the Nagas substituted the regiment for their tribe. Many of the soldierly values were already inscribed in tribal culture.[51] Major-General (retired) K.K Ganguly, an officer who had served with the Nagas sidestepped my question and held that while the Nagas joined due to lack of employment opportunities, the army instilled among them the values of patriotism at various training centres.[52] While one would not pass a judgement on Sharma's controversial views it is worth pondering as to whether regiments fill the task of absorbing recruits at odds with the grand narrative of Indian nationalism. A perusal of the regimental ethos somewhere else would illustrate this point better.

Coming to the major theme discussed here, one can argue that the apolitical nature of the Indian Army is, contrary to popular wisdom, a break from the past. I argue that the legacy the colonial army bequeathed was in fact of a political nature despite the best efforts of the colonial government. What then does it tell us about the moot question discussed here? How colonial was the post-colonial army? There is no clear-cut answer. While the colonial legacy is very strong and remains invariably a force to be reckoned with, the Indian state is prepared to confront this formidable legacy if it does not agree with its state-building agenda. While there is admiration for the colonial baggage, there is no blind adherence to the colonial policies pertaining to the army. Continuity is favoured as long as it suits the purposes of the Indian state. If on the contrary it proves to be a stumbling block, the baggage is firmly sought to be shed. Therein lies possibly the arguable success of the incipient Indian state in building an army firmly tailored to its overall agenda. While there were hiccups, fate and hindsight ensured that a critical boiling point was never reached. While problems still exist, the army has become today a truly national army in the sense that while it may be dissatisfied with the system and the current state of affairs in a manner akin to civil society, the system is taken as given and is considered the best possible alternative despite its many aberrations. Despite possible occasional loose talk the focus is on reform and not the overthrow of the system, an attitude congruent with major sections of civil society.

NOTES

1. See Stephen P. Cohen, *The Indian Army: Its Contribution to the Development of a Nation* (1971, rpt., New Delhi: Oxford University Press, 1991), p. 74.
2. Edwin S. Montagu and Venetia Montagu (eds.), *An Indian Diary* (London: Heinemann), 1930 pp. 201, 352.
3. See M. Muthanna, *General Cariappa* (Mysore: Usha Press, 1964).
4. Cohen, *The Indian Army*, p. 75.
5. General B.M. Kaul, *The Untold Story* (Bombay: Allied, 1967), pp. 9-11, 62, 70-1, 74-5.
6. See Apurba Kundu, *Militarism in India: The Army and Civil Society in Consensus* (New Delhi: Viva Books, 1998), p. 84.
7. Lieutenant-General K.M. Cariappa, 'Speech to Indian Officers, 8 October 1947', reprinted in *United Service Institution Journal* (*JUSII*), vol. LXXVIII, no. 330 (January 1948), p. 4. Major-General Cardozo argues that Cariappa set the tone by insisting that officers should have no truck with politics. Cariappa was apparently echoing the thoughts of other senior officers in the army. See Major-General Ian Cardozo, 'Ethos, Values, Training', in Ian Cardozo (ed.), *The Indian Army: A Brief History* (New Delhi: United Service Institution), 2005, p. 259.
8. See 'Military Rule Under President's Control', *The Mail*, Monday, 22 June 1970, sr. no. 6, Cariappa Papers, p. 4, National Archives of India (NAI), New Delhi.
9. Ibid.
10. Ibid.
11. Humphry Evans, *Thimayya of India* (New York: Harcourt Brace), 1960, p. 124.
12. Ibid., p. 116.
13. Harbakhsh Singh, *In the Line of Duty: A Soldier Remembers* (New Delhi: Lancer), 2000, pp. 8-9.
14. Ibid., pp. 28-33.
15. Ibid., p. 111.
16. Ibid., p. 135.
17. Lieutenant-Colonel Rajendra Singh, 'Is Scientific Selection Successful?', *JUSII*, vol. LXXVI, no. 325 (October 1946), p. 335.

18. Major Gurbachan Singh, 'The Right Type and Some Thoughts on Indianization', *JUSII*, vol. LXXVI, no. 324 (July 1946), p. 44; Lieutenant-Colonel B.L. Raina, 'Leadership', *JUSII*, vol. LXXX, nos. 334-5 (January-April 1949), p. 18.
19. Cohen, *The Indian Army*, p. 96.
20. 'Sir Roy Bucher's Speech to the Staff and Cadets, 28 May 1948', reprinted in *JUSII*, vol. LXXVIII, no. 333 (October 1948), p. 331.
21. Kaul, *The Untold Story*, pp. 30-1. Kaul was appreciative of his tenor at Sandhurst. He said that he learnt a code of conduct, a sense of discipline, a sense of honour, apart from learning the basic techniques of the military craft. It seems his instructor forgot to educate him on the need of the military man to keep aloof from politics.
22. One of the interviewees who argued strongly about the legacy of the British was General Shankar Roy Choudhury. I interviewed him at his house in Kolkata on 26 December 2005.
23. S.E. Finer, *The Man On Horseback: The Role of the Military in Politics* (London, 1962).
24. See Brigadier S.K. Sinha, 'Career Prospects for Officers in Armed Forces', *JUSII*, vol. LXXXVIII, no. 412 (July-Sptember 1968), pp. 263-9. For an excellent discussion on the subject see Cohen, *The Indian Army*, pp. 173-80; Kundu, *Militarism in India*.
25. Jawaharlal Nehru, *Jawaharlal Nehru: An Autobiography* (1936, rpt., New Delhi: Oxford University Press, 1980), pp. 3-4.
26. Jawaharlal Nehru, 'The Defence of India', *Young India*, 1 October 1931, pp. 284-5.
27. For a first hand account of the bad blood arising out of the Menon-Kaul nexus see D.K. Palit, *War in High Himalaya: The Indian Army in Crisis, 1962* (New Delhi: Lancer), 1991. Also see Air Marshal P.C. Lal, *My Years with the IAF* (1986, rpt., New Delhi: Lancer, 1987) and Kundu, *Militarism in India*, Ch. 6.
28. Military officers in freewheeling conversations blame the bureaucracy for many of their travails. My interview with Lieutenant-General Pratap Narayan (on 3 March 2006 in New Delhi) reveals the officer exhibiting this tendency. While a lot of the blame for the mismanagement of army affairs could be laid at their door, the bureaucracy also absorbed whatever punches the military landed. So, its transience in matters military also proved to be a safety valve for the political class.
29. General K.S. Thimayya Papers, Speeches and Writings, sr. no. 7, Speech by K.S. Thimayya to the Gentleman Cadets of the IMA, Dehradun, 13 December 1958, Nehru Memorial Museum & Library, New Delhi.
30. Written answers to questionnaire, Lieutenant-General S.N. Sharma, 25 February 2006.
31. Palit, *War in High Himalaya*, p. 74.
32. See Table 1 in Kundu, *Militarism in India*, p. 6.

33. *Annual Report of the Ministry of Defence: 2006-2007*, New Delhi, p. 193.
34. Ibid., p. 196.
35. Kaushik Roy, 'Recruitment Doctrines of the Colonial Indian Army', *Indian Economic and Social History Review*, vol. 34, no. 3 (1997), p. 331.
36. *Papers Connected with the Reorganization of the Army in India, Supplementary to the Report of the Army Commission*, London, 1859, House of Commons, C 2541, pp. 4-33, 45-66, 71-2, 308-9, 312-13.
37. Ibid., pp. 14, 27, 29-31.
38. Field Marshall Earl Roberts of Kandahar, *Forty One Years in India: From Subaltern to Commander in-Chief* (London: Richard Bentley and Sons, 1897), pp. 499, 530, 532, 534.
39. Ibid., pp. 499, 532.
40. Major George F. Macmunn, *The Armies of India* (1911, rpt., New Delhi: Heritage Publishers, 1991), pp. 129-31.
41. Roy, 'Recruitment Doctrines of the Colonial Indian Army', p. 325.
42. Cohen, *The Indian Army*, pp. 32-57.
43. A.L Venkateswaran, *Defense Organization in India* (New Delhi: Publications Division, 1969), p. 188.
44. Cardozo, *The Indian Army*, p. 261.
45. Lieutenant-General S.K. Sinha, 'Class Composition of the Army', *Indian Defence Review*, no. 1 (July 1986), p. 83.
46. Major-General Joginder Singh, *Behind the Scene* (New Delhi, 1995), pp. xii, 5.
47. Written replies to questionnaire, Lieutenant-General (retired) Mathew Thomas, PVSM, AVSM, VSM, 31 March 2006, Meerut, reply to question no. 10.
48. Written replies to questionnaire, Lieutenant-General S.N. Sharma, 25 February 2006, reply to question no. 9.
49. *Annual Report of the Ministry of Defence, 2007-2008*, New Delhi, p. 196.
50. Written reply by Mathew Thomas, to the questionnaire 31 March 2006.
51. Written response of S.N. Sharma to questionnaire, answer to question no. 4.
52. Oral interview with Major-General (retired) K.K. Ganguly, 28 December 2006.

CHAPTER 14

The *Dharam Yudh* or Just War in Sikhism

TORKEL BREKKE

This paper attempts to explore the concept of *dharam yudh* in Sikhism. Just war is the most obvious and probably the most common English translation of the Punjabi (Gurmukhi), *dharam yudh* the equivalent of *dharmayuddha* in Sanskrit. Hindu traditions about *dharmayuddha* are rich and complex and can be found in many of the most important texts in classical Hinduism, such as the great epics.[1] However, Sikh ideologies of just war seem to be quite distinct from ideas found in Hindu texts, although some of the vocabulary is similar.

The Sikhs put great emphasis on the violent aspects of the history of their community. They are often conscious of the violent nature of the relations between their own group and the Indian state, whether under the great Mughal Aurangzeb in the early 1700s, or under Prime Minister Indira Gandhi, assassinated by her Sikh bodyguards after ordering a bloody assault on the Golden Temple in 1984. Indeed, Sikhs may be more aware of their 'history' than most other groups. W.H. McLeod has pointed to the significance of the distinction between academic history and tradition, i.e. popular perceptions of history, in the study of Sikhism.

As scholars interested in the history of religious communities we need to understand both history as it really was—*wie es eigentlich gewesen*, in the words of Leopold Von Ranke—and history as it is handed down to new generations through popular historical writing in Punjabi, through the telling of stories in Sikh homes and in gurdwaras, as well as through graphic representations in popular art, such as the posters and calendars that are often seen on walls when one visits Sikh homes in Punjab. When trying to make sense of the political side of modern Sikhism, it is not sufficient to analyse the immediate political background to the conflict with the Indian state. In McLeod's words: 'What we must also understand is the perception of Sikh history (the tradition or the myth) that lies behind current claims to justice and to the due recognition of Sikh rights.'[2]

Taking a cue from this distinction, the present article will focus on how Sikhs have thought about and discussed the ethics of war, and *dharam yudh* in particular, during conflicts with the Indian state in modern times. Thus, my focus is on the period from 1980 until today. First, however, I want to look at some important documents from the eighteenth century. The period after the death of the tenth guru, Guru Gobind Singh in 1708, is the formative and troubled period of the Sikh Khalsa. This was 'a period obscure, yet very important' in the words of McLeod.[3] The early Khalsa was persecuted by the Mughal Army and leading Sikhs saw their struggle as a *dharam yudh*. But, what did this concept actually mean to them? Did they have clear notions of right authority or just cause, the themes we are used to looking for in the Christian justification of resort to arms, the *jus ad bellum*? Did they formulate a warrior code for right behaviour on the field of battle?

I should stress that this article has no ambition of arriving at new insights about the history of Sikhism during the eighteenth century. I rely on published sources translated into English, most importantly the *Dasam Granth* and the *Rahit-namas*, and on works by other scholars. Much work remains to be done on the sources for the life of Guru Gobind Singh, as Gurinder Singh Mann has pointed out recently, and we may expect a more complete picture of the life and times of the tenth guru to appear in the future.[4] In fact, Mann has suggested that the attribution of compositions in the *Dasam Granth* to the guru is based on a twentieth-century misinterpretation of the title *Sri Mukhvak Patishai 10* found in the compositions.[5] However, the point here is to briefly relate the key events of the history of the guru the way it has been understood by modern Sikhs thus providing a necessary background for a presentation of the ethics of war in the last few decades.

Secondly, I want to look at how Sant Jarnail Bhindranwale, the charismatic militant leader of the early 1980s, thought about *dharam yudh*. In particular, this article tries to explore how modern militant Sikhs use examples from the formative period of the Khalsa as models for their own conception of *dharam yudh*. Thirdly, I intend to carry out an analysis of 24 issues (two years' publication: 2005/06) of *Sikh Shahadat*, which is a magazine in Punjabi devoted to ideas about martyrdom and struggle for Sikh rights. One might object to this choice of historical foci: what has the early history of the Khalsa to do with the modern militant rhetoric of my two contemporary examples? And how can one possibly discuss ideologies of war in Sikhism without a decent treatment of sources from the time of Ranjit Singh's empire of the first four decades of the 1800s, when the Sikhs created a powerful political entity through military expansion?

Apart from the limitations of space, there is one main reason for my choice here. Initially, I wanted to understand the legitimization of violence among modern militant activists and writers; hence my choice of Sant Jarnail Singh Bhindranwale and *Sikh Shahadat*. However, after working on these sources for a brief period, and after talking to a number of people in Punjab who were among the readers of *Sikh Shahadat*, or similar journals, it became clear to me that the life of Guru Gobind Singh, and particularly the Battle of Chamkaur, is the point of departure for their reasoning about violence in modern times. In other words, it is hard to follow modern thoughts about just war in Sikhism without an idea about this historical point of reference. Thus, in my exploration of the just war in the minds of contemporary Sikhs, their readings of the life of Guru Gobind Singh will be a primary focus.

THE *DHARAM YUDH* OF THE EARLY KHALSA

Through his detailed study of ideals of martyrdom in the Sikh tradition, Louis Fenech has made a very important contribution to wider questions concerning Sikh ideas about warrior ethos and justification for violence. He makes interesting observations about transformations in Sikh ideas regarding *dharam yudh* when discussing the notion of martyr (*shahid*) in the Sikh lexicon by Kahn Singh Nabha called *Gur-shabad Ratanakar Mahan Kosh*, compiled in the second and third decades of the twentieth century. Kahn Singh defines *dharam yudh* this as: '(i) A war which is fought while keeping the principles of *dharma* foremost [in one's mind and heart]. A war in which deception, betrayal and falsehoods are not used; (ii) A war fought in order to protect the principles of *dharma*.'[6]

Fenech discusses the definition by looking at the many different meanings of the word *dharma* and pointing out that the Sikh concept of *dharma* is close to that found among Hindus. Kahn Singh's definition of *dharam yudh*, then, implies that it is a war fought to safeguard all those ideals that a community should treasure. However, while emphasizing the meaning of *dharma* and the intention and ideals behind wars fought by Sikh martyrs, it seems that Fenech ignores the other aspect of Kahn Singh's definition, i.e. the *way* in which wars are fought. In the study of the ethics of war in comparative perspective, we are interested in both issues—of resort to war (*jus ad bellum*) and means of fighting (*jus in bello*). Fenech's discussion is focused exclusively on the first because he is exploring the links between concepts of martyrdom and the concept of *dharam yudh*; matters of *jus in bello* are less relevant to him. In his discussion, he uses Kahn Singh's authoritative definitions to make a surprising logical connection: a *shahid* is someone who dies heroically in the defence, maintenance or establishment of *dharma*. Therefore, any war where there are Sikhs who die the death of a *shahid* is a *dharam yudh*. This disconnection of the *dharam yudh* from the actual rules of combat is symptomatic for most of the writing on the place of war in Sikh tradition, both strictly academic writing and popular historical writing.

The just war as it is discussed in classical Hindu texts like the *Mahabharata*, emphasize *jus in bello*. In the long passages relating stories about the war at Kurukshetra the driving force is the long list of transgressions of the warrior code laying down the rules for right behaviour in war. For instance, Bhima breaks the warrior code and commits a terrible sin when he strikes below the navel and crushes Duryodhana's thighs during a mace-duel. Can we find anything comparable in Sikh literature about war? Are there traces of a Sikh warrior code?

The *Dasam Granth* must be a primary source in an understanding of the ethics of war of Guru Gobind Singh and the people around him. However, as already noted, the life and times of Guru Gobind Singh are still somewhat in the dark and I go into these issues here mainly to provide a background for an understanding of modern ideas about violence and war. In the *Dasam Granth*, special significance must be given to texts like the *Bachitar Natak*, traditionally regarded as the guru's autobiography, and the *Zafarnama*, the letter thought to have been sent by the guru to the Mughal Emperor Aurangzeb after the Battle of Chamkaur. The battles described in the *Bachitar Natak* follow the model of the heroic battles found in Hindu epics and the author refers to Hindu gods and heroes. The brave and mighty warriors are the leading characters in this tale. They are skilled in the use of bows and swords and are fearless in facing enemies. These kinds of battle-scenes become ethically relevant in the books on *Mahabharata* when heroes break rules in one way or another and the rules become the subject of discussion between God and human heroes. It would be deeply unfair to compare the *Bachitar Natak*, a work composed by one author, with a great literary tradition that is the *Mahabharata*. However, the fact is that the *Bachitar Natak* does not touch on moral questions except when the author explains the terrible consequences of abandoning the guru when the strong Mughal Army is closing in. Those who turn away from the feet of the guru will be ridiculed in this world and end up in hell, as the text explains over a number of verses.[7] The rest of the *Dasam Granth* is also silent on the questions of rules of combat; the lack of ethically relevant considerations about war and warfare in the three texts about the goddess Chandi in the *Dasam Granth*, i.e. *Chandi Charitra Ukti Bilas*, the *Chandi Charitra* and the *Var Sri Bhagauti Ji Ki*, is understandable; these texts are gory poems about the struggle between gods and demons.

The more realistic *Zafarnama* is supposedly a letter written by Guru Gobind Singh to Aurangzeb shortly after the Battle at Chamkaur. It was written in Persian and it has been translated a number of times.[8] The *Zafarnama* is an interesting document when we want to understand the situation that the tenth guru and his followers were in around the time of creation of the Khalsa. One of the main contentions of the guru against the emperor in the letter is that the latter is a breaker of promise and therefore not a truly religious man. The guru repeats this claim throughout his letter showing his great disappointment that the guarantees given by the generals were not honoured by the Mughal forces. Aurangzeb's generals had promised the guru and his followers safe passage once they left Anandpur but the Mughal forces attacked the Sikhs once they were outside the walls of the town. In *Zafarnama*, Guru Gobind Singh famously asks the emperor what he could do when his forty famished Sikh warriors were engaged by a huge Mughal Army of 100,000 men:

> I knew not that the man (Aurangzeb) is a promise-breaker and Mammon worshipper . . . and has no faith. You lack faith in God and religion. Neither you do follow the creed, nor believe in Muhammad. Whosoever is true to faith, never breaks promises, once made. I do not trust such a man, who does not know that God is one and also does not value the oath on the *Quran*.[9]

The intention of the letter from the guru to Aurangzeb is to make the emperor realize that his generals have fought an unjust battle against the Sikhs and to convince the emperor to enter into talks with the Sikhs on an equal footing. These were high ambitions for a leader in Guru Gobind Singh's position after the battle, to say the least. As evidence for the 'unjust' conduct of the Mughal generals, the guru offered to send the emperor letters from the generals in which they granted the guru safe passage from Anandpur Sahib. The guru insists that according to the religious ideals of Islam, the Mughal ruler must now himself travel to Guru Gobind Singh's place of refuge in order to have talks. He urges the emperor to come to the village of Kangra to give them an opportunity to meet. To quote the *Zafarnama*: 'You should come here to Dina Kangra village and have face-to-face talks with me. This will give us an opportunity to converse. I may show some kindness towards you.'[10]

The guru writes that he is in a position to guarantee the safety of the emperor, if he should come to the village, as the people of the area, the Brars, are his subjects and accept his word. Although Guru Gobind Singh is first and foremost a great and righteous warrior in Sikh tradition, his life is also tied to the theme of martyrdom. All his four sons were killed at Chamkaur or in the aftermath of the battle.

Guru Gobind Singh was the son of the ninth guru, the martyr Tegh Bahadur, and he was born in 1666 in Patna. 'When the Lord commanded I took birth in this dark age (*kaliyuga*)', Gobind states in the *Bachitar Natak*.[11] In other words, he was born to carry out Gods plan on earth, i.e. spread the *dharma* and resist tyranny and false religion. He was taken to Punjab, properly trained and he assumed guruship at his capital Anandpur. From the point of view of Aurangzeb in Delhi, the Sikh leader was one among a number of rebellious hill chieftains. In 1699, on *Baisakhi* day, Gobind invited every Sikh to join him and he founded the Khalsa. The founding of the Khalsa meant the abandoning of earlier authority of the *masands*, administrative deputies, and the creation of a new relationship between the guru and his followers. It also meant the introduction of a code of conduct and belief, the *rahit*. It was one of the most crucial events in Sikh history. At the same time, the Mughals decided that Guru Gobind Singh was creating too much trouble and planned to get rid of him. This resulted in the siege of Anandpur in 1701 and again in 1704 the Battle at Chamkaur. We will return to the Battle at Chamkaur later in the article because this is the key historical focus for modern Sikh discussion on the *dharam yudh*, the just war.

The many descriptions of campaigns and battles in the *Dasam Granth* do not really contain much information about practical rules regulating warfare. In other words, we look in vain for what we refer to as *jus in bello* when talking about the just war in the Christian tradition. On the other hand, it could be argued that the cultural background of the *Dasam Granth* is so close to Hinduism, and particularly to the heroic culture of the epics, that we may assume that the same Hindu warrior code (*kshatradharma*) would apply for Sikh warriors of the early 1700s. On the other hand, the realities confronting the Khalsa in the decades following its founding in 1699 bore no resemblance to the idealized heroic world of epic Hinduism. The Khalsa was resisting oppression from Delhi and this was the fundamental context of their warrior-ethos.

RULES IN BATTLE

If we are looking for rules about the behaviour of Khalsa Sikhs on the field of battle we might instead turn to the code of belief and conduct, i.e. the *rahit*, laid out in the class of texts known as the *rahit-nama*. If we want to use the *rahit* to understand early Sikh views on war, the only sensible strategy is to study the groundbreaking work of W.H. McLeod on these sources. In particular, it is relevant here to look at McLeod's work in the six *rahit-namas* of the eighteenth century. These are the *Tanakhah-nama*, the *Prahilad Rai Rahit-nama*, the *Sakhi Rahit ki*, the *Chaupa Singh Rahit-nama*, the *Desa Singh Rahit-nama* and the *Daya Singh Rahit-nama.* Of these texts, the *Tanakhah-nama* is the oldest and simplest in form; it may have been composed in the second decade of the eighteenth century. To get an idea of the ideology concerning war in the *rahit* we can look at some relevant extracts from the *Tanakhah-nama* attributed to Nand Lal:

> He is a *Khalsa* who in fighting never turns his back. He is a *Khalsa* who slays Muslims. He is a *Khalsa* who fights face to face. He is a *Khalsa* who destroys the oppressor. He is a *Khalsa* who fights his enemy. He is a *Khalsa* who is always fighting battles. He is a *Khalsa* who carries weapons.[12]

Clearly, Nand Lal has not given much thought to moral dilemmas arising in battle. Let us look at relevant passages from the *Rahit-nama* of Bhai Desa Singh:

> Intoxicating liquor may be taken before battle, but there should be no mention of it at other times. In battle the Singh should roar [like a lion]. Fighting them face to face he defeats the Muslims. In battle let [the *Khalsa*] never be defeated. He should forget sleep [remaining ever alert] to fight the Muslim. Let him with determination do the deeds of a Kshatriya, crying 'Kill! Kill!' as he fights in battle. Fear not, for many are fearlessly fighting. Sustain the spirit [which declares]: 'I shall kill the enemy!' Those [*Khalsa*] who die in the course of a battle shall certainly go to paradise. He who defeats the enemy in battle will find his glory resounding the whole world over. Stand firm, therefore, in the fiercest conflicts. Never turn and flee from the field of battle. If anyone fears fighting in battles let him earn his sustenance by such activities as agriculture.[13]

Desa Singh gives a few more details than Nand Lal but again we meet a simple view of war, where the enemy is obvious and the righteousness of the violent struggle against the Mughal/Muslim oppressors is never questioned. If we turn to another early and important *Rahit-nama*, the Chaupa Singh *Rahit-nama*, we will find some rules pertaining to weapons and warfare and the Khalsa Sikhs' reverence for swords and other weapons is emphasized in several passages. There are also a few references to ideas of legitimization: 'The right to rule is won and sustained by the sword. Arms should only be used, however, when there is good cause for doing so.'[14] Here, Chaupa Singh points to the need for limiting the use of deadly force to the times and circumstances when there is good cause for doing so but, unfortunately, he does not give any details about what such good causes might be. From the historical context we know that resistance against oppression of the Mughals was the legitimate cause for war *par excellence* and most of the old *Rahit-namas* contain extremely negative view of Mughal rule and, by implication, of Muslims in general, referred to as *malecchans* (Sanskrit: *mleccha* = barbarian) or *turak* (i.e. Turk). Chaupa Singh also gives some brief comments on rules for the battlefield but they are so few and lacking in detail that they hardly constitute an attempt to define a *jus in bello*. He writes: 'The battlefield: A Singh should never turn his back in battle. Always aid a wounded, disabled or exhausted Sikh on the battlefield. Always have slain Sikhs cremated on the battlefield if possible.'[15]

The *Prem Sumarag* is a slightly different *Rahit-nama* because it is a *Sanatan* work, which means that it is less concerned with distinguishing Sikhs from Hindus. The *Prem Sumarag* is hard to date but seems to be from the mid- or late-eighteenth century. In the *Prem Sumarag*, the Sikhs are told that if they are attacked because of their religion (*dharam*) they should curse the master or prophet (i.e. Muhammad) of the attackers. Again, the only possible attacker in this work is the Mughal Empire and its servants. 'The time (for the destruction) of the Muslims (*malecchan*) is near', the text states.[16] Thus, the *Prem Sumarag*, like the other *Rahit-namas*, offers a picture of the Sikh community in the eighteenth century where the main religious distinction is made *vis-à-vis* the Muslims who are associated with the political establishment of the day. On the other hand, the boundaries against Hindu society seem less clear in the *Prem Sumarag*; this is implied when the text is called a *Sanatan* work by modern scholars.

One of the most interesting aspects of the *Prem Sumarag* is its strong apocalyptic tones. It starts its first chapter by describing the fallen state of the world. God has established a *panth* where *dharma* has its abode to serve as witness to the world, it says.

The true message has been delivered through the gurus, but evil remains. According to the *Prem Sumarag*, this is the time of *Kaliyuga*, i.e. the last end evil period in the classical Hindu scheme of the cycle of world-ages. All other *panth*s of the time will submit to the evil *panth* (*malechh panth*). The evil *panth* is the Muslim *panth* and the central factor of the present scene is the coming into power of the Muslim *panth*. '[The people of] the Muslim *panth* have become enmeshed in their own beliefs, forgetful [of the truth]. To them I now declare my purpose.'[17] In other words, the *Prem Sumarag* was written at the end of a great historical cycle where the evil have conquered much of the world and where the conditions are continuously getting worse. All other *panth*s must draw together to defend themselves against the Muslim *panth* and the Muslim *panth* will eventually be destroyed and everyone associated with it will go to hell.

On the other hand, those who enter the Khalsa *panth*, i.e. those who become Sikhs, will be able to retain their religion or righteousness (*dharam*).[18] *Kaliyuga* will eventually come to an end and the good *Satiyuga* will replace it. But at present, in the last days of the *Kaliyuga*, the *Akal Purakh* (God) will protect the Khalsa against the growing strength of the Muslims. The transformation will start in the town of Sambhal, where a disciple of a warrior will stand forth and destroy the Muslims and hoist the banner of religion or righteousness (*dharam*). Many false gurus will appear but they will not be accepted. However, when the right person appears on the scene, all who belong to false *panth*s, Hindus, Muslims and others, will be destroyed.[19]

Let us sum up this brief discussion on the idea about *just war* in relevant Sikh documents from the eighteenth century, i.e. the first decades after the founding of the Khalsa by Guru Gobind Singh. We can reasonably say that there is no trace of a detailed warrior code for righteous behaviour on the field of battle in any of our sources. We get a picture of a nascent community of faithful followers of the guru who perceive themselves as oppressed and persecuted by the formidable armies of the Mughal power in Delhi. There is a sense of apocalyptic urgency and a consciousness of defending righteousness in dark times. The *Rahit* requires members of the Khalsa to bear arms and be ready to use them. The kind of war that the *Rahit* refers to is really armed resistance against the Mughal authority. In this situation of supreme emergency and fear of apocalyptise, we search in vain for a *jus in bello*.

The lack of any reference to matters of *jus ad bellum*, like just cause or legitimate authority, is perhaps attributable to the same causes. The ever-present just cause is the constant resistance against Mughal power. Legitimate authority has been given to the Khalsa as a community by Guru Gobind Singh. The Khalsa as a family has inherited the authority of the gurus and this entails the authority to use force against enemies. In matters of war and warfare, the Khalsa as a whole has the right authority to wage war and this is the reason why we cannot find many details in the *Rahit-namas* of matters of authority. As McLeod notes in the introduction to his study of the history of the Khalsa *Rahit*, it was a vital function of the Khalsa and the duty of the Khalsa Sikhs to engage in *dharam yudh* against the Mughals.

THE *DHARAM YUDH* OF BHINDRANWALE

My question from this point concerns the relationship between, on the one hand, the conception and ideology of war identified in the material at my disposal from the formative period of Sikh history, and, on the other hand, the conception and ideology of war found in crucial modern documents. Can we see any important indications of continuity or change in the ethics of war in the Sikh tradition over this large span of time? I want to start by looking at the ideology of war found in the statements of the most famous of the modern militant Sikh leaders: Sant Jarnail Singh Bhindranwale. Bhindranwale emerged as a charismatic militant leader during the militancy in Punjab from the late 1970s till he was killed during the storming of the Golden Temple by the Indian Army in June 1984. The troubles in Punjab had their roots in political demands made by Akali Dal politicians well before the wave of violence broke out and the then president of the Sikh political party Akali Dal, Harchand Singh Longowal, formulated the declaration of a *dharam yudh* against the Government of India in 1982 on behalf of the Akali Dal. In a pamphlet entitled *Why This Holy War?* he wrote: '[t]he Sikhs have been forced to launch a holy war against the Government of India . . .' because they were treated as second-class citizens.[20]

Still, Bhindranwale became the most important figure in the

fight against the authorities because of his considerable religious authority and his uncompromising and simple rhetoric against both the state and other religious communities, like the sect of Nirankaris and the Hindus.

Bhindranwale saw the militants' struggle against the Indian government as a *just war.* For instance, in a speech on 29 April 1983 he said: 'Regarding the *Dharam Yudh Morcha* that is going on, we should proceed with resoluteness according to [the Guru's words]. Resolutely may I ensure my victory.'[21] But to Bhindranwale, the *Dharam Yudh Morcha*, the campaign that is a just struggle or a war according to religion, is really the name of more than simply an armed response to the perceived and expected hostilities from the government or from sections of the Hindu population. In a speech on 13 May 1983 Bhindranwale talked about the *Dharam Yudh Morcha* as the general struggle to secure the rights of Sikhs, a struggle that was started a long time ago.[22] On 2 July 1983 he talked about the *Dharam Yudh Morcha* as the struggle of all the different organizations and individuals who have come together on one platform to remove the yoke of slavery from the necks of the Sikhs and to secure benefits for all Punjabis.[23] It is clear that for Bhindranwale a great struggle for justice was going on in which violent means is one necessary element.

Bhindranwale made similar statements on a number of occasions. In a speech on 13 April 1983, he asked all Sikhs to be prepared for the attack from the government that he was expecting. He exhorted the members of the congregation to act on their own without consulting himself or his fellow leader Longowal Sahib.

> In that eventuality, at various places, in the villages as in the cities, wherever there is a critic of *Satguru Guru Granth Sahib*, or one who dishonored sisters and daughters, or one who rejoices in this dishonor, or one who creates rifts among us, kill them wherever they are, punish them. This is my firm request to you. You must not hesitate at that time. So long as they hold back from such action, we shall hold back too. Beyond Mecca, everything is wilderness to a Muslim, after Harmandar Sahib, for the Sikh all is wilderness.[24]

In fact, Bhindranwale often expressed the view that physical attacks on the Sikhs should be suffered without retaliation. But, there is a definite limit at the borders of the Golden Temple. The

Golden Temple is the source of order and sanctity for the Sikh community. Beyond the Temple, or without the Temple as the source of order, everything is wilderness for the Sikhs, in his view. Therefore, an attack by the government on the Temple was seen by Bhindranwale, and certainly by many Sikhs, as an attempt to destroy the source of communal survival and vitality of the Sikhs.

Bhindranwale seems to have believed that violence was legitimate only when the Sikhs were attacked on their own holy ground, inside the Harmandir Sahib. In a speech on 27 March 1983 he appealed to his followers to stay peaceful until the Golden Temple itself was invaded by the government. He said: 'We have to maintain peace. We have to maintain order. So long as the Government confines its activities to outside [of Harmandir Sahib], definitely stay cool because we are engaged in the *Dharam Yudh Morcha*.'[25]

In this speech at least, Bhindranwale has rather clear ideas of when and under what circumstances violence is just. The Sikhs are engaged in a *Dharam Yudh Morcha*, a war according to the tenets of *dharma*. He is keen to point out that none must act to any provocation and not to false alarm. He continues:

> However, when at any time, on any day, the Government enters the boundary of this complex to destroy its sanctity, let me appeal most strongly to the entire Sikh congregation—to all of you who live in villages, towns and in the entire country—that when you learn that they have entered the boundary of the complex and attacked then it will be your responsibility everywhere to kill every critic of the Guru and every enemy of the Sikh Nation. At that time there should be no hesitation on your part.[26]

Bhindranwale gave very clear instructions about the responsibility of all Sikhs to fight and destroy anyone who insults the most holy symbol of the Sikhs. If there is anyone who insults the *Granth Sahib*, the holy text of the Sikhs, a Sikh should kill him and then take shelter in the Golden Temple, according to Bhindranwale. He does not make any distinction between types of offenders or different classes of opponents in the struggle. The obligation falls on everybody. In a speech on 18 May 1983: 'Remember this about this Harmandar Sahib that whether one is a woman, or a child, or a Singh of the Guru, he/she will be cut to pieces rather than tolerate insult to the Harmandar Sahib.'[27] In other words,

Bhindranwale was of the opinion that the community as a whole had a collective responsibility to resist oppression from the government and to use force to repel attacks on the most holy place of the Sikhs and the central symbols of Sikhism.

In preparation to defend the Golden Temple, Bhindranwale insisted time and again that the Sikhs should acquire weapons. But the acquisition of weapons was more than a practical preparation for the defense of the Golden Temple, in Bhindranwale's world-view. He believed that Sikhs were obliged to carry weapons because the gurus had said so and his idea of carrying weapons as a religious obligation was expressed in many ways. He often quoted the gurus or referred to their examples, especially that of Guru Gobind Singh, when he defended the right and obligation of Sikhs to carry weapons. This obligation included not only the *kirpaan*, or the knife, one of the five Ks. It also included modern weapons because the religious and the functional aspects of the weapons are hard to separate in Bhindranwale's statements about arms. He stressed that Guru Gobind Singh said that it was a sin for a Sikh not to bear arms, and it was a sin not to be able to defend sisters and daughters against attackers. 'Without hair and weapons a man is but a sheep and can be led anywhere', he would say. Arms are much more than religious symbols to Bhindranwale; they must enable a Sikh to defend himself and his fellows and his faith.

Honour is an interesting focus for the study of ideas of just war in a South Asian context and it is clearly a topic that transgresses a number of religious boundaries. Threats against honour have been a source of motivation for the use of force among Sikhs, as among Hindus and Muslims of the subcontinent. One reason that Bhindranwale often gave for using deadly force was to avenge or defend girls or women who were molested. In a number of speeches, he told his listeners to kill anyone who would molest or dishonour sisters and daughters. In a speech on 13 April 1983 he referred to a case where a policeman pulled an earring of a Sikh girl and took it home. This is the kind of behaviour where Sikhs have a duty to punish the offender with death. Anyone who dishonours the holy text or a Sikh woman should be killed, was his message.[28]

I think this is a crucial point for an understanding of just war in the Sikh tradition. Honour remains a principal motivation for the use of force both in what we would call the private and the public sphere. The study of just war traditions often takes for granted that other cultures have developed a clear distinction between private and public violence. However, just war in the strictest sense presupposes a public sphere and it presupposes authority in that sphere that is competent to decide on the use of force for the whole political community, i.e. the state in most contexts. This distinction, both in actual historical terms and in philosophical terms, is a precondition for the development of the just war tradition in early and modern Europe. However, in the Sikh tradition this distinction has not developed, at least not along the same lines. In other words, for militant Sikhs the whole religious/ethnic group is seen as one community and a breach of the honour of the group may be conceptualized, it may even be felt to be, a breach of the honour of a family. The preoccupation of Bhindranwale and other militants with the honour of the girls and women of the community as part of the legitimization of armed resistance indicates that there is no real distinction between the private matters of Sikh men and women and the public matters of the relationship between the Sikhs and the Indian state. Violence by private persons is seen to be legitimate when the honour of the family is harmed by others and, by extension, violence by militant groups is seen to be legitimate when the honour of Sikh women is harmed by police or army personnel, or by members of other communities.

To take this line of reasoning one step further, we could explore the details about the particular sense of community and belonging among militant Sikhs. In a long interview I did with the most seasoned and famous of the Akali leaders associated with militancy in Punjab, Simranjit Singh Mann, on 22 July 2006, Sikh nationalism was one of the most important issues. Mann stressed again and again that Sikhs are a nation, and have a right to nationhood on the basis of their religion, language, script and territorial belonging in Punjab. Bhindranwale, too, stressed that Sikhs are a separate nation. For instance, in a speech on 27 March 1983:

> Sikhs are a separate nation. My educated brothers must have been delighted at the news that appeared some three or four days back. In connection with the turban there has been a decision in England. My educated brothers know better. It is a weighty matter. The Parliament

in England has decided that not only are the Sikhs a separate nation, they are a separate race. Sikhs are a separate race. Yes, a race. Yes, a separate race.[29]

To Bhindranwale this was an issue that was settled a long time ago. He did not need official decrees to know that he belonged to a separate nation. The separateness of the Sikhs from other communities of India was, in his eyes, self-evident from the textual and cultural heritage of the Sikhs:

The British Parliament has issued a decision now regarding Sikhs being a separate nation. Bhai Gurdas Ji decided this four hundred years back; the Tenth King [Guru Gobind Singh] decided this three hundred years back; and Satguru Nanak Sahib decided this five hundred years back. We learn about this if we read the religious texts.[30]

I think it is reasonable to say that the ethics of war espoused by Bhindranwale display features that places it quite close to that found in the formative period of the Khalsa in the early 1700s. Bhindranwale and his fellow militants were engaged in a *dharam yudh* in the defence of the Sikh community, in defence of its rights and honour, and against the oppression by the Indian state. Their just war was fought on behalf of the Sikh community both against security forces and against sections of the Sikh society that did not realize the urgency and necessity of the struggle. The honor of Sikh women and the communal honour were of great importance in Bhindranwale's world-view and the breach of honour by members outside the Sikh community was the primary cause for a *dharam yudh*. An attack on the Golden Temple in Amritsar was seen as the most serious attack on the symbol of collective honour and identity and was therefore the ultimate cause for counter-attacks on anyone involved in hostilities against the Sikhs.

DHARAM YUDH IN THE *SIKH SHAHADAT*

In the last part of this paper, I intend to use a contemporary Sikh publication, a magazine called the *Sikh Shahadat*, to illustrate how the *just war* or *dharam yudh* is discussed among Sikhs today and how the *just war* of the eighteenth century is the key script for the understanding of the concept in a contemporary setting.

What kind of magazine is the *Sikh Shahadat* ? First, of all it should be noted that this is one magazine among a number of publications in Punjabi taking an assertive, and sometimes militant stance on issues concerning Sikh rights and the relation between Sikhs and other communities in India. In the March volume of 2005, the editor of the magazine, Harjinder Singh, gives a summary of the ideas and aims after five years of publication. Five years ago the *Sikh Shahadat* started as a journal for a new war (*jang*) of ideas, he wrote. The aim of the journal was to bring forward the voices of Sikhs who are more honest to religion than their own lives.

Needless to say, martyrdom is crucial to the *Sikh Shahadat*, the name of the journal meaning *Sikh martyrdom. Sikh Shahadat* is a magazine that is devoted to the great martyrs, those who 'faced great martyrdoms to save the honour of mother earth Punjab', according to the editor. Many historically and culturally conscious Sikhs attach great moral significance to the martyrdom of certain leaders in Sikh history, especially Guru Arjan and Guru Tegh Bahadur. The discussion of martyrs is not only about ancient history. For instance, an important recent martyr was Harjinder Singh Jindha, who killed an Indian general in revenge for his role in Operation Blue Star, i.e. the attack on the Golden Temple in 1984. Harjinder Singh Jindha is lauded as a martyr in several articles of the magazine, like the one called 'Harjinder Singh Jindha in the Window of Memory' (October 2006). The leader of the Khalistan movement is also discussed in the pages of *Sikh Shahadat*, as in an article called 'The Martyrdom of Sant Bhindranvale: A Long Conspiracy Regarding the Historical Happening' (October 2005).

Clearly, the publication of the *Sikh Shahadat* is meant to create a stronger awareness of Sikh identity and a greater willingness to sacrifice and suffer for the community. There are strong forces that want to slowly destroy the Sikh community, in the opinion of the editor and many of the contributors. The most important force is the Government of India, which calls itself secular but is really, in the minds of militant Sikhs, a Hindu-chauvinist state:

> The publication of *Sikh Shahadat* intends to recreate the dream of the martyrs and their principles. The Sikh martyrs show the way of how to stay awake to the robbery in the community. . . . Through *Sikh Shahadat* we stand against those who want to silence the voice of the community.

On the communal stage we present the brave stories of community martyrs [*panth di shahidan*].[31]

Shahadat (martyrdom) and its relevance today is the key issue in the journal. The editor and writers of the *Sikh Shahadat* see themselves as the group who are loyal to Guru Gobind Singh at Chamkaur. At the Battle of Chamkaur, a few Sikh warriors would stand by the guru in the face of certain death while many fled or betrayed their leader. The spirit of loyalty unto death is seen to be on the decline in the modern Sikh community. It is of particular importance to make the youth understand the spirit of giving up one's life for the sake of the community and the *Sikh Shahadat* wants to impart this spirit. To this end, the journal has devoted the first page of every issue to the presentation of a fallen Sikh fighter. The title of this page is simply 'Martyr Hero of the Sikh Struggle' (*Sikh sangarsh de shahid yode*). On this page, the magazine always presents two fallen martyrs, with colour pictures and short biographies. The martyr heroes who are presented are chosen on the basis of the date when they died (mostly during the early 1980s); the date of the issue corresponds to the date of the current month's fallen martyrs. In several places, the articles of the *Sikh Shahadat* mention or discuss the *dharam yudh* and these passages reveal valuable information about the perception of just war among Sikhs today. However, *yudh* (Sanskrit: *yuddha*) is not the only word for *war* in contemporary Punjabi Sikh literature. In fact, the most common Punjabi word for armed conflict is *jang*.

The Battle of Chamkaur (*chamkaur di jang*) is one of the key episodes referred to by many contributors to the articles in *Sikh Shahadat*. In a long article entitled 'The War at Chamkaur: Spiritual Colours and Historical Connections', one Karamjit Singh discusses the significance of the war to the Sikh community. The author starts by stating that the war of Chamkaur is one of the important events in Sikh history, always refreshing the soul of the Khalsa. He asks why that event is important for the Sikh history and replies that 'it spreads flower of *gurbani* [i.e. the word of the guru], the duty of *Khalsa* and speech of *gurbani*'. There was something special about the combat at Chamkaur because the motivation behind the war set it apart from all other wars in world history, according to the author. Most great wars are fought for the same reasons, the author claims. 'The reason for all wars is to expand geographical areas.' But, the war of Chamkaur was different. 'That war became a heritage of the whole of humanity.' One of the proofs of this universal character of the war is the testimony of the Muslim poet Hakim Mirza Aula Jarkhan, who has written about the tragedy of the princes, i.e. the sons of Guru Gobind Singh. He described in his poem the words of the princes before the court of Wazir Khan as the words of God [*khuda ki zaban* = the tongue/words of God]. According to the author, the fact that Muslim poets can write such things shows that the experience of the Sikh community carried a message to the whole of humanity. The article goes on to describe, page after page, how the guru was betrayed, how a few loyal friends and servants remained by his side, how he escaped from the fort at Anandpur, how a handful of Sikh warriors clashed with the numerically superior Mughal Army. The author also explains how all this had a larger meaning as the guru now placed guruship on the Khalsa, the community, as a whole, and how he pointed out that the execution of his sons had no significance in the larger life of the Khalsa. All this is the stuff of popular Sikh history; most Sikhs know the details by heart. Then, the author asks whether the war at Chamkaur was a just war and answers his own question in the affirmative:

> Was the war of Chamkaur *dharm yudh*? Yes, it was *dharm yudh*. If *dharm yudh* is when politics is removed and the principles of the Guru are present then it is *dharm yudh*. But what is the meaning of war? What is the real definition? *Britannica Encyclopaedia* says that war is a fight between political groups. If politics is removed it will be a small fight between thieves and smaller groups. The only aim of that war is to rob. But here Guru Gobind Singh is fighting that kind of *dharm yudh* which is totally different from the world wars. At the same time he created *Khalsa* in order to reach the aim of *dharm yudh*.[32]

The January volume of 2005 is dedicated to the 300th anniversary of the martyrdom of the princes and the war at Chamkaur. The two young sons of the tenth guru were martyred at Sirhind in 1705. In one of the battles against Guru Gobind Singh's city of Anandpur, the Sikh guru was besieged in the city. The guru's mother, Mata Gujri, and the guru's two youngest sons, Zorawar Singh and Fateh Singh, escaped from the besieged town but they were betrayed and captured at Sirhind. The young boys were urged to convert to Islam but they refused. They were executed by being bricked alive into a wall by the Mughal governor of Sirhind, Wazir Khan. The *Sikh Shahadat* for January 2005 carries articles with titles like: 'The Martyrdom of the Small Princes'; 'The Martyrdom of the Small Princes and the Horizon of a New Future'; 'The Matchless Martyrdom of the Princes and its Message'; Importance of the Martyrdom of the Princes in World History and 'The Martyrdom of the Princes and the Present Situation of Sikh Youth'.

In the same January 2005 volume there is an article by one Harnam Singh Shah with the title 'The Horizon of Martyrdom in the Sikh Religion'. This article illuminates some key points in the modern and popular conceptions of the just war among Sikhs. Moreover, it also offers an opportunity to discuss the connections between Sikh ideas about just war and martyrdom. The first part of the article briefly mentions the meaning of the word martyr in Persian, English and Arabic and then continues to explain that the present understanding of a martyr in Christianity relates to a believer who has died in a fight with non-believers [*kafar*]. According to the author, the Arabic word for martyr, *shahid*, was influenced by Christianity and refers to someone who gives his life in the battle with non-believers and to confirm his belief. In the Sikh religion, the word *shahid* is derived from Islam since the word was introduced to Indian society by Muslims, he writes. In Hinduism, Jainism and the traditions of China and Japan, the word *shahid* is not used with this meaning but instead signifies rituals and sacrifices in religious worship, according to the author. As opposed to this, Christianity, Islam and Sikhism have many stories about great martyrs, such as the Jewish teacher Akiba, the Persian philosopher Mukrat, the crucified Jesus Christ, the Shia Muslim Hussein and the Sufi master Mansor.

The article then explains that in Sikhism the word *martyr* was very important. It was connected to the idea of people's rights. Sikh martyrs wanted to save religion and righteousness in general, not only their own community. The writer explains that Guru Arjan followed this principle. He died by having hot sand poured over his body and by being placed in boiling water. With bravery he accepted this cruelty and followed what God wanted. The emperor forced him to convert to Islam but he did not accept it. Guru Arjan did this to save other people's rights and to go against the Mughals who forced people to convert to Islam. The author writes that the spirit (*jot*) of Guru Nanak and the other gurus are present in the *Guru Granth Sahib* and then clarifies what signifies a martyr in the Sikh tradition. Those saints who were killed for the protection of people, for truth and saving religious rights, to go against cruelty, to stop that and to get people's basic rights. They did it without fear and without greediness, not for their own liberation and not for their family, not for country and community, but for protecting the rights of whole humanity and the right of religion. After this the author mentions stories about the martyrdom of Guru Tegh Bahadur, Guru Gobind Singh and his sons. Towards the end of the article, the author returns to a comparison between different concepts of martyr in world religions. He repeats how martyrdom in the Sikh tradition is different from martyrdom in other religions. According to the Parsi religion, *shahid* means a person who dies for religion and belief. According to the Jewish community, the martyr is the one who dies knowing the danger of preaching religious books. According to Christianity, the martyr is one who believes in Christianity and dies for it. According to Islam, a martyr is one who dies in religious war (*dharam yudh*) or is killed for injustices. Sikh martyrs, on the other hand, have given their lives for protecting the rights of humanity and all religious communities.

The ideas in the article just discussed are highly representative for a modern idea of the just war among Sikhs. In fact, many contributors to the *Sikh Shahadat* believe their tradition occupies a very special place among world religions because the Sikhs recognized the rights of non-Sikh communities, too, and they were willing to die to protect these rights. We have just seen how one author traced this tradition to Guru Arjun (1563-1606). However, the popular history of the ninth guru, Tegh Bahadur (1621-75), is an equally important source for the idea of martyrdom and just war as the fight to protect universal rights, rather than the rights of Sikhs. Guru Tegh Bahadur is said to have given his life in the fight to protect the rights of Hindus. This idea is traced to the text *Bachitar Natak* of the *Dasam Granth*, where Guru Gobind Singh, the tenth guru, writes about his father, the ninth guru, that he performed the supreme sacrifice in order to protect the sacred thread and the frontal mark. For most Sikhs and popular Sikh historians this is taken to mean that Guru Tegh

Bahadur was martyred protecting the religious rights of people of all religious communities (except the Muslims) and not just the Sikhs. For instance, the popular history by Khushwant Singh renders this story thus.[33] 'One of the main problems in academic historical writings about the Sikhs in recent years has been the fixing of a clear and unified Sikh identity in the nineteenth century. Harjot Oberoi's work has been important for the argument that the modern definition of Sikhism and the modern perception of Sikh history were formulated by *Tat Khalsa* Sikhs from the 1880s and implied the suppression of diversity within the Sikh religion'.[34] Not everyone agrees that the *Tat Khalsa* view of Sikh historical identity is a complete revision. However, following these lines of thinking about religious identities and boundaries it is clear that thc Sikh perception of the universality of their just war tradition may be challenged. Thus, Louis Fenech points out that the assumption that the sacred thread and the frontal marks were not part of Sikh culture around 1700 'is an understanding clearly rooted in later *Tat Khalsa* thinking'.[35] In other words, it is not correct to see Tegh Bahadur stand for the struggle for universal human rights.

The conception of the Sikh *dharam yudh* as a just war for universal rights and protection of people, rather than for the protection of just but narrow Sikh political causes, is found in a number of articles in the *Sikh Shahadat.* For instance, in volume 11 of 2005, an anonymous author wrote an article about the just war of the tenth guru. The article had the title 'Historical Page: The Dharm Yudh of Guru Gobind Singh'. Like the author discussed above, the message of the anonymous contributor is that the just wars of Guru Gobind Singh was fundamentally different from all other wars because they were fought out of love and care for the whole of humanity and not in order to secure the interests of the Sikhs in particular. I quote the article at some length because it is representative of a view of the Sikh *dharam yudh* in the magazine:

> There was only love and love in the *dharm yudh* of Guru Gobind Singh. There was no sign of revenge and hatred in his wars. That is why the Guru Sahib never gave the name violence to his wars. Violence is related to hatred and anger and is for revenge. Those Mogul emperors who martyred Guru Arjan Sahib, cut the head of Guru Tegh Bahadur, martyred many Sikhs, the four *sahibzadian*, the five *panj pyare*, the

forty *mukhtian* and Beant Singh, their hands were full of blood. Even if this happened Guru Gobind Singh agreed to have a conversation with the emperor king Aurangzeb and that proves clearly that the Guru *ji* had no hatred, anger or desire to take revenge in his heart. In the war of Bhangani he fought against the mountain kings and then in the war of Nadaun he helped them. That proves that Guru *ji* never fought with anger and hatred but with good behaviour [*asul di*]. Whoever suffered cruelty, he fought for them. When someone complained about Bhai Ghanaia for giving water to the enemy and the wounded then Guru *ji* called him in to ask for a reason. He answered 'Maharaj I did not see any enemy. According to your teaching I see God's soul in all.' The true Guru praised him and embraced him. He gave medicine and told him to not only give water but also give aid.[36]

A number of important issues are touched on in this extract. First of all, the author uses the story about the wars of the tenth guru to discuss the nature of the warfare of the most important of all Sikh political leaders. Most importantly, the wars were not fought out of a sense of anger or a thirst for revenge and this is proved by the guru's willingness to talk to the cruel Mughal emperor whose generals had recently killed the guru's four sons. There was no hatred in the guru's heart, the author states. Indeed, Gobind never fought with hatred but always with good behaviour. But, more importantly the whole spirit of the article is an attempt to show that the wars of the tenth guru were in fact wars fought on behalf of humanity under the banner of human rights.

There are a number of interesting points to be made from the contents of the 24 issues of the *Sikh Shahadat* that I have analysed here. The main lesson is the importance of popular history, as I mentioned in the introduction. In the world-view of the editors and contributors of the *Sikh Shahadat* (and probably of some of similar publications), there are certain crucial moments in the history of the Sikhs that serve as examples for behaviour today. In the words of G.A. Almond, R.S. Appleby and E. Sivan, the past is inextricably entwined with its behavioural consequences in the enclave cultures of religious fundamentalisms.[37] Certain moments of the past are highlighted and are experienced by the members of the community with a sense of urgency and contemporaneity. Important such events are the martyrdom of gurus like Arjan and Tegh Bahadur, the wars of Guru Gobind Singh, and the martyrdom of his sons. However, certain modern events are also about to gain an important status in the historical consciousness of the Sikhs, not least the genocidal attacks on Sikhs in the aftermath of the assassination of Indira Gandhi in 1984. We lack the space to discuss this here although these events are a major issue in the articles of the *Sikh Shahadat*. Another main lesson is the striking universality with which the history is seen by many of the contributors. In their view, Sikhism certainly needs to be strengthened and revived, but at the same time they see Sikhism as a world religion on par with other religions and they perceive the highlighted episodes of Sikh history as struggles fought and martyrdoms suffered for the sake of human rights in general. This point could lead us into discussions of Sikh identity in an age of globalization but that would take us far away from our main subject, the ethics of war.[38] The main conclusion at this point must be that *dharam yudh* in the *Sikh Shahadat* has a slightly different meaning than it had both in the formative period of the Khalsa and in the world-view of Bhindranwale and his militant followers. It is now time to draw some larger lessons.

CONCLUSION

I have attempted to explore the concept of *dharam yudh* in the Sikh tradition by looking at relevant documents from three points in the history of Sikhism. Firstly, I looked at ideas about war in the 1700s in the *Dasam Granth* and the *Rahit-namas*. With all the uncertainties surrounding the composition and authorship of these texts, it has not been my intention to give anything like an authoritative overview of the ideas of war in the times of Guru Gobind Singh and the following decades of the 1700s. On the contrary, I simply wanted to give a brief introduction to key themes from the guru's times as perceived and used by modern Sikhs when they think about the legitimization of violence and war. Secondly, I looked at the concept of *dharam yudh* in the world-view of Sant Jarnail Singh Bhindranwale in the early 1980s. Finally, I analysed articles and editorials from the magazine *Sikh Shahadat* in 2005-6. We have seen that the Battle at Chamkaur provides a very significant, indeed *the* most significant point of reference for modern Sikhs reasoning about the justness of war.

In my opinion, the Sikh tradition presents us with an interesting example of tradition for the comparative study of the ethics of war. For one thing, it is a tradition of ethical reasoning that is still alive and relevant to a number of people. The Sikh tradition of *dharam yudh* has been kept alive by the Sikhs' keen sense of being a special people with a distinct identity and by recurrent conflicts with the central authorities in Delhi. The violent conflicts between the Khalistan movement and the Indian state in the early 1980s made questions of legitimate rule and violence relevant for many Sikhs because the conflicts affected so many people directly.

However, the main conclusion to be drawn from my analysis is the relative *lack* of ethical reasoning about war in the Sikh tradition. From the early period of Khalsa history, we can find a number of texts with opinions or brief comments about war, but they rarely comment on matters of ethics. There is no systematic attempt to discuss under what circumstances, under what authority and for what causes war is legitimate or just. In other words, there is no clear statement of a *jus ad bellum*. Moreover, we can find no traces of a warrior code like the one that is so often referred to in the *Mahabharata* (by the word *kshatradharma*) and some other Hindu texts. In other words, there is no trace of a *jus in bello*, i.e. a systematic thinking about the rules of war and combat typically revolving around issues such as the distinction between combatants and civilians, and proportionality of means and ends in war. The lack of interest in rules for the restraining of war, combined with the clear sense of religious righteousness being on their side, makes it necessary to return to the comparative aspects of our topic. Is the *dharam yudh* really an equivalent to the just war-tradition of Christian Europe, formulated by Thomas Aquinas and developed by theologians and lawyers from the 1500s? Or is it closer to the tradition of Holy War that we find in the books of the Old Testament, and in unrestrained war of later holy warriors?

Perhaps the lack of interest in the morality or the means of war can be explained, in part at least, by the fact that the sources used here are from writers who live(d) in times when the Sikhs tended to perceive themselves as persecuted and politically weak relative to the dominant power in northern India. This has made systematic thinking about restraint in war superfluous; the Sikhs were always on the defensive. Defending livelihood and religion against tyranny does not seem to have created the need for a systematic exploration of the rights to resist political authority, like the long philosophical debate about tyrannicide in Europe, for instance. Having pointed to the lack of ethical reasoning about war in our sources one could easily object that there are a number of other sources from other periods that should be consulted to get a more complete picture of the history of the *dharam yudh* in Sikhism. In particular, it would be of great interest to explore the ideology of war in times when the Sikhs were not on the defensive, as during Ranjit Singh's empire of the first four decades of the 1800s. How did Sikhs think about *dharam yudh* in this period of political dominance? What were legitimate causes for war? What were the ideal righteous—or *dharmic*—intentions of Ranjit Singh's generals and soldiers? And how were they expected to behave in battle? These would be fascinating questions for further studies in the ideology of war in the Sikh tradition.

ACKNOWLEDGEMENTS

This chapter was written as part of a research project about the ethics of war at the Peace Research Institute of Oslo (PRIO). I thank all the other participants in the project for their generous cooperation and comments. Thanks to Professor Gurinder Singh Mann for his help in guiding me to some of the relevant material. Thanks also to Professor W.H. McLeod for his helpful advice and to Lou Fenech for very helpful comments on an earlier version. For readings of the material in Punjabi, I have relied on invaluable help from Dr. Kristina Myrvold.

NOTES

1. Torkel Brekke, 'Between Prudence and Heroism: Ethics of War in the Hindu Tradition', in Brekke (ed.), *The Ethics of War in Asian Civilizations* (London: Routledge, 2005), pp. 113-44.
2. W.H. McLeod, 'The Sikh Struggle in the Eighteenth Century and its Relevance for Today', *History of Religions*, vol. 31, no. 4 (1992), p. 347.

3. Hew McLeod, *Sikhism* (London: Penguin, 1997), p. 62.
4. See in particular his two chapters on the sources for the study of Guru Gobind Singh's life and times in *Journal of Punjab Studies*, Special Issue on Guru Gobind Singh, vol. 15, nos. 1 and 2 (2008), pp. 227f.
5. Gurinder Singh Mann, in *Journal of Punjab Studies*, Special Issue on Guru Gobind Singh, vol. 15, nos. 1 and 2 (2008), p. 258 fn. 141.
6. Louis Fenech, *Martyrdom in the Sikh Tradition: Playing the 'Game of Love'* (New Delhi: Oxford University Press, 2005), pp. 146-7.
7. *Dasam Granth*, text and translation by Jodh Singh and Dharam Singh (Patiala: Heritage Publications, 1999), vol. 1, pp. 198-202.
8. It is challenging to choose the right translation for somebody who does not know Persian, like myself. I have consulted the following three translations: *Zafarnama*, English translation by Sardar Darshan Singh with a foreword by Shri I.K. Gujral (New Delhi: ABC Publishing House, 2000). *Zafarnama*, translated into Punjabi and English by Naranjan Singh Noor (Patiala: Punjabi University, 2000). *The Zafarnama of Guru Gobind Singh*, rendered into English from original Persian by Jasbir Kaur Ahuja (Mumbai: Bharatiya Vidya Bhavan, 1996).
9. *The Zafarnama of Guru Gobind Singh*, rendered into English from original Persian by Jasbir Kaur Ahuja (Mumbai: Bharatiya Vidya Bhavan, 1996).
10. Darshan Singh's translation, op. cit.
11. *Dasam Granth*, vol. 1, p. 153.
12. W.H. McLeod, *Sikhs of the Khalsa: A History of the Khalsa Rahit* (New Delhi: Oxford University Press, 2005), pp. 284-5.
13. Ibid., p. 300.
14. Chaupa Singh (8, a, IV), in W.H. McLeod, *The Chaupa Singh Rahit-Nama* (Dunedin: University of Otago Press, 1987).
15. Chaupa Singh (8, d, I-III) in McLeod, *The Chaupa Singh Rahit-Nama.*
16. *Prem Sumarag: The Testimony of a Sanatan Sikh*, tr. with an introduction by W.H. McLeod (Delhi: Oxford University Press, 2006), p. 19.
17. Ibid., p. 11.
18. Ibid.
19. Ibid.
20. Kushwant Singh, *History of the Sikhs*, vol. 2, *1839-2004* (Delhi: Oxford University Press, 2005), p. 345.
21. *Struggle for Justice, Speeches and Conversations of Sant Jarnail Singh Khalsa Bhindranwale*, tr. from Punjabi, audio and video recordings by Ranbir Singh Sandhu (Dublin, Ohio: Sikh Educational and Religious Foundation, 1999), p. 101.
22. *Struggle for Justice*, p. 127.
23. Ibid., pp. 161-2.
24. Ibid., p. 72.

25. Ibid.
26. Ibid.
27. Ibid., p.121.
28. Ibid., p. 88.
29. Ibid., p. 74.
30. Ibid., p. 78.
31. *Sikh Shahadat*, March 2005, vol. 1, p. 3.
32. Ibid., January 2005, vol. 11, p. 13.
33. Kushwant Singh, *History of the Sikhs*, vol. 1, *1469-1839* (New Delhi: Oxford University Press; 2005), pp. 71-2.
34. Harjot Oberoi, *The Construction of Religious Boundaries* (New Delhi: Oxford University Press, 1997).
35. Fenech, *Martyrdom in the Sikh Tradition: Playing the 'Game of Love'*, p. 152.
36. *Sikh Shahadat*, vol. 11, 2005, p. 45.
37. Gabriel A.R. Almond, Scott Appleby and Emmanuel Sivan, *Strong Religion: The Rise of Fundamentalisms Around the World* (Chicago: Chicago University Press, 2003), p. 61.
38. Many of these issues are discussed in a recent collection of articles: Verne A. Dusenbery, *Sikhs at Large: Religion, Culture, and Politics in Global Perspective* (New Delhi: Oxford University Press, 2008).

CHAPTER 15

The Strategic Revolutionary Phases of the Maoist Insurgency in Nepal

SCOTT GATES
JASON MIKLIAN

After receiving a majority of votes in the Nepalese Constituent Assembly election, the Maoists formed the government in August 2008. Considering that the Maoists were only founded in 1994 and had fought an insurgency for over a decade, the event was indeed momentous. This article traces the path of the Maoist ascent to power from the start of the civil war to its end. In the pattern of several other movements inspired by the success of Mao's 1948 takeover of the Chinese state, the Maoists of Nepal tactically fought their war in stages, consisting of creation, expansion, and consolidation. Re-examining the lifecycle of Nepal's civil war in this context allows us to better understand: the course of events and the timing of significant episodes; the decisions that Maoist leadership made; and the techniques the Maoists used in order to make in-roads during each phase of the conflict.

Nepal is one of the poorest countries in the world. A country of 27 million people, Nepal ranks at the bottom quartile of most development indicators. Social, economic, and political inequalities cleave castes, ethnic groups, and geographic regions. Rooted in both its colonial legacy and decades of heavy-handed autocratic rule, Nepal suffers from institutionalized discriminatory practices. Nepal's 1996-2006 civil war between the Communist Party of Nepal-Maoist (or Maoists) and the Government of Nepal that left 13,000 dead was rooted in some of these disparities.

The Maoists started with a small base of troops in the mid-western hill region, adopting a hybrid of communist and republican ideologies for their recruitment efforts and tactical decision-making. They successfully incorporated Mao's strategies of guerrilla warfare, lessons learned from other revolutions, and tailored a strategy exploiting local grievances in rural areas of Nepal for their own benefit.

As the war progressed, the Maoists expanded their territorial control over Nepal. When the Maoists and the Royal Nepal Army reached a period of stalemate, the Maoists merged with Nepal's seven major political parties to spearhead peaceful protests that successfully pressured King Gyanendra to abdicate power on 24 April 2006. And in 2006, a Comprehensive Peace Agreement was signed, an interim constitution adopted, and national elections held. After the elections, the Maoists emerged victorious, and led a national coalition government headed by former leader and now Prime Minister Pushpa Kamal Dahal *a.k.a.* Prachanda. The peace was short-lived, however, as new conflicts arose in their place. Further, hardliners within the Maoists have called the revolution 'incomplete', and wish to return to armed struggle in order to finish the takeover of the state.

True to their name, the Maoists have based their strategic struggle against the Nepalese state on Mao's theory of revolution and guerrilla warfare. Their basic aim has been to capture state power and establish a 'new democracy'.[1] Echoing Mao's justification of violence that 'it was necessary to bring about a brief reign of terror in every rural area. . . . To right a wrong it is necessary to exceed the proper limit, and the wrong cannot be righted with the proper limit being exceeded',[2] Nepalese Maoist policy on armed struggle adopted in March 1995 states: 'It is an incontrovertible fact the Nepalese people have been waging violent struggle for their rights since . . . historical times. Till today whatever general reforms have been achieved by the Nepalese people, behind them there was the force of violent and illegal struggle of the people.'[3] This view is reinforced by Prachanda,[4] who asserted that: 'People have not obtained even the least of gains without waging violent struggles. Today, the Nepalese society has arrived at such a point of crisis under the existing political system that there is no alternative on the part of the people other than to smash it.'[5]

Throughout the military campaign, the Maoists also engaged citizens as a political party and broader social movement.

Obtaining and securing political power has always taken precedence over all other strategic goals. This reflects Mao's conceptualization of an insurgency as a political undertaking: 'The problem of political mobilization of the army and the people is indeed of the most importance . . . political mobilization is the most fundamental condition for winning the war.'[6] Indeed, the Maoist insurgency in Nepal is best characterized as a political movement with a military wing.

Beyond accepting Mao's call for violent revolution, the Nepalese Maoists also adopted Mao's theory of revolutionary guerrilla warfare. Mao's theory specifies three phases, involving different sets of strategies that are designed to maximize the potential for forward momentum over the course of a revolutionary war.[7] These can be categorized as follows:

Phase 1 (organization, consolidation, and preservation): The primary objective in the first phase is to build political strength, primarily by persuading as many people as possible to convert to the movement. Initially, agitators are to work to enlist the support of peasants to supply food, recruits, and information to the Maoists. Through a slow, clandestine and methodical effort the objective is to gradually build a 'mass' following. Military action is sporadic, selective and limited. Political assassinations and kidnappings are employed for propaganda purposes. Informers and collaborators are killed as means of protecting the revolution.

Phase 2 (progressive expansion): The goal is to gain strength and consolidate control of base areas. Administrative control of contested areas begins and 'reactionary elements' are liquidated. Political indoctrination efforts increase in peripheral districts where the reach of the state is limited. Direct action against the state also increases. Police outposts are attacked and vulnerable military forces ambushed and infrastructure destroyed. Procuring arms, ammunition, in addition to other valuable supplies plays a role in these attacks. Guerrilla warfare is employed to make it difficult for governments to maintain military presence.

Phase 3 (mobile war, strategic offensive): The goal of this phase is to destroy the enemy and take control of the state. Guerrilla forces are transformed into conventional armies. Direct attacks against the state increase. Conventional military forces and tactics are employed against the government's army. The broader strategic aim is to expand territorial control and force the government to retreat to the major cities.

This essay proceeds by viewing the Maoist insurgency through the lens of Mao's three phases of guerrilla warfare. We are not the first to do this. Nihar Nayak also examines the changing tactics of the Maoists through the lens of Mao's theory of guerrilla warfare.[8] We differ markedly from Nayak, however, in that we feature the Nepalese Maoists as a political movement foremost, emphasizing the political, social, and economic aspects of the Maoist movement as opposed to Nayak's focus on military tactics and strategy.

Mao's recommendations for a progression of strategies were followed by the Nepalese Maoists fairly closely over the first two phases of the conflict. The first phase of the war (organization, consolidation, and preservation) took place from 1994 to 2001. The Maoists (CPN-M) participated in the first Nepalese election of 1991, but by 1994 had abandoned democratic participation. The actual armed conflict began in 1996 with an attack on a police post in Rolpa district. This first phase begins before the actual war begins as it accounts for the early activities of the Nepalese Maoist movement. Phase two (progressive expansion) took place from 2001 to 2005. This phase included the build-up to explosions of violence that took place after 2002. The war ended with a negotiated settlement in 2006 and after a two-year transition period, Prachanda was elected as prime minister. The third phase never culminated as Mao specifies. Nonetheless, aspects of the early stages of Phase 3 were evident just before the peace agreement was signed.

PHASE ONE (ORGANIZATION, CONSOLIDATION, AND PRESERVATION): 1994-2001

The Maoists began their struggle in the remote hill districts of Rolpa and Rukum in 1994.[9] Figure 15.1 indicates the location of these first confrontations. The geopolitical environment for revolution was favourable at this time, as worsening abuse of citizenry by the government reinforced the entrenched popular opinion of an urban bias at the centre, which viewed rural lives and livelihoods as expendable. Rural concerns about absolute deprivation gave way to concerns over relative deprivation *vis-*

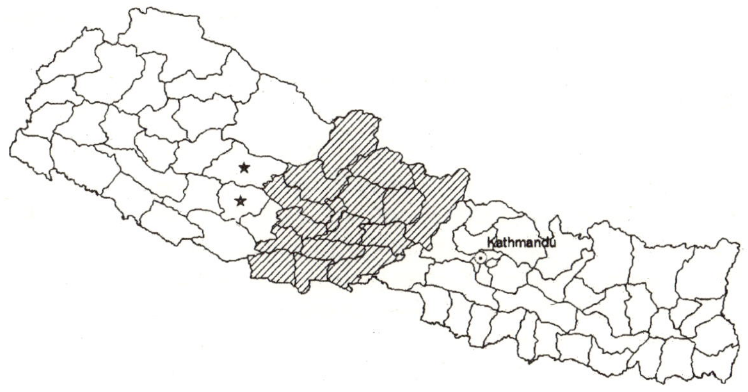

Initial conflict area 1996-2000

★ Initial conflict districts and Maoist strongholds during the conflict

▨ Conflict areas 1996-2000

FIGURE 15.1: PHASE ONE OF THE WAR

à-vis Kathmandu.[10] Mid-western Nepal in particular was plagued with bad governance, endemic corruption, non-existent rule of law, violence, thuggery, repression by local police, poor infrastructure, widespread poverty, exploitation of manual labourers, poor distribution of wealth, and lack of access to education.[11] When buttressed with policies from the government that were anti-poor, anti-rural and anti-agrarian, there was fertile ground for the creation and initial expansion of the Maoist insurgency.

The Maoists promised a wide variety of improvements, including universal education, health care, electricity, jobs, abolition of the monarchy in favour of a republic, the ending of police and military oppression, elimination of corruption, and a significant transfer of wealth from Kathmandu to Rolpa.[12] Combatants joined primarily to end the pervasive corruption of government officials and the brutality of Nepal Police.[13] Operating in areas with little to no infrastructure, the insurgency slowly spread through both persuasion and coercion. One long-time resident of Rolpa district encapsulated the local security situation during the movement's inception period thus:

The government brutally oppressed us in the two or three years before the Maoists came. Had they not been so brutal, the Maoist movement never would have started. . . . In the early 1990s, the police forced bribes from everyone—the old, women, and even children. One time they rounded up everyone in the school, and kept them there while they looted the houses and broke our safe boxes, and even took our old coins and precious family heirlooms. These actions made many join the Maoists.[14]

For two years, the Maoists undertook of minor but increasingly successful village-level retaliatory attacks on police, government representatives, and other political leaders of the area. This period allowed the Maoists to learn which tactics were likely to be successful if the conflict was expanded, and which messages to the grassroots carried the greatest resonance. The crystallization of this strategy was unveiled in 1996, when the Maoist People's War against the state was launched after the government rejected a Maoist 40-point list of demands. The 40-point list largely concerned change at the Kathmandu level, and was not addressed for or presented to rural audiences for the most part. Prachanda allayed potential elite fears through talk of prosperity, growth and market economics if the Maoists should come to power, downplaying more revolutionary demands such as expropriation of assets by the state. Indeed, most rural dwellers even in Maoist strongholds could not name a single item on the 40-point demand list when asked, and few knew of its existence at all.[15]

Figure 15.1 shows the location of the initial Maoist attacks as indicated by the stars. Most of the fighting during this early phase of the war, however, occurred in the districts located between Rolpa, Rukum and Kathmandu. The Maoists were able to quickly secure control of a number of provinces in the Middle West and Far West as indicated in Figure 15.2, which shows the relative levels of Maoist influence across districts in the first phase of the war. It should also be noted that the casualty figures during the first phase of the war were generally low. In addition, for the most part, those districts with the greatest degree of Maoist influence were the districts most disadvantaged *vis-à-vis* Kathmandu.

Several academic studies have concluded that ethno-religious, caste divisions, and other measures of inequality correlate positively with the location of armed conflicts, leading several academic works that examined the conflict at the sub-national level to propose that the conflict was driven by horizontal (or inter-group) inequalities rooted in ethnicity and/or the caste system.[16] These results (especially Murshed and Gates) reflect a

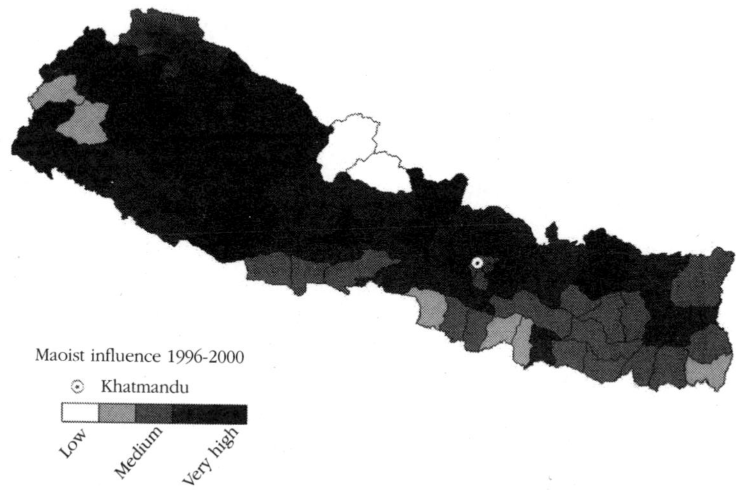

FIGURE 15.2: EXTENT OF MAOIST INFLUENCE IN PHASE ONE OF THE WAR

relationship between horizontal inequality and deadly violence.[17] These statistics do not regard the origins of the conflict, nor do they provide information as to what factors motivated people to support the Maoists. To the extent that horizontal inequalities coloured the initial mobilization in Rolpa and Rukum, they were not ethnic, religious, or cultural, but were instead urban-rural in nature, or at least centre-periphery conflicts with the outlying regions taking on Kathmandu. Other sub-national quantitative analyses tend to agree that ethnic/caste distinctions were not factors.[18] Rural dwellers did not see those in Kathmandu as belonging to a different culture or ethnicity, but more as criminals or 'fat cats' who reaped the spoils from pro-urban exploitation of Nepal's resources and policies, suggesting that vertical inequalities played a larger role.[19]

Inequalities are rooted in Nepal's caste system, but Nepal's patchwork of hundreds of overlapping castes, sub-castes, religious groups and ethnic groups blur attempts to claim that the conflict is rooted in the discrimination of any one group. In addition, most individuals in Nepal perceive as having more than one 'identity', and caste divisions are not as rigid or monolithic as the 4+1 caste hierarchy (Brahmin, Chhetri, Vaishya, Shudra, and Dalit or untouchable) would indicate. Each caste has hundreds of sub-castes, fragmented spatially, ethnically and economically. For example, three major communities in Nepal (Newar, Khas, and Maithili) each have members from all four castes who also sub-divide themselves by region, their degree of Indian *versus* Nepali origin, and other criteria. The notion of a singular 'Brahmin identity' or a 'Newari policy' is for all intents and purposes non-existent. To the extent kinship networks based upon caste do exist, they are generally applicable to a specific sub-caste only, if at all.

Ethno-religious cleavages were not a driving factor during the early stages of the conflict. Maoist leaders belonged to Brahmin/Bahun/Chhetri 'ruling elite' classes, as were many combatants. Maoist supporters[20] came from different castes and ethnicities, but Maoist leaders played the caste card sparingly, speaking the language of Dalit empowerment through an ideological class-inequity lens.[21] In other words, the Maoists may have been pro-Dalit, but they were not a Dalit organization. Rolpa and Rukum combatants and supporters from the 1994-6 period, joined primarily to protect themselves and their families from a brutal police presence that operated with impunity.[22] As observed by Paul Richards in Liberia and Sierra Leone, local factors as much as grand ideological claims play a significant role in explaining the onset of the armed conflict. 'War is bound up with abuses associated with rural custom. Chieftaincy, 'customary' courts and traditional bride service bear down unfairly upon impoverished rural youth';[23] a fact as true in Nepal as it was in western Africa. Indeed, civil wars in general tend to be fought in the periphery where the reach of the state is limited and where local factors dominate.[24]

An examination of public statements, official Maoist documents, and interviews of original supporters from the period suggests that the Maoists did not overtly use caste or ethnicity as a mobilizing agent against the state during the initial period of insurgency.[25] Neither high-caste Brahmins nor Newars were singled out for attacks (although a disproportionate share were extorted from them due to their status as landowners), and there is no evidence to suggest that the Maoists attempted to create any kind of 'us *vs.* them' dichotomy against the cultural composition of those in power in Kathmandu during this period. Former combatants, supporters, and non-supporters originally from the two initial districts of Rolpa and Rukum agree that the ultimate goal of the movement as presented to them was not about overthrowing a different ethnic group or Brahmin elite, but was about removing the king, and in his place creating a republic ostensibly based on equality.[26]

Official statements of Maoist leaders and interviews conducted with Maoist supporters suggest that exploitation of vertical inequalities (classic inequality between classes) played a significant role inducing individuals to join the rebellion.[27] The Maoists spoke a language of rectifying vertical inequalities, made possible through the hybridization of communist and capitalist ideology that underpinned the essence of the 'land to the tiller' campaign. This departure from traditional communist ideology allowed grassroots supporters to individualize their potential gains if they gave their allegiance. Karen Mancours argues that increasing inequality over time between socio-economic groups may marginalize and motivate those left lagging behind enough to mobilize, but it is unclear how increasing group inequality motivates individual actions.[28] In addition, her analysis fails to account for how increasing inequality is endogenously induced by the conflict itself and the statistical model is most likely misspecified.

Although vertical inequalities are generally less likely to cause civil conflict, there may be a point at which a continually degrading situation also becomes untenable, as happened in Nepal. Although caste/ethnic horizontal inequalities were likely not the primary driver of the conflict at its inception, there are substantial (and worsening) horizontal inequalities in Nepal that played essential roles in the growth and sustainment of the conflict in other regions, as seen in later phases of the war.

MAOIST WAR, PHASE TWO: PROGRESSIVE EXPANSION, 2001-2003

The Maoists made an abrupt tactical shift at what is now recognized as the mid-point of the conflict. As the Maoists consolidated their stronghold hill territories, they looked to expand their influence into other parts of the country.[29] Thus, a February 2001 Maoist plenary session meeting ended with an agreement that the Maoists would continue to consolidate their power in rural areas, but shift their primary focus to urban theatres of conflict, tailoring the 'progressive expansion' stage to the geopolitical characteristics of Nepal. This new doctrine was called 'Prachanda Path', to highlight both the similarity and unique elements *vis-à-vis* previous communist revolutions.

Maoist gains from the period of progressive expansion comprised a central component to their rise to power, but there were as many 'push' factors for citizens to join the Maoists from government actions as 'pull' incentives from the Maoists themselves. The massacre of almost the entire royal family in 2001 by Crown Prince Dipendra, and Gyanendra's subsequent rise to power as king was never accepted as legitimate by many Nepalese citizens. Further, King Gyanendra's incremental steps towards authoritarian rule exacerbated public anger, particularly when he banned political parties and withdrew media freedoms in the name of security. These actions stoked resentment among the middle and upper classes of Nepali society, turning popular opinion decisively against the king for the first time. As the king became increasingly authoritarian, Maoist calls for a republic became increasingly palpable and therefore the 'lesser of two evils' in the minds of many citizens. With declining legitimacy of the king came an expansion of the conflict with the Maoists. Phase Two of the war occurred from 2001 to 2003 with an intensification of fighting and a geographic spreading of the conflict.

Figures 15.3-15.5 show the districts experiencing over 100 battle-related casualties aggregated over the respective years, 2001, 2002, and 2003 (Phase 2 of the war).[30] Note that in contrast to Figure 15.1, these maps do not show conflict areas with lower levels of violence. Figure 15.3 shows that while the war had intensified, involving many more deaths than in Phase One, its geographic scope was limited. The conflict was concentrated close to the location of the origin of armed conflict (as indicated in Figure 15.1). In 2001, the war was limited to five districts in the midwestern region, remote from Kathmandu.

By 2002, the number of districts affected by fairly serious conflict (exceeding 100 battle-related casualties) expanded considerably. The war in the western districts had spread in the mid-western and far western regions, most notably to the Western Terai (the broad plains to the south of Nepal, near the Indian border). In 2002, intense Maoist fighting occurred in

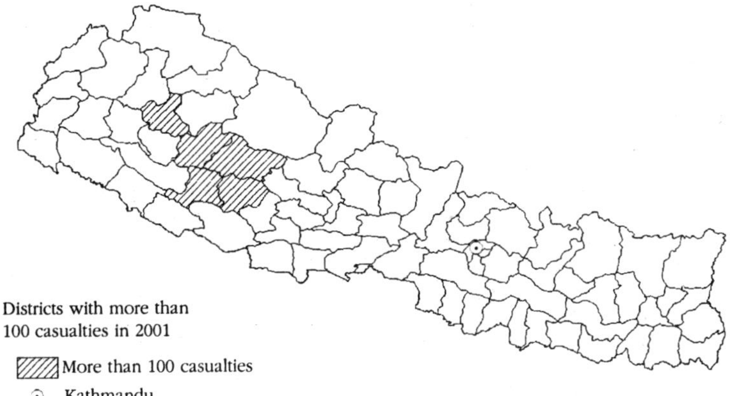

FIGURE 15.3: THE GEOGRAPHIC SCOPE OF THE WAR IN 2001

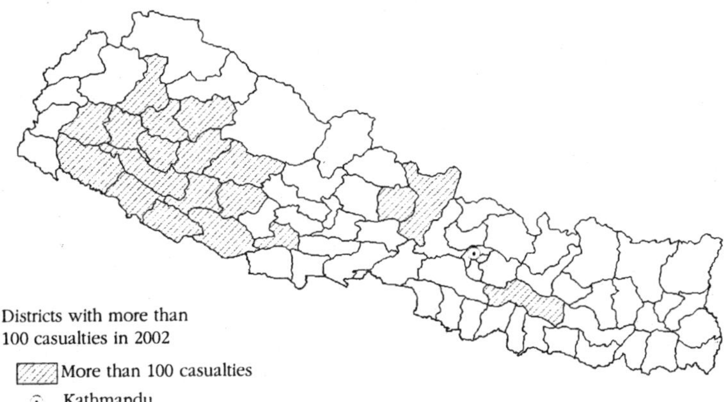

FIGURE 15.4: THE GEOGRAPHIC SCOPE OF THE WAR IN 2002

districts a long way away from rebel headquarters. Two districts in the western region and one district in the central region experienced more than 100 battle casualties. The scope of the war had expanded considerably.

By 2003, the war had spread throughout much of the far western and mid-western regions. It had also spread throughout the mountains and hill districts in the central region. The war was now getting close to Kathmandu.

A key aspect of the second phase of the war was to bring the

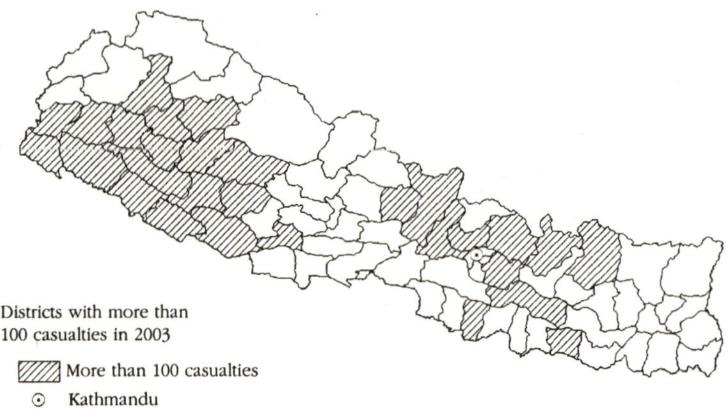

FIGURE 15.5: THE GEOGRAPHIC SCOPE OF THE WAR IN 2003

war to new parts of Nepal. In order to effectively implement their new strategy, Prachanda and Bhattarai realized that they needed to contextualize their revolutionary message within the varied demands of local actors if they were to gain support in areas of Nepal where, economic and political dynamics were quite different from that of the western hill regions. The Maoists considered the belt of southern plains districts of Nepal that border India as the most important target for expansion. These districts, collectively known as the Terai, house over half of Nepal's population and 90 per cent of its agriculture. Figure 15.6 shows the population distribution in Nepal. The concentration of the people of the Terai is quite evident, shown as the dark belt along the border with India. Although many concerns of Terai citizens were similar to those in the hills, including police oppression, poor infrastructure, and corruption, the Maoists hoped to exploit the grievances of the Madhesi ethnic community in particular to their benefit. In this regard, the Madheshi Janadikhar Forum (MJF) was targeted by Maoist leadership due to their extensive grassroots mobilization networks, respect among the Madhesi community, and overall influence within Nepal's second-largest city of Biratnagar.

The Terai has long been a marginalized area of Nepal.[31] The state has attempted to assimilate the 100-plus ethnicities of Nepal into a pan-Nepali identity through language, schooling, and legal

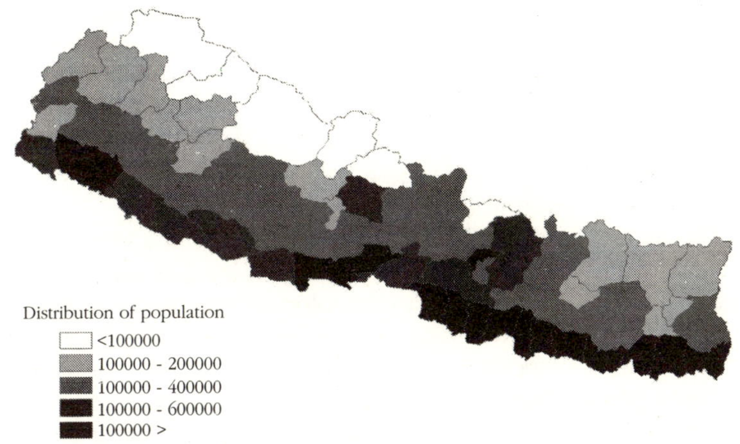

FIGURE 15.6: POPULATION DISTRIBUTION IN NEPAL

directives. These policies codified the cultures of the largely Kathmandu-based citizens with lineage ties to the hill regions (also known as *Pahadis*), legalizing systematic discrimination and under-representation in the government of any Nepalese who did not have this lineage, making even public discussion of ethnic difference or inequality a jailable offence.[32] Spurred into action by these discriminations, several organizations were formed in the 1990s using ethnic markers for cohesion. The largest of these, the MJF, was formed in 1997. Originally designed as an academic exchange by which Madhesis could air grievances, the MJF quickly became ethnically radicalized after its leadership developed a working relationship with the Maoists around 2002.[33] The MJF leaders (including Upendra Yadav and Matrika Yadav) were given senior regional leadership positions as MJF cadres were incorporated into the Maoist rank and file. In exchange, the Maoists ramped up talk of ethnic equality and regional autonomy for ethnic minorities. The Maoists were thus able to penetrate a new 'conflict market' quickly using the grassroots legitimacy that MJF leaders had built over the previous five years. The MJF leaders believed that a Maoist victory would assure them the regional autonomy they craved.[34] And grassroots supporters were promised 'scientific land reform' requiring large landowners to give their land to those who farm it; the forced return of all *Pahadi* groups to the hills and abandonment of the land they had acquired during the waves of settlement to the Terai; Madhesi, Tharu, and other communities historically disadvantaged by *Pahadi* elites were promised equal recognition and opportunities in the civil bureaucracy and army; and the district map of Nepal would be redrawn to give the Terai greater autonomy.[35] Those who elected to join were given weapons and increased status within their communities.

The inclusion of Madhesi communities was a key step to the growth of the Maoist movement. Insurgent leaders successfully channelled rural anger into a neo-colonization discourse while sidestepping the fact that large landowners and Madhesi political leaders often worked in collusion against Tharu and other minority groups for mutual gain. Maoist party manifestos and supporting documents[36] focus on discrimination by other ethnic groups against Madhesis, but do not address the discrimination within the Terai by Madhesis against Tharus and other minorities. Regardless, Madhesi grassroots support coupled with increasing authoritarianism from the centre quickly turned the tide of popular support in the Terai to the Maoists.

PHASE THREE, 2004-2006: AN 'INCOMPLETE' REVOLUTION?

By the end of 2004, the conflict had spread to most corners of the country. A 2004 ceasefire attempt was aborted, and the Maoists ruled over most of Nepal's rural districts despite increased international assistance to King Gyanendra's Royal Nepal Army in the form of weapons and funding. Figure 15.7 shows the scope of the conflict in 2004. The extent of Maoist activities remained largely unchanged in 2005 and 2006 and for this reason maps for these years are not included.

In 2005, the Maoists strengthened their control over the districts they had taken over. By early 2006, the conflict had devolved into a military stalemate as the Royal Nepal Army had abandoned attempts to win back rural areas, but still maintained an iron grip over Kathmandu and the surrounding valley. However, King Gyanendra's increasingly draconian attempts to control the conflict by lashing out at the political parties, media organizations, and civil society groups that were heretofore neutral or even pro-palace created the lever the Maoists needed to gain access to the capital. After two weeks of peaceful mass

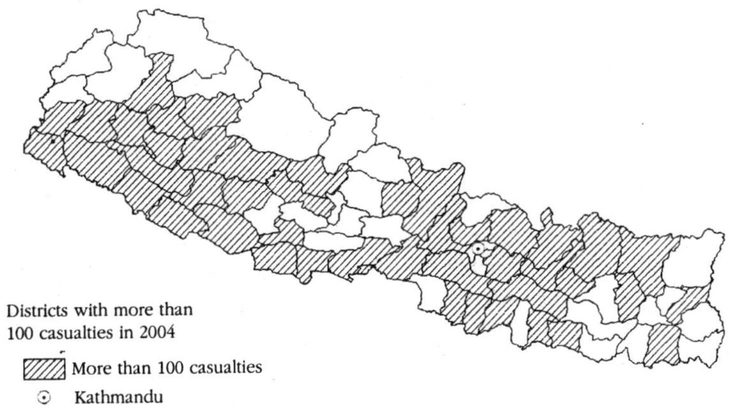

FIGURE 15.7: THE GEOGRAPHIC SCOPE OF THE WAR IN 2004

protests by a combination of Maoists, political parties, and civil society groups in Kathmandu, King Gyanendra stepped down from power in April 2006, for all intents and purposes ending the civil war. Shortly thereafter, an alliance of the Maoists and seven major political parties (SPA) drafted a Comprehensive Peace Agreement (CPA).

Up until this point, the Maoists had fought a 'textbook' war. Maoist control of nearly all rural areas and the surrounding of the Kathmandu valley in Phase Three of the war all follow Mao's prescription. Where the path diverges is that the Maoists never fought the war conventionally, nor did they push the war further to achieve complete victory. The alliance with the political parties composing the SPA constituted a profound shift in strategy.

Like the alliance with the MJF, the Maoist arrangement with the SPA was mutually beneficial at the time of initiation. The Maoists desperately wanted to develop an air of legitimacy in the capital in order to overcome the Army's superior military capabilities. Likewise, the SPA was searching for a way to regain the political power taken away from them by the king. Even though most SPA members did not fully trust Maoist motives, a deal was struck to protest peacefully together. Seizing the opportunity, negotiators recognized the urgency of the 2006 peace negotiations as perhaps the best opportunity to secure a settlement. The shared belief by both the SPA and Maoists at the time was that 'if we fail here, the King wins'.[37]

Maoist recruitment rhetoric after 2001 was steeped in wealth redistribution and land reform, but negotiators were surprised to find that the Maoists did not address these or any other demands 'of the people' during CPA talks; instead, they were primarily interested in maximizing power at the top, and expressed little interest in the details of the negotiations or specific policies.[38] A jointly-drafted interim constitution in 2007 also showed a similar lack of reform momentum. Maoist promises of wealth distribution and redistribution at that time encompassed vague plans for more government jobs and 'giving more' to the poor through increased development, but specifics never materialized. What on the surface would seem to illustrate a perplexing lack of accountability by top-level Maoists to their supporters becomes clear if one is to view the peace process in light of the greater Maoist strategic objective.

The MJF leaders viewed these documents as a sign that Maoist leaders were not serious about their demands, exposing the rockiness of the Maoist/MJF bargain. Increased tensions over perceived Maoist discrimination led to public disillusionment and intra-group violence in the Terai, most visibly manifested when MJF supporters killed 29 Maoists in the border village of Gaur in March 2007. The renewed violence took many by surprise, due to the Kathmandu-centric focus of both politicians and the international community during the peace process,[39] and the fact that the Terai was not the primary battleground during the conflict.

While the Maoists and MJF fought a common enemy (Kathmandu elites), differences between their goals and broader objectives were brushed aside. However, after the peace agreements were struck, these differences served as wedge issues which were used by the MJF and other Terai actors to foment splintering and violence. Most Maoist leaders were *Pahadi* themselves, and although they promoted MJF leaders to Terai-based positions, they did not promote any to senior positions.[40] This slight was followed by rapid exodus of Maoist commanders and troops of Terai origin to splinter groups defined on the basis of ethnicity. Over the past two years, the splinter groups have battled each other, the Maoists, other ethnic groups, and the state, as all attempt to control the Terai. A post-conflict political vacuum of power, increased numbers of violent actors, communalized radicalization of violence, and generalized lawlessness left the Terai as dangerous and more violent in 2006-8 than during the civil war period, and raised the sceptre of secession upon the minds of many violent actors, an option never on the agenda during the war between the Maoists and the king.

The rise of Terai-based groups proved perplexing for the Maoists, who had difficulty in crafting a coherent response to the demands that did not undermine their greater goal of state supremacy. Both during and after the conflict, Maoist leadership was ambivalent towards addressing the struggles of many Terai groups, despite their use of Terai grievances in order to build support during the conflict. In addition, the post-peace agreement period was characterized by a fundamental lack of respect for Terai-specific issues amongst Maoist leaders. One Maoist leader who contested in a Terai district in the 2008 elections, and subsequently lost to an MJF candidate, lamented the situation thus: 'The root cause of the Maoist decline in the Terai is the lack of political sense of the people here. There is a disconnect between high-level thought (Maoists) and what people in the villages (Madheshis and Tharus) say. We should be sensitizing people at the village level.'[41]

The rise of regional ethnic conflict cleavages was also poorly predicted by many scholars who emphasized the primacy of geographic indicators to explain the locality and severity of conflict in Nepal.[42] Almost all post-2006 violence has taken place in not the hill or mountain districts, but in the flat Terai plains. In addition, violent events have increasingly been committed in or near urban centres, the opposite of what many Nepal conflict models predicted.[43] However, Murshed and Gates' hypothesis of horizontal inequality drivers for conflict is highly relevant and their four key aspects of horizontal inequality are all visible, namely inter-group asset inequality, unequal access to public employment, unequal access to public services and over-taxation, and economic mismanagement.[44] Let us examine each briefly in the case of the Terai:

Inter-group asset inequality: Per capita income in the Terai and hill regions are roughly equal (about $1,400 per capita) despite the formidable barriers to rural wealth creation in the hills, including problems of infrastructure, arable land, and geography.[45]

Further, the numbers obscure the fact that much of the wealth generated in the Terai (two-thirds of the country's GDP) is actually created by large numbers of landowning *Pahadis*, who extract the wealth as soon as it is created. In contrast, Terai citizens own negligible amounts of hill land.

Unequal access to public employment: Government positions are highly prized in Nepal as they constitute some of the only stable, well-paid jobs in the country. But, the aforementioned historical legacies of 'Pahadicization' of Nepal's government decimated Terai representation in both the military and bureaucracy. Terai citizens constitute 50 per cent of Nepal's population, but only 1-4 per cent of the Nepal Army,[46] with similar proportions of the foreign service, civil service and police.

Unequal access to public services and over-taxation: As in other poor rural locales, the lack of documentation has been a barrier to successful tenancy challenges by the poor against either rich landlords or the state. Infrastructure is poor (worse than Kathmandu, but not worse than hill areas). The 'Pahadicization' of Nepal society left Terai citizens at a distinct disadvantage. Many Terai citizens have to pay bribes just to take school and professional entrance exams.[47] Health care is poor, and Terai citizens are routinely denied service based upon their ethnicity. One villager said that the hospital staff 'take one look at our faces', and deny help.[48]

Economic mismanagement: Nepal ranks poorly on most development indicators, ranking at or below the bottom quartile on all of the World Bank's 2008 governance indicators, and ranks 124th in the Human Development Index. Further, the financial sector is in a state of complete disrepair.[49] The relationship between landlessness and poverty in Nepal is robust,[50] and the previously mentioned growth figures also illustrate inequality between the Terai and hill regions.

Phase Three of the war in Nepal follows Mao's prescription quite closely. Moreover, the nature of expansion of Maoist influence and the adaptation of the conflict to suit the special conditions of the Terai also fit Mao's writings on guerrilla warfare. Even the alliance with the political parties of the SPA exhibits some superficial similarities with the Communist-Nationalist (KMT) alliance in China against the Japanese. Ultimately, the most significant difference between the nature of the war in Nepal and other Maoist movements is the lack of conventional warfare in the final phase of war. General Vo Nguyen Giap in Vietnam strayed from strict adherence to Mao's strategy for the Phase Three of the war by integrating conventional and guerrilla warfare together. Giap never abandoned guerrilla warfare. In Nepal, however, the Maoist Army never engaged in conventional warfare.

IS HALF A REVOLUTION BETTER THAN NONE? ANGER FROM ABOVE AND BELOW

Perhaps the most unorthodox element of the Nepalese Maoist strategy was how the Maoists integrated their demand for a republic and direct elections into Maoist discourse, which traditionally calls for one-party rule. Nonetheless, this hybrid solution, while effective from a strategic standpoint to gain followers and support amongst key segments of Nepal society during the conflict, has consequences that continue to reverberate after official hostilities have ceased. Long-time grassroots Maoist supporters and hardliners within their own party have clamoured for fulfilment of ideological promises made during the conflict; promises locked in *stasis* during the Maoists' further transition to an institutionalized political party. While the Maoists publicly profess to uphold respect for constitutional democracy, they blame this same system as the reason for inaction when speaking to party faithfuls.

The decision to enter multi-party politics was contested internally by Maoist leaders throughout the course of the conflict and peace agreement, and remains a divisive issue. Prachanda and second-in-command Baburam Bhattarai have feuded publicly several times since 1990; at one point, Bhattarai was even expelled from the party. They now work together as joint mediators within their own party, asking their hardline factions and PLA (Maoist army) troops for patience while at the same time assuring grassroots supporters and political actors in Kathmandu that they are not as radical as portrayed. The balancing act has taken its toll on party unity, as a once cohesive unit has developed several internal factions that threaten to split the party.

Promised change from peace and subsequent elections have resulted in few tangible gains in the 2006-9 period for the people of Nepal. The Maoists continue to comb schools for recruitment in stronghold areas, human rights violations from the conflict remain unaddressed, and no perpetrators of abuses from either the Maoists or Nepal Army have been tried for their crimes. About 100,000 conflict refugees from hill districts currently live in the Terai, placing extra pressure on an already overpopulated area in terms of land rights issues and ethnic stereotyping. In addition, violent youth wings of many of the major political parties engage in proxy turf wars as their parent organizations have condemned them in public but funded them in private.

Without giving Maoist leadership too much credit by claiming that their policies and strategies have unfolded exactly as they may have hoped, it is worthwhile to view the creation, expansion, and integration of the movement into the political mainstream more as a strategic framework of decisions rooted in the successful (and failed) revolutions of the past than as a function of impromptu decision-making and good luck. Further, reflection of Maoist strategy in this regard may contextualize future decisions that the organization takes, as well as expose the likelihood of intra-group tension if leadership has lost its thirst for revolution. Miklian has previously called for an honest appreciation of the Maoists' efforts to transit from an insurgent group to a mainstream political party,[51] but their continued foot dragging in the military integration and lack of implementation of worthwhile policies during their year of leadership in government suggest that the honeymoon period has come and gone, with populist rhetoric of republicanism proving too difficult or undesirable to prioritize over the allure of power in the face of ideologies indoctrinated over the previous decade.

ACKNOWLEDGEMENT

The authors give special thanks to Siri Aas Rustad for creating the maps used in this chapter.

NOTES

1. The concept of 'new democracy' is traceable directly to Mao himself and his essay, 'On the New Democracy' (1940), reproduced in Dan N. Jacobs and Hans H. Baerwald (eds.), *Chinese Communism: Selected Documents* (New York: Harper & Row, 1963), pp. 66-77.

2. Mao Zedong, 'Report of an Investigation into the Peasant Movement in Hunan to the Central Committee of the Chinese Communist Party' (February 1927), reprinted in Jacobs and Baerwald (eds.), *Chinese Communism*, p. 23.
3. 'Strategy and Tactics of Armed Struggle in Nepal', document adopted by the third plenum of the CPN (Maoist), March 1995 (International Crisis Group, Asia Report No. 104, 27 October 2005), p. 20.
4. Literally, Prachanda means 'the ferocious one'.
5. Prachanda, 'War Policy of the Nepalese New Democratic Revolution in the Context of Historical Development', in *Problems and Prospects of Revolution in Nepal* (Janadisha Publications, 2004).
6. Mao, *On Protracted War* (19xx: 137), cited in Colonel Thomas X. Hammes, USMC, *The Sling and the Stone: On War in the 21st Century* (St. Paul, MN: Zenith Press, 2006).
7. See Mao Tse-tung, *On Guerrilla Warfare* (1937), translated and introduction by Samuel B. Griffith (Urbana, IL: University of Illinois Press, 1961), pp. 20-2.
8. Nihar Nayak, 'The Maoist Movement in Nepal and Its Tactical Digressions: A Study of Strategic Revolutionary Phases, and Future Implications', *Strategic Analysis*, vol. 31, no. 6 (November 2007), pp. 915-42, also examines the Nepalese Maoist movement through the lens of Mao's theory of guerrilla warfare.
9. Both senior Maoist leaders Prachanda (now prime minister) and second-in-command Baburam Bhattarai have been involved in politics since 1990. See Nayak, 'The Maoist Movement in Nepal and Its Tactical Digressions: A Study of Strategic Revolutionary Phases, and Future Implications', pp. 915-42.
10. Sonali Deraniyagala, 'The Political Economy of Civil Conflict in Nepal', *Oxford Development Studies*, vol. 33, no. 1 (2005), pp. 47-62; Ted Gurr, *Minorities at Risk: A Global View of Ethnopolitical Conflict* (Washington DC: United States Institute for Peace Press, 1993).
11. Author interviews, Maoist supporters, former supporters, and non-supporters, Rolpa district, October 2008.
12. Author interviews, Maoist supporters, former supporters, and non-supporters, Simal Banh and Jhankot, 2008.
13. Author interviews, Maoist combatants, former supporters, and non-supporters, Rolpa district, October 2008.
14. Authors' interview, farmer, Rolpa district, November 2008.
15. Authors' interviews, farmers and Maoist supporter farmers, Libang area rural villages, Rolpa, October 2008.
16. Mansoob S. Murshed and Scott Gates, 'Spatial-Horizontal Inequality and the Maoist Insurgency in Nepal', *Review of Development Economics*, vol. 9, no. 1 (2003), pp. 121-34; Karen Macours, 'Relative Deprivation and Civil Conflict in Nepal', SAIS-Johns Hopkins University working

paper, 2006; Graham Brown, Stewart Frances and Arnim Langer, 'The Implications of Horizontal Inequality for Aid', United Nations University Research Paper No. 2007/51.

17. Moreover, Murshed's and Gates' analysis only regarded Phase One of the war.
18. Alok K. Bohara, Neil J. Mitchell and Nepal Mani, 'Opportunity, Democracy and Political Violence: A Sub-National Analysis of Conflict in Nepal', Nepal Study Center, University of New Mexico. NSC Working Paper #1, 2004; Quy-Toan Do and Lakshmi Iyer, 'Poverty, Social Divisions and Conflict in Nepal', World Bank Working Paper 07-065, 2007.
19. These perceptions had roots in many criminal and financial scandals that went unpunished. Jason Miklian, *Illicit Trading in Nepal: Fuelling South Asian Terrorism*, PRIO South Asia Policy Report No. 3. Oslo: International Peace Research Institute, Oslo, 2009.
20. In guerrilla wars, the line between 'civilian', 'supporter' and 'combatant' is often blurred, as happened in Nepal. Further, support was coercive in many areas of strongest Maoist support.
21. Authors' interviews, Rolpa, April/October 2008. Also see Rajendra Pradhan and Ava Shrestha, 'Ethnic and Caste Diversity: Implications for Development', *Asian Development Bank, Nepal Resident Mission Working Paper No. 4* (2005), pp. 12-14.
22. Authors' interviews, Jhankot and Marichaur villages, Rolpa district, November 2008.
23. Paul Richards, 'To Fight or Farm? Agrarian Dimensions of the Mano River Conflicts (Liberia and Sierra Leone)', *African Affairs*, vol. 104, no. 47 (2005), pp. 571-90.
24. Halvard Buhaug and Scott Gates, 'The Geography of Civil War', *Journal of Peace Research*, vol. 39, no. 4 (2002), pp. 417-33.
25. Author interviews, former combatants, Jhankot and Marichaur villages, Rolpa district, October 2008.
26. Authors' interviews, supporters, Kathmandu and Libang (Rolpa), April/October 2008.
27. Authors' interviews, Jhankot and Marichaur villages, Rolpa district, November 2008.
28. Karen Macours, 'Relative Deprivation and Civil Conflict in Nepal', SAIS-Johns Hopkins University working paper, 2006.
29. Nayak, 'The Maoist Movement in Nepal and Its Tactical Digressions: A Study of Strategic Revolutionary Phases, and Future Implications', pp. 915-42.
30. These maps were constructed using data from Quy-Toan Do and Lakshmi Iyer, 'Poverty, Social Divisions and Conflict in Nepal', World Bank Working Paper 07-065, 2007.
31. For a comprehensive overview of Terai marginalization and the Madhesi

movement, see Jason Miklian, *Nepal's Terai: Constructing an Ethnic Conflict*, PRIO South Asia Policy Report No. 1. Oslo: International Peace Research Institute, Oslo, 2008.

32. For example, the 1964 Reform Act and the 1963 Citizenship Act forbade Terai citizens to own land due to their supposed status as 'conquered people' (Tharus and other indigenous communities), Indian immigrants (Madhesis) and 'inauthentic' Nepalese citizens (Muslims), beliefs still held by many in Kathmandu. Several economic programmes were implemented during the 1990s and 2000s in an attempt to reduce the disparities, but they were ineffective and at times even counterproductive as large landowners used liberalization policies to further solidify holdings. See Miklian, *Nepal's Terai: Constructing an Ethnic Conflict*, PRIO South Asia Policy Report No. 1, Oslo: International Peace Research Institute, Oslo.
33. Miklian, *Nepal's Terai: Constructing an Ethnic Conflict*, PRIO South Asia Policy Report No. 1, Oslo: International Peace Research Institute, Oslo.
34. Authors' interview, Upendra Yadav, November 2007.
35. Authors' interviews, farmers, Morang and Sunsari districts, October/November 2008. For more information, see Jason Miklian, 'Regional Aspects of Land Reform and Conflict in Nepal', Mimeo, International Peace Research Institute, Oslo (PRIO) (forthcoming).
36. See for example, Upendra Yadav, *Conspiracy Against Madhesh: Subversion of Democracy, Racial Discrimination, Internal Colonization, Aggression* (Madhesi People's Rights Forum Party Manifesto, Nepal, 2005).
37. Authors' interview, Mohan Banjade, Law Reform Commission, Peace and Reconstruction Ministry.
38. Authors' interviews, Iswor Pokharel (CPN-UML representative to CPA drafting team), Dr. P.S. Mahat (Nepali Congress representative to CPA drafting team), Mohan Banjade (independent expert on Law Reform Commission and CPA framer), Kathmandu, April 2008. See Jason Miklian, *Power Sharing in Nepal*, PRIO South Asia Policy Report No. 2. Oslo: International Peace Research Institute, Oslo, 2008.
39. 'Nepal's Maoists: Their Aims, Structure, and Strategy', International Crisis Group (Crisis Group), Asia Report #104, 27 October 2005; Jason Miklian, 'International Media's Role on U.S.-Small State Relations: The Case of Nepal', *Foreign Policy Analysis*, vol. 4, no. 4 (2008), pp. 399-418.
40. Authors' interview, Upendra Yadav, November 2007.
41. Authors' interview, Siddnarayan Mandal, April 2008. Mandal contested from Eastern Morang.
42. Alok K. Bohara, Neil J. Mitchell and Mani Nepal, 'Opportunity, Democracy and Political Violence: A Sub-National Analysis of Conflict in Nepal', Nepal Study Center, University of New Mexico. NSC Working

Paper #1, 2004; Quy-Toan Do and Lakshmi Iyer, 'Poverty, Social Divisions and Conflict in Nepal', World Bank Working Paper 07-065, 2007; Stathis N. Kalyvas, 'Promises and Pitfalls of an Emerging Research Program: The Microdynamics of Civil War', in Stathis N. Kalyvas, Ian Shapiro and Tarek Masoud (eds.), *Order, Violence and Conflict* (Cambridge: Cambridge University Press, 2008), pp. 397-421.

43. But, explained well by Richards' 'agrarian revolt' model. Paul Richards, 'To Fight or Farm? Agrarian Dimensions of the Mano River Conflicts (Liberia and Sierra Leone)', *African Affairs*, vol. 104, no. 47 (200), pp. 571-90.
44. Murshed and Gates, 'Spatial-Horizontal Inequality and the Maoist Insurgency in Nepal', pp. 121-34.
45. *Resilience Amid Conflict: An Assessment of Poverty in Nepal (1995-96 and 2003-04)*, World Bank Publication No. 34834-NP, 2006.
46. Govinda Neupane, *Nepalko Jatiya Prashna: Samajik Banot ra Sajhedariko Sambhawana* (Kathmandu: Center for Development Studies, 2000).
47. Authors' interviews, Biratnagar and Morang citizens, November 2008.
48. Authors' interviews, Sunsari village, April 2008. Some Terai citizens look physically different from *Pahadi*s owing to their Indian heritage.
49. Jason Miklian, *Illicit Trading in Nepal: Fueling South Asian Terrorism*, PRIO South Asia Policy Report No. 3, Oslo: International Peace Research Institute, Oslo, 2009.
50. Chandra Adhikari and Paul Chatfield, 'The Role of Land Reform in Reducing Poverty across Nepal', Paper presented for the Third Annual Himalayan Policy Research Conference, Nepal Study Centre, Madison, Wisconsin, 16 October 2008.
51. Jason Miklian, 'International Media's Role on U.S.-Small State Relations: The Case of Nepal', *Foreign Policy Analysis*, vol. 4, no. 4 (2008), pp. 399-418.

Index